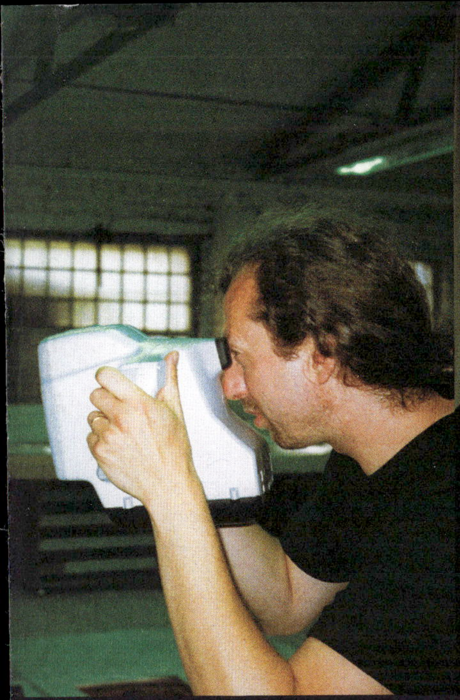

ologne

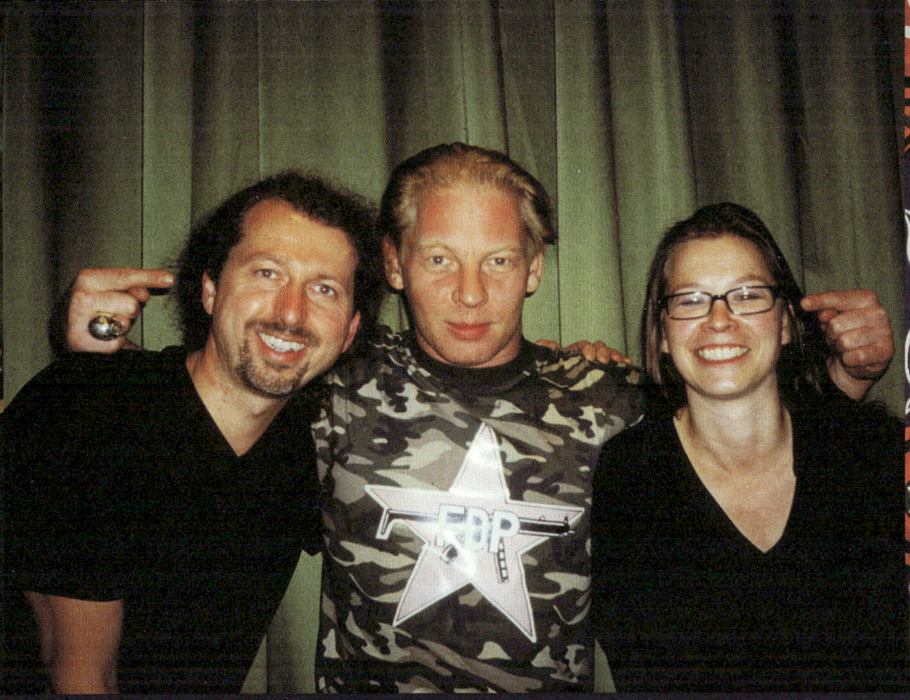

Ben Becker, Berlin

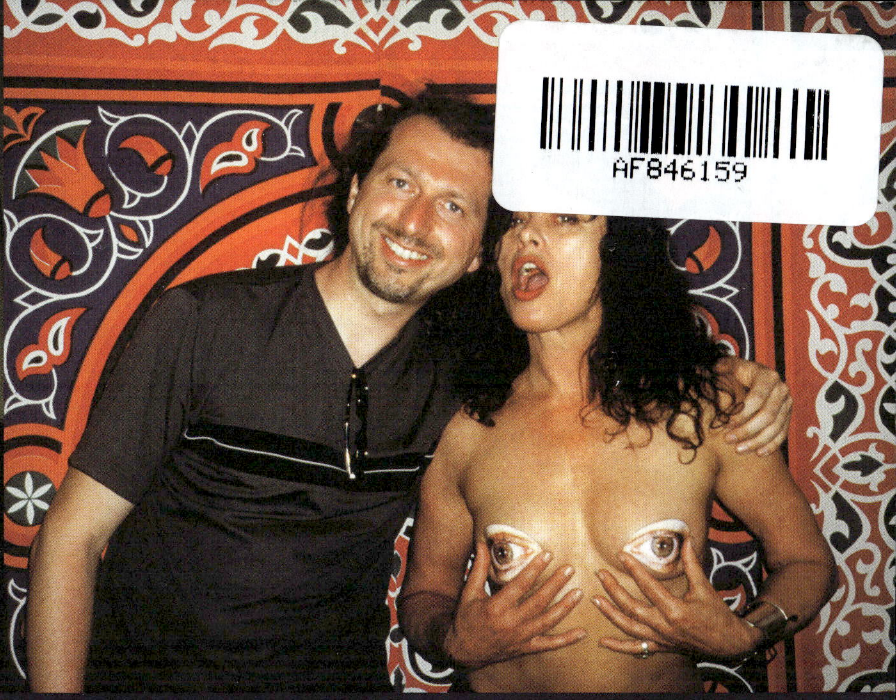

Uschi Obermaier, Los Angelos

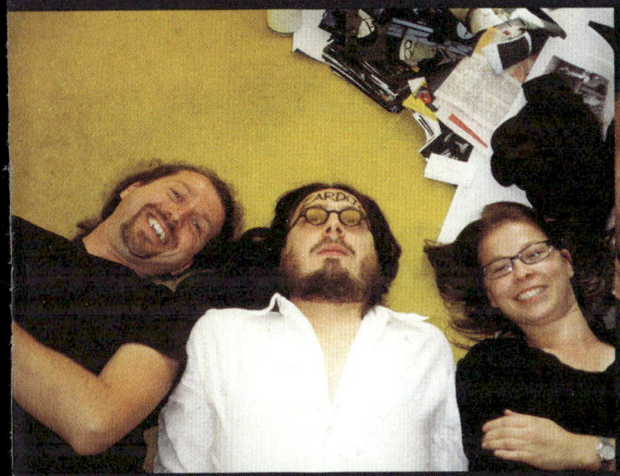

Jonathan Meese, Berlin

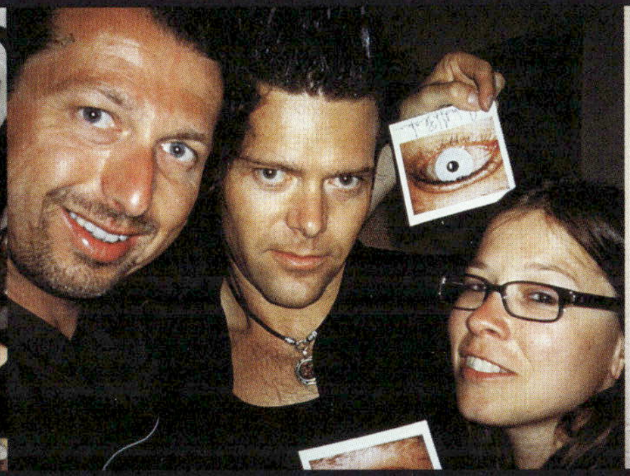

Richard of Rammstein, Berlin

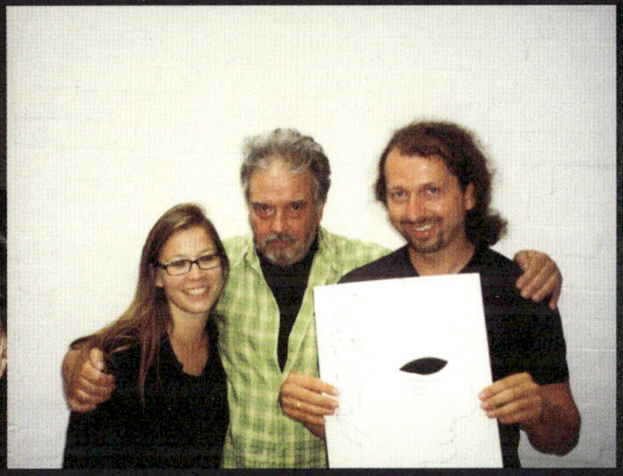

David Bailey, London

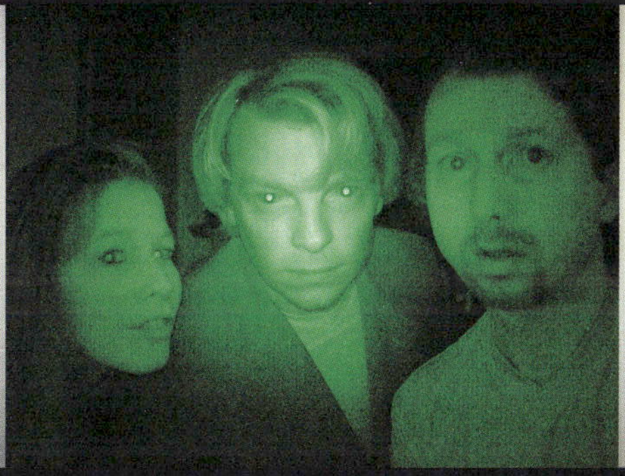

André Eisermann, Hamburg

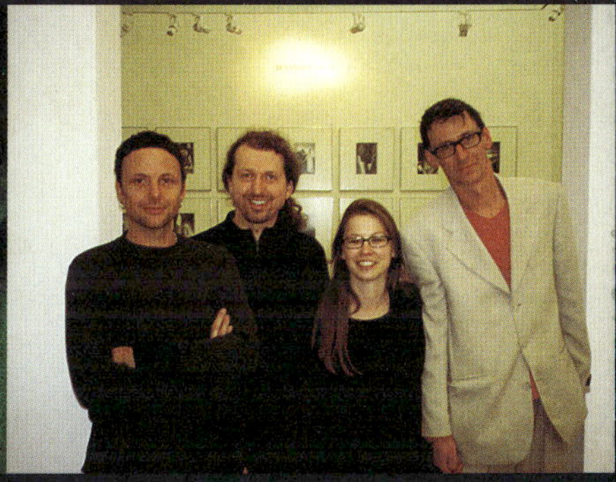

Paul & Flake of Rammstein, Berlin

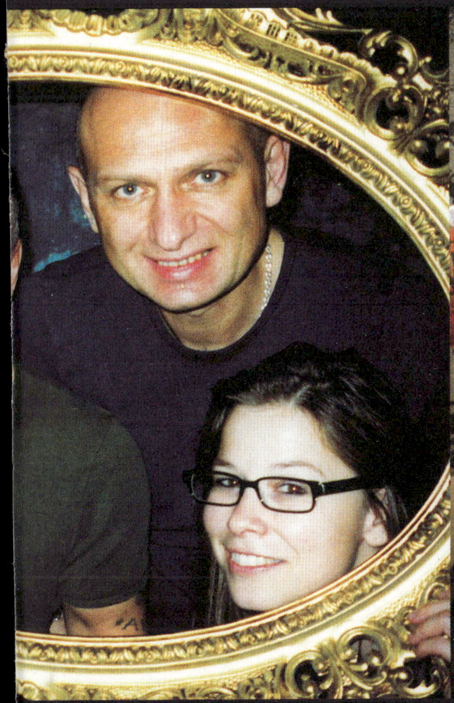

atthias Reuss, Paris

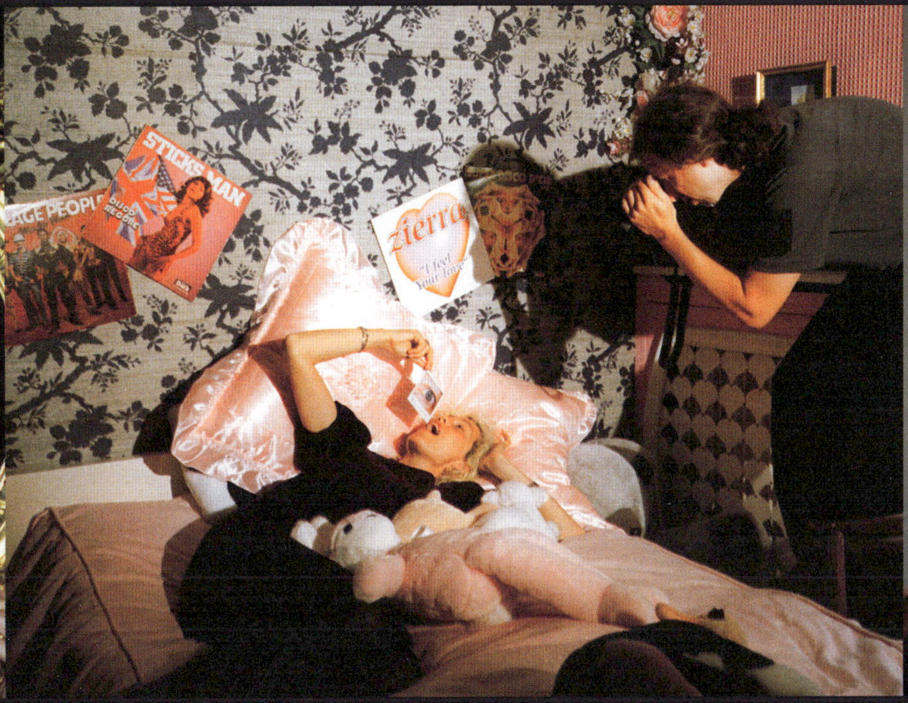

Ellen von Unwerth, Paris

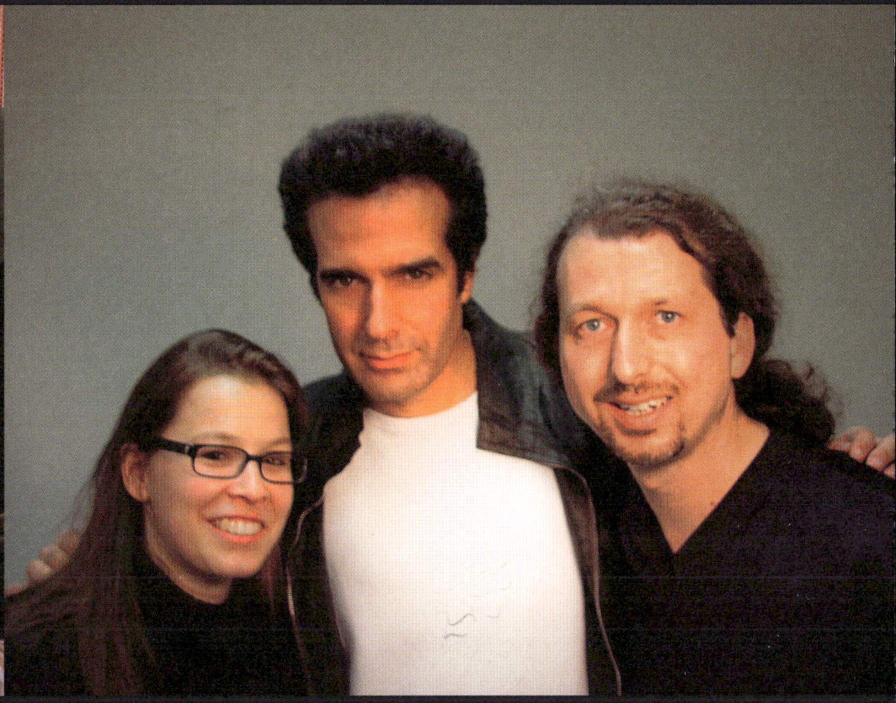

David Copperfield, Frankfurt

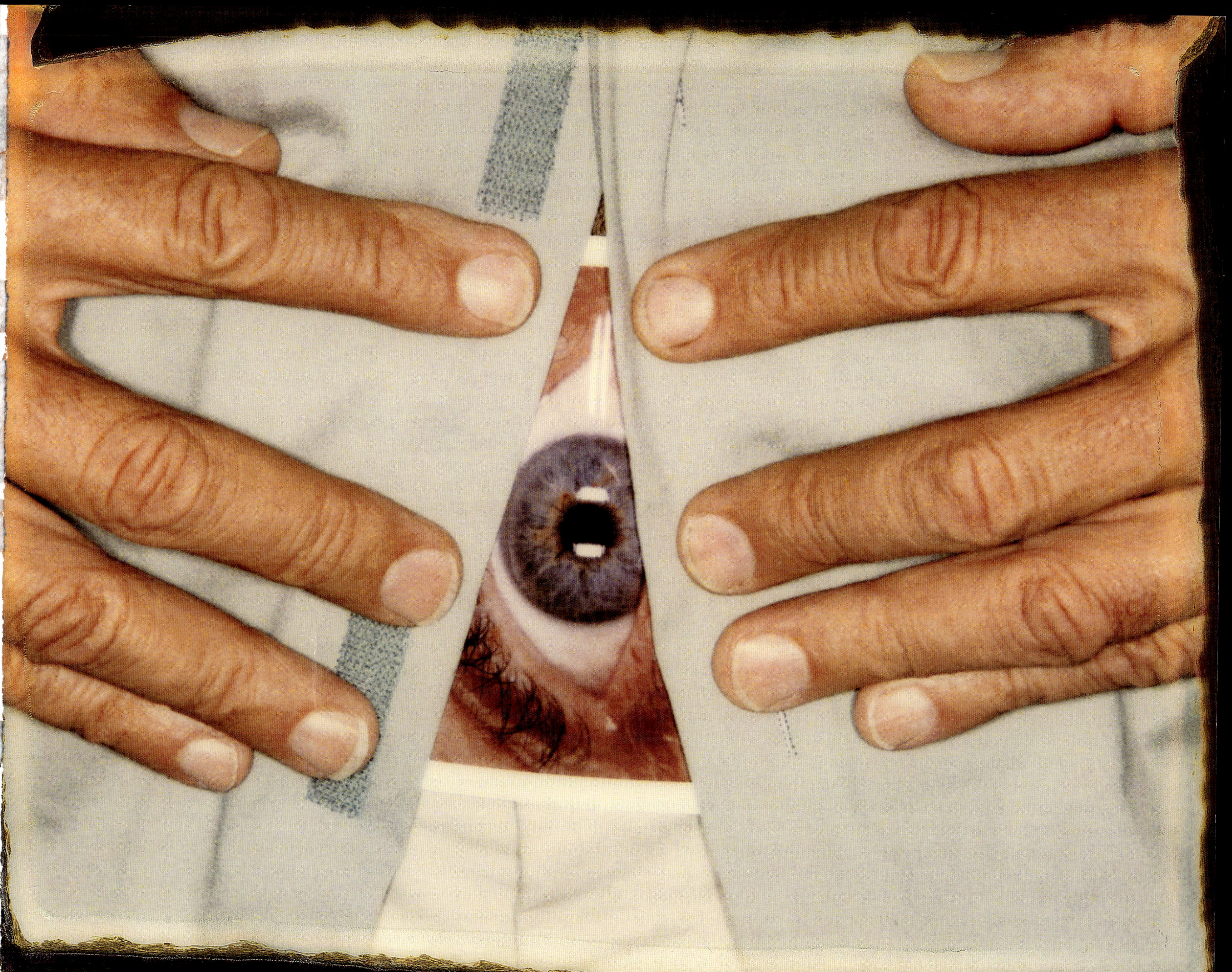

Ralph Gibson, photographer

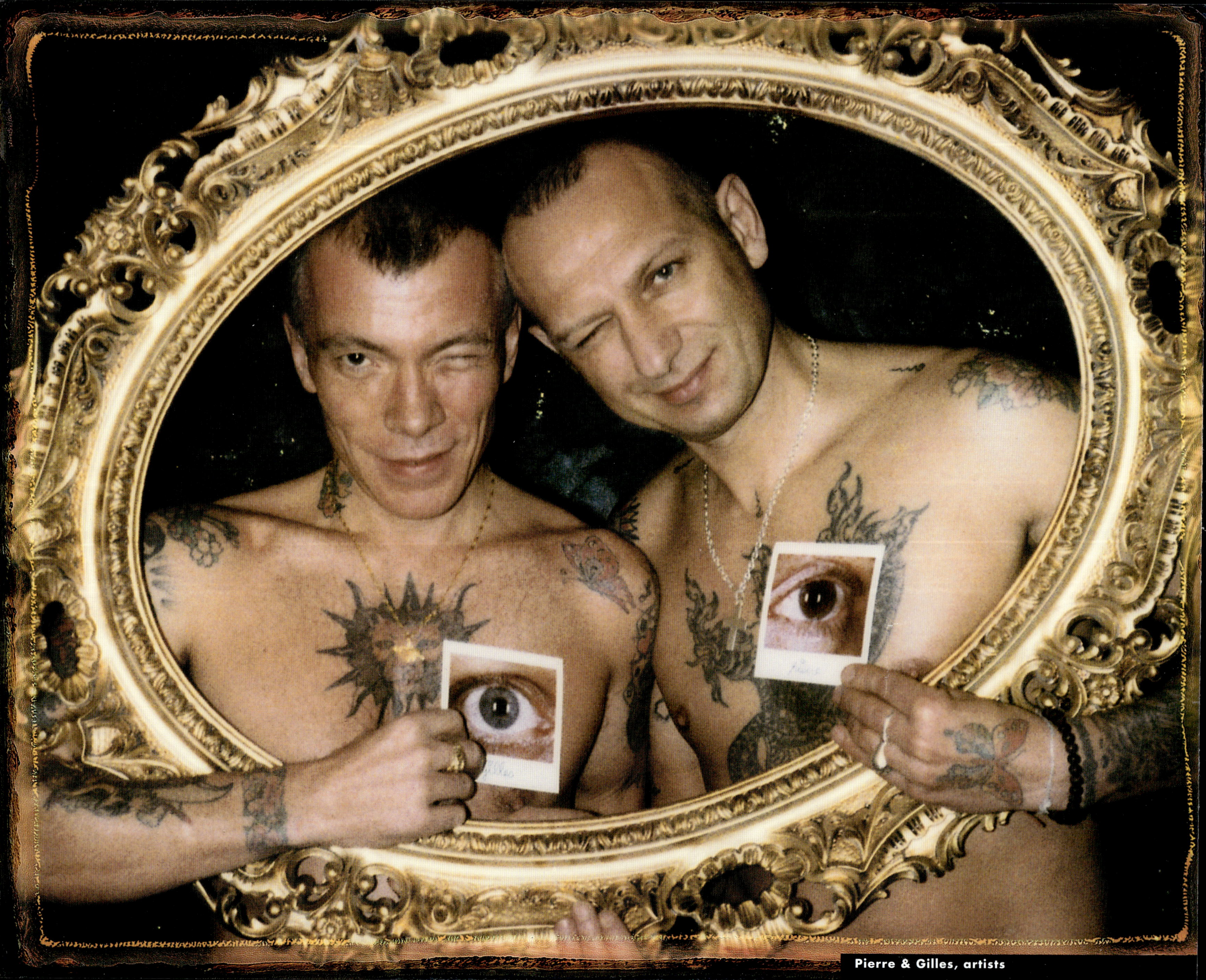

Pierre & Gilles, artists

FAMOUS EYES

DANIEL AND GEO FUCHS

EDITION REUSS

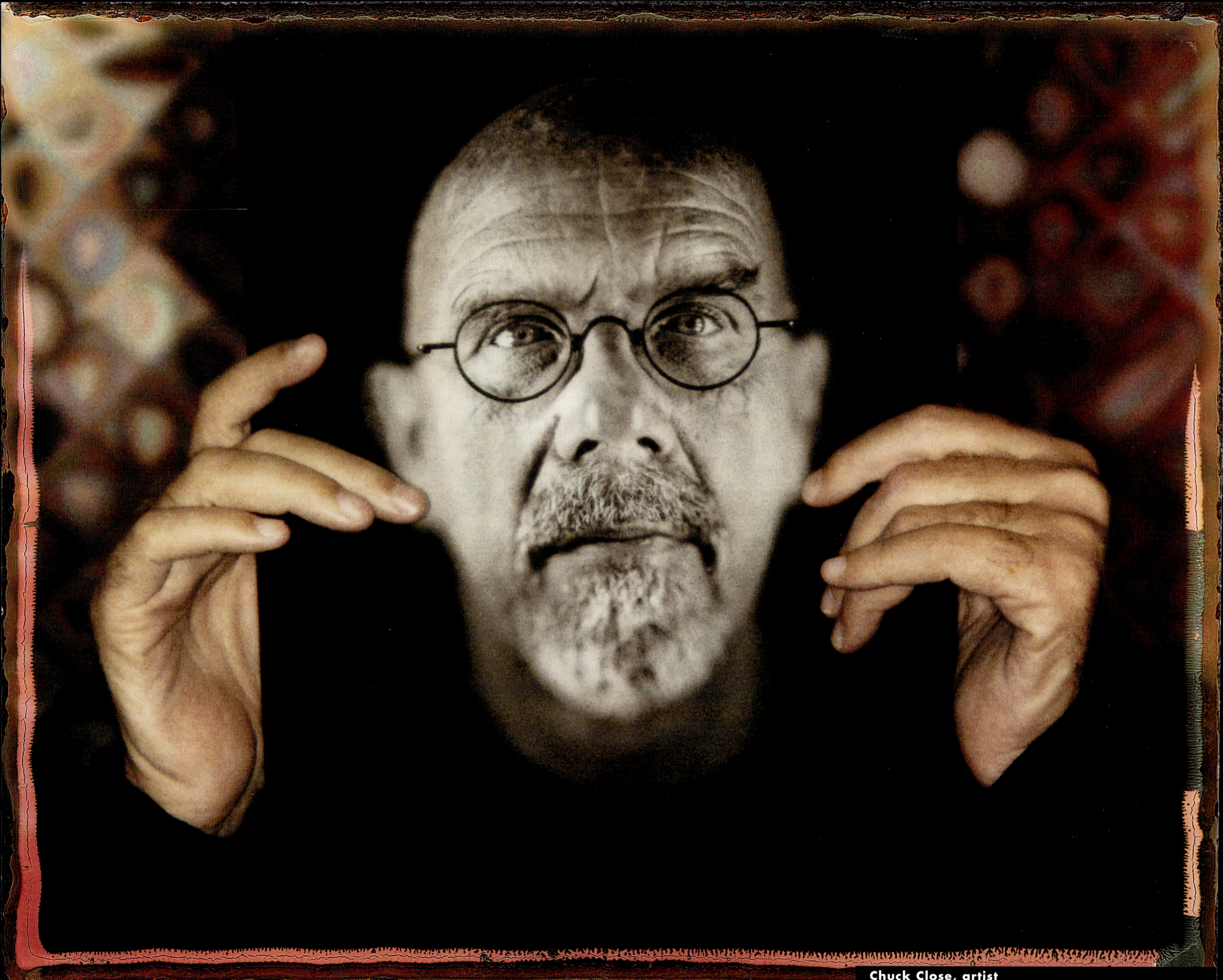

Chuck Close, artist

CONTENTS

IRISIERENDE BLICKWECHSEL

Wär' nicht das Auge sonnenhaft,
Wie könnten wir das Licht erblicken?
Goethe

Je größer der Pupillendurchmesser,
desto stärker der Appetit auf das Erblickte.
Detlef B. Linke

Man schaut einem Gegenüber in die Augen, um zu erfahren, mit wem man es zu tun hat, Freund oder Feind. Indem man blickend den eigenen Blick preisgibt, stiftet man den anderen Menschen dadurch seinerseits zum verräterischen Augenspiel an: Blicke werden getauscht. Und da Blicke einem oft mehr als tausend Worte sagen, ging es mir beim Künstlerpaar Fuchs so, dass ich schon im Bruchteil einer Sekunde wusste, woran ich war: sympathisch auf Anhieb! Das lag keineswegs am Blau von Daniels und auch nicht am Braungrün von Geos Augenfarbe, an die ich mich übrigens später gar nicht mehr genau hätte erinnern können. Diese Erinnerungslücke hat keineswegs etwas mit einem bloß flüchtigen Blick zu tun. In einer Welt voller Farbigkeit ist es, paradox genug, eher Schwarz-Weißem vorbehalten, die Gedankenbewegungen unseres Gegenübers spürbar mitzuverfolgen: Stets liegt das Wesentliche des Blicks im kontrastierenden Wechselspiel von weißer Bindehaut und schwarzer Pupille. Und die Vergrößerung der Pupille kann unter Umständen so viel Schwarz in die Augenmitte rücken, dass die eigentliche Augenfarbe auf einmal eine untergeordnete Rolle spielt (1). Genau das aber ist bei Daniel und Geo meist der Fall: Mit großen, neugierigen Augen stellen sie ihre Pupillen auf Durchlass. Und je feuriger und ungebremster sie sich in ein Projekt hineinsteigern, desto intensiver wirkt das Weiß ihrer Augapfelbewegungen im Kontrast zur Pupillenschwärzung, wenn sie versuchen, es einem nahe zu bringen. Wie seinerzeit bei der Trilogie Conserving (2), ihrem letzten, mehrfach preisgekrönten und wie nun bei Famous Eyes, ihrem allerneuesten Projekt.

Wie der Umstand, dass ich sie direkt beim Vornamen nenne, deutlich macht, sind die Fotokünstler Daniel und Geo Fuchs längst gute Freunde von mir, und es fasziniert mich von Mal zu Mal mehr, wie sie gerade der fachlichen Seite der Fotografie stets neuartige schöpferische Impulse für ungewohnte Bilderwelten abzugewinnen vermögen. Bei Conserving war es die wohl wichtigste Wesensart des Mediums Fotografie – dass es Gesehenes konserviert –, die sie sich zur Arbeitsmetapher und schließlich zum Sujet erkoren. Dafür fotografierten sie in etlichen anatomischen und naturwissenschaftlichen Sammlungen in Europa in Alkohol oder Formalin eingelegte Präparate. Was dabei herauskam, grenzt an ein Wunder: Mensch und Tier wirken nun so, als träumten sie nur und wären jeden Moment imstande, aus ihrem vermeintlich ewigen Schlaf zu erwachen.

Bei Famous Eyes dagegen richten Daniel und Geo ihr Augenmerk zwar auf die altbekannte Analogie von Auge und Kamera, wissen sie allerdings für ihr Vorgehen völlig neu zu deuten – nämlich als Metapher und als Motiv zugleich. Daraus ergibt sich ein Gedankenspiel, das in mehrere Richtungen funktioniert und so genügend Raum lässt für Überraschungen. Wie man weiß, dient ja die Iris, rein technisch gesehen, zur Regulation einfallenden Lichts, zur Steuerung der Lichtmenge. Und hinter der zentralen Öffnung der Iris, der Pupille, befindet sich die Augenlinse, deren Brennweite wie bei der Kamera veränderlich ist. Selbst die „Blende", um im Bilde zu bleiben, kann beim menschlichen Auge variiert werden: Die Pupille erweitert oder verengt sich je nach Helligkeit der Umgebung, je nach Gemütsverfassung. Und schließlich entsteht auf der Netzhaut – wiederum analog zur Kamera – ein Bild, das Aufgenommenes seitenverkehrt und auf dem Kopf stehend wiedergibt (2). Natürlich hört die Analogie damit auf, dass man als Mensch über ein Paar Augen verfügt, wohingegen die Kamera „einäugig" sieht. Von dieser Differenz lebt die Fotoaktion Famous Eyes: Polaroid für Polaroid sollten möglichst viele prominente Augenaufschläge in Szene gesetzt werden. Dabei war Famous Eyes zunächst mal ein Versprechen, das der nun vorliegende Fotoband überhaupt erst einlöst. Und dies selbst bei so weltberühmten Bürgern des globalen Dorfes wie Louise Bourgeois oder Christo, deren Hervorbringungen längst als Kulturerbe der Menschheit bewundert werden, deren Augen jedoch den meisten von uns keineswegs geläufig gewesen sind.

Es freut mich, dass auch ich etliche Berühmtheiten zu diesem Projekt beisteuern konnte. Wer immer jedenfalls den Weg des umtriebigen Künstlerpaares Geo und Daniel Fuchs kreuzte, der wurde von ihnen mit geradezu naturwissenschaftlich protokollierendem Blick, als ginge es um Augendiagnostik, konsequent kühl fotografiert. Was nun angesichts solcher medialer Kälte fehlte, war Warmherzigkeit. Für die aber sorgten die beiden, indem sie alles Weitere – beim Fotografiertwerden sonst durchaus nicht üblich – zu einem denkbar offenen Vorgang gemeinschaftlichen Herausfindens und Gestaltens machten. Denn wie das zum Augenporträt dazugehörige Konterpartfoto am Ende jeweils aussah, das probierten das Künstlerpaar Fuchs und der oder die betreffende fotografierte Person so lange aus, bis die stimmige Konfiguration schließlich gefunden war, und zwar auf sehr spielerische, auf sehr schöpferische Weise, ich hab's am eigenen Leib erlebt. Etwas Bestimmtes ins Auge zu fassen, hat immer den Charakter von Zuwendung, weil man dabei aus dem visuellen Umfeld eine ganz persönliche Auswahl vornimmt. Beim hier stattfindenden verschmelzenden Wechselspiel von Fremd- und Selbstdarstellung könnte dies die Garantie dafür sein, dass hinter der nicht fassbaren Berühmtheit ein winziges Stück vom eigentlichen Menschen sichtbar wird.

Anke Degenhard

(1) Detlef B. Linke: Kunst und Gehirn. Die Eroberung des Unsichtbaren. Rowohlt Verlag, Reinbek bei Hamburg 2001, S.31

(2) Daniel & Geo Fuchs, Conserving, Edition Reuss, München 2000

AN IRIDESCENT CHANGE OF PERSPECTIVE

If the eye weren't like the sun,
how else would we be able to see the light?
Goethe

The wider the diameter of the pupil,
the greater the appetite for that which is seen.
Detlef B. Linke

We look a person in the eye to find out whether we're dealing with a friend or a foe. As we look at another person, we also surrender our gaze to that person and thus participate in the revealing game of the eyes, in the exchange of glances. Because a gaze often says more than a thousand words, my first glance at this artist couple convinced me: Daniel and Geo Fuchs instantly won my confidence! The reason for this had nothing to do with the blue of Daniel's eyes nor with the brownish green of Geo's eyes – a color, incidentally, whose hue I couldn't recall precisely sometime afterwards. This gap in my memory wasn't due to the brief duration of the glances we exchanged. In a world full of colors, it's paradoxical but true that black and white are the colors of the organs which enable us palpably to pursue the movements of the thoughts in the mind of the person we're looking at. The essence of a person's gaze always seems to lie in the contrast and interplay between the white conjunctiva and the black pupil. Under certain conditions, the dilation of the pupil can put so much black into the center of the eye that the color of the iris comes to play a subordinate role.[1] Precisely this is what happens most often in Daniel's and Geo's eyes. Wide open and agape with curiosity, their pupils dilate to let the world in. The more ardently and unconditionally they devote themselves to a particular project, the more intense do the movements of the whites of their eyes become, the more starkly they contrast with the black of their pupils as these artists explain their work to another individual. This was the case with their previous and prize-winning *Conserving*[2] trilogy, and it is again the situation here in this newest project.

As the fact that I call them by their first names suggests, photographic artists Daniel and Geo Fuchs have long since become close friends of mine, and it fascinates me more each time to see how they repeatedly succeed in finding in the technical side of photography innovative and creative impulses for unconventional worlds of imagery. The theme explored in *Conserving* was the most important characteristic of the medium of photography, i.e. the genre's ability to conserve and eternalize a fleeting image. This was the aspect which they chose as their working metaphor and ultimately as the subject of their work. Visiting numerous anatomical and scientific collections throughout Europe, they photographed hundreds of specimens which had been preserved in alcohol or formaldehyde. The results of their research border on the miraculous. These preserved specimens of humans and animals look as though they're dreaming, as though at a moment's notice they might awaken from a slumber which was meant to last forever.

In *Famous Eyes* on the other hand, Daniel and Geo turn their gaze to the familiar analogy between the eye and the camera, albeit in a manner which they reinterpret for this project in an entirely new fashion: namely, as both a metaphor and a motif. This reinterpretation results in a play of thoughts which functions in numerous directions and leaves plenty of room for surprises. From a technical point of view, it's well known that the iris serves as a regulator of incident light by controlling the amount of light which it allows into the eye. Behind the pupil, which is the central opening of the iris, lies the eye's lens, which can vary its focal length much the way focal length can be adjusted in a camera. To continue with our metaphor of the camera, the human eye can even adjust its aperture setting or "f-number": the pupil dilates or contracts in accord with the brightness of the surroundings and in response to an individual's moods. Finally, and again in a process which seems closely analogous to what goes on inside a camera, an image appears on the retina which represents the object that is being seen, but in which left and right, up and down are reversed so that the object appears as a mirror image standing on its head.

Of course, we can stretch this metaphor only so far before it snaps. Unlike the Cyclops of the camera, we humans are blessed with two eyes, not one. And it's this crucial difference that enlivens the photographic activity presented to us in *Famous Eyes*. One Polaroid after another, Daniel and Geo Fuchs capture an extraordinarily large number of celebrities' eyes. The resulting volume is the fulfillment of a promise which is ultimately kept by the book itself. And fulfilled it is, most ably indeed, in the eyes of such world-famous citizens of our Global Village as Louise Bourgeois or Christo, two artists whose creations have long since become part of humankind's cultural heritage, but whose eyes have seldom been seen, even by people who know and admire their artworks. I'm especially pleased that I too was able to contribute a few famous people to this project. All who chanced to cross paths with the indefatigable artist couple of Geo and Daniel Fuchs soon found themselves scrutinized with a nearly scientific, documentary gaze, almost as though they were being subjected to an ophthalmic examination, then photographed with commensurate coolness. What was lacking in this cool medium, of course, was warmheartedness. But this human warmth was provided by the two artists as they pursued their joint project of openly discovering and designing everything else, a process which is by no means typical when a person is being photographed! The ultimate appearance of the counterpart photos which augment each of the eye portraits was determined in a lengthy process of collaboration between Daniel and Geo Fuchs and the various people they photographed. This collaborative search was continued in a very playful and creative fashion for as long as necessary, and it didn't come to an end until precisely the right configuration had been found. I know that this was the case because I experienced it firsthand. To fix one's gaze on some specific object always has the character of concerned attention because one opts for a highly personal selection from the wide spectrum of the visual environment. In the interplay between depiction by others on the one hand and self-presentation on the other, perhaps it is this attention which guarantees that behind the celebrity who eludes our grasp, a tiny piece of the real person always becomes visible.

Anke Degenhard

1. Detlef B. Linke: Kunst und Gehirn. Die Eroberung des Unsichtbaren. Rowohlt Publishers, Reinbek bei Hamburg, 2001, p. 31.

2: Daniel & Geo Fuchs, Conserving, Edition Reuss, Munich, 2000.

REGARDS IRISÉS

Si le soleil n'était dans l'œil,
comment pourrions-nous capter la lumière ?
Goethe

Plus le diamètre de la pupille est important,
plus notre soif de découverte est grande.
Detlef B. Linke

Quand nous regardons l'autre dans les yeux, c'est pour savoir à qui l'on à affaire, ami ou ennemi. En livrant son propre regard, on incite l'autre à se trahir à son tour : les regards s'échangent. Ces derniers en disant souvent plus que maints mots, il m'a fallu moins d'une fraction de seconde pour savoir que penser des Fuchs. Le couple d'artistes me fut d'emblée sympathique ! Un fait qui ne revient nullement au bleu des yeux de Daniel ou au bleu-vert de ceux de Geo, couleurs dont je n'aurais pu me souvenir précisément par la suite. Une défaillance de ma mémoire qui ne s'explique pas davantage par l'aspect fugitif de leurs regards. Dans le monde de couleurs qui est le nôtre, il est en effet paradoxal qu'il soit plutôt réservé au noir et blanc de suivre les mouvements de la pensée de notre vis-à-vis. L'essentiel du regard se situe dans le contraste qu'offrent le blanc de l'œil et la pupille noire. L'agrandissement de celle-ci noircit le centre de l'œil parfois au point de conférer à sa véritable couleur un rôle secondaire. (note 1)

Voilà ce que nous montrent Daniel et Geo : ouvrant de grands yeux avides, ils laissent pénétrer leurs pupilles. Et plus ils s'exaltent dans un projet, avec ardeur et sans limite, plus le contraste du blanc des mouvements du globe oculaire avec le noir des pupilles se révèle intense lorsqu'ils tentent de le rendre perceptible à quelqu'un. Tout comme lors de la réalisation de leur trilogie Conserving (note 2), leur dernier ouvrage, objet de plusieurs prix, et de Famous Eyes, leur tout nouveau projet.

Les artistes Daniel et Geo Fuchs, que je nomme par leurs prénoms, comptent depuis longtemps parmi mes amis. Or leur capacité, d'œuvre en œuvre, à enrichir le monde de la photographie grâce à des idées toujours novatrices et créatrices pour des univers d'images insolites me fascine chaque fois un peu plus. Dans Conserving, il s'agit de la nature principale du médium qu'est la photographie chargé de conserver ce qui est vu, métaphore, puis sujet choisis par les auteurs. Ainsi les auteurs ont-ils photographié des êtres conservés dans l'alcool ou le formol dans nombre de musées européens d'anatomie et de sciences naturelles pour donner vie à ce que j'appellerais presque un miracle dans lequel êtres humains et animaux semblent rêver et prêts à sortir à tout moment d'un sommeil éternel feint.

Dans Famous Eyes en revanche, Daniel et Geo Fuchs convergent leurs regards sur l'analogie de l'œil et de la caméra certes bien connue, expliquant toutefois leur approche d'une façon résolument nouvelle, à la fois métaphore et sujet. Il en ressort un jeu de pensées fonctionnant dans plusieurs directions et laissant suffisamment de place à l'effet de surprise. Comme chacun le sait, l'iris, d'un point de vu purement technique, sert à diriger la quantité de lumière. Derrière son orifice central, la pupille, se trouve le cristallin, dont la longueur focale est, à l'instar de la caméra, variable. Même le diaphragme, pour rester dans le registre de la photographie, peut varier dans l'œil humain. La pupille s'agrandit ou rétrécit selon la lumière présente, selon notre état d'âme. A l'instar de la caméra apparaît alors sur la rétine une image inversée, tête en bas. La comparaison entre l'être humain doté de deux yeux et la caméra " borgne " s'achève naturellement ici. Une différence dont se nourrit Famous Eyes qui, au fil des polaroïds, met en scène les yeux d'une pléiade de personnalités éminentes, tenant ainsi sa promesse du départ. Cela au travers de personnes connues dans le monde entier, comme Louise Bourgeois ou Christo, dont les œuvres font désormais figure d'héritage culturel de l'humanité, et dont pourtant la plupart d'entre nous ne connaît pas les yeux.

Je suis heureuse d'avoir pu convaincre un grand nombre de célébrités à contribuer à ce projet. Tous ceux et celles qui ont été photographiés par le couple toujours en action que forment Geo et Daniel Fuchs montrent un regard quasi scientifique et protocolaire, comme s'il s'agissait d'un diagnostique ophtalmologique qui n'aurait eu la moindre chaleur humaine si les artistes n'en avaient fait une démarche ouverte de quête et de conception commune, fait peu commun dans la photographie. Car, comme en témoignent à la fin les photos correspondant aux portraits des yeux, les artistes, ainsi que la personne photographiée, ont fait maints essais avant de trouver la configuration idéale, suivant une voie extrêmement ludique et créative dont je ne peux que témoigner.

Regarder quelque chose de précis a toujours un caractère affectif, puisque l'on opère alors un choix tout personnel dans le contexte visuel. Dans le jeu fusionnel de la représentation de soi et de l'autre ici illustré, cela pourrait être la garantie que derrière la célébrité inimaginable apparaît une bonne part de l'être humain, le vrai.

Anke Degenhard

Note 1 :Detlef B. Linke : Kunst und Gehirn. Die Eroberung des Unsichtbaren (Art et cerveau. La conquête de l'invisible) Rowolt verlag. Reinbeck, près de Hambourg 2001, page 31

Note 2 : Daniel & Geo Fuchs, Conserving, édition Reuss, Munich 2000

MAKING OF – EINIGE DER SCHÖNSTEN BEGEGNUNGEN

„I need a gun, I need a gun!" Oliviero Toscani

Wir sind mit Oliviero Toscani im Loft des Talk Magazins verabredet. Er hat eine imposante Statur und versprüht von Anfang an eine unglaubliche Warmherzigkeit. Konzentriert sieht er sich die bisher entstandenen Famous-Eyes-Bilder an, dann springt er unvermittelt auf und ruft: „I need a gun, I need a gun!" Seine spontane Idee ist, sein Augen-Polaroid mit einer echten Waffe zu durchschießen. Aber weder im Büro lässt sich eine Pistole auftreiben, noch haben die Wachleute eine, und sein Assistent kommt nach einiger Zeit nur mit einer Plastikpistole zurück. Plötzlich nimmt er einen Cutter und sticht es durch sein Augenfoto.

Mitten im Niemandsland. Jack Pierson

Jack Pierson lebt seit einiger Zeit in einem Haus in der Wüste Kaliforniens. Von Los Angeles sind es ein paar Stunden mit dem Auto bis nach Twentynine Palms. Er wohnt weit außerhalb, mitten in der Wüste, und trotz der vierstelligen Hausnummer gibt es in dieser Straße nur ein Haus, das von Jack Pierson. Hier tankt er Energie, „in the middle of nowhere". Wir wollen das Bild im Freien machen, umgeben von der einzigartigen Wüstenatmosphäre. Es weht ein starker Wind. Der Sand bläst uns ins Gesicht, während Jack wie gekreuzigt auf der Erde liegt. Seine zwei kleinen Hunde hüpfen durchs Bild und buddeln Löcher, sie freuen sich über die Abwechslung. In dieser fast unwirklichen Atmosphäre entstehen tolle Bilder, die uns beinahe davongeflogen wären.

Wie in einem Bienenstock. David LaChapelle

Als wir bei David eintreffen, ist er noch mitten in einem Cover Shooting für die japanische Vogue. In seinem Studio geht es zu wie in einem Bienenstock, an die 20 fleißige Assistenten sind pausenlos im Einsatz. Es ist spannend, David bei der Arbeit zuzusehen. Kurz bevor er zu fotografieren beginnt, dreht er seine Lieblingsmusik bis zum Anschlag auf, dann legt er los. Danach durchstöbern wir seinen Fundus und stoßen auf eine Admiralsjacke. Er zieht sie an, und alle meinen, er sähe darin aus wie Michael Jackson. Wir fotografieren ihn von einer hohen Leiter aus, unter der amerikanischen Flagge. Sein ganzes Team hilft. Schließlich sind alle von der Aufnahme begeistert.

„In der Kunst bin ich ja noch ein junges Talent." Udo Lindenberg

In der schummrigen Hotelbar des Hamburger Atlantik erwartet uns mit tief ins Gesicht gezogenem Hut und Bodyguard eine deutsche Rocklegende: Udo Lindenberg. Sein Auge zu fotografieren, gestaltet sich schwierig. Das diffuse Licht in der Bar mag älteren Damen schmeicheln, ist aber ungeeignet, um gute Fotos zu machen. Dann bleibt auch noch die kastige Polaroidkamera ständig an seinem Hut hängen, der irgendwie mit Udo verwachsen sein muss. Schließlich philosophieren wir über die Kunst, und Udo meint: „In der Kunst bin ich ja noch ein junges Talent." Die Zeit vergeht, und Udo muss zu einem weiteren Termin, vielleicht an einer anderen Hotelbar ...

Schließlich fallen die Hüllen! Uschi Obermaier

Es ist eine landschaftlich beeindruckende Fahrt vom Santa Monica in die Topanga Canyons – hier lebt Uschi Obermaier. Sie begrüßt uns herzlich mit ihrer angenehm tiefen Stimme und ihrem liebenswerten Bayerisch. Ihr Hund beobachtet uns scheu aus sicherer Entfernung. Wir überlegen lange, probieren viel aus, und schließlich fallen doch die Hüllen: Uschi ist in ihrem Element! Nacktheit ist für sie etwas so Natürliches, das sie schon fast wieder angezogen wirkt. Wer käme bei diesem kessen, fast jugendlichen Bild auf die Idee, dass unzählige Männer schon in den 70er Jahren von ihr geträumt haben? Wir trinken noch einen frischen Pfefferminztee in ihrem Garten und genießen einen wunderschönen kalifornischen Sonnenuntergang.

Eine wirklich gespenstische Atmosphäre. Rammstein

Die Musiker der deutschen Band Rammstein sind von unserem Projekt „Conserving" begeistert. Kein Wunder, sind sie doch berühmt-berüchtigt für ihren außergewöhnlichen Geschmack. Für ihr neues Album „Mutter" sollen wir das Artwork im Stile von „Conserving" fotografieren. Alle Musiker müssen in ein riesiges, wassergefülltes Bassin abtauchen und sich dann wie tote, in Formalin konservierte

Körper treiben lassen. Eine wirklich gespenstische Atmosphäre. Das ist natürlich auch eine wunderbare Situation für „Famous Eyes", denn Rammstein unter Wasser wird es wohl so schnell nicht wieder geben.

„Rani, Rani!" Fredy und Franco Knie

Fredy und Franco aus der weltbekannten Zirkusfamilie Knie treffen wir während ihrer Tournee durch die Schweiz. Franco stellt sich für die Aufnahme mit einer eleganten Elefantendame in Pose. Mit „Rani, Rani" wird die Elefantendame aufgefordert, ihren Rüssel zu kringeln. Die restlichen Elefanten stehen Spalier, nur etwa einen Meter hinter Daniel. Prinzessin Stéphanie ist gut gelaunt und scherzt unterdessen mit den Artisten. Auch wenn Elefanten ein überaus gutmütiges Wesen haben: Wenn man so nahe mit dem Rücken vor ihnen steht, bekommt man schon etwas Angst. Aber alles geht gut. Dann wird Fredys stolzer Palominohengst hereingeführt. Zunächst ist er sehr nervös, wird aber zunehmend ruhiger. Drinnen übt die Zirkuskapelle, es wird geklatscht und gelacht an diesem herrlich warmen Nachmittag in Bern.

Seine stahlblauen Augen funkeln im Morgenlicht. Jürgen Prochnow

Diffuses Morgenlicht, Nebelschwaden ziehen über das Meer, und die ersten Möwen fliegen kreischend über das Pier. Die Luft ist kalt. Alles ist perfekt geplant. Wir blicken erwartungsvoll auf das dunkle Meer – und tatsächlich, zur verabredeten Stunde taucht in sicherer Entfernung DAS BOOT auf. Ein schwarzes Beiboot nähert sich dem Pier. Mit jedem Ruderschlag werden wir nervöser. Plötzlich reißt uns der Wecker aus unseren Träumen. Nur ein Traum? Nein, nicht ganz – um 11 Uhr treffen wir Jürgen Prochnow am Pier von Santa Monica. Seine stahlblauen Augen funkeln im Morgenlicht ...

„Bunte Lämpchen und goldene Altare" Pierre & Gilles

Gemeinsam mit unserem Verleger Matthias Reuss tauchen wir staunend ein in das wundervolle Reich von Pierres und Gilles. Räume, prall gefüllt mit kitschigen Kunstwerken, bunten Lämpchen und goldenen Altaren, entführen uns in eine andere Welt. Die beiden Künstler strahlen eine unglaubliche Harmonie aus. Mit einem Riesenspaß durchforsten wir die Wohnung nach dem genialen Motiv und werden schließlich im Atelier fündig: Ein wuchtiger, neobarocker Goldrahmen bildet den Kontrast zu den wilden Tätowierungen von Pierres und Gilles.

Wir hatten schon Angst, er erstickt. Ben Becker

Ben Becker treffen wir in Berlin. Wir probieren alles Mögliche aus, und plötzlich kommt dem extremen Schaupieler und Musiker die zündende Idee. Er greift sich eine Rolle dickes Klebeband und verschwindet damit im Bad. Als er nach ein paar Minuten zurückkommt, hat er sich das Klebeband um den ganzen Kopf gewickelt, auch über Mund und Nase. Nun muss alles schnell gehen, denn er bekommt allmählich keine Luft mehr. Er zerrt noch sein T-Shirt über den Kopf, und während er sich das Band fast panisch wieder herunterreißt, entstehen die neun Polaroids für das Tableau. Ben schnappt tief nach Luft – wir hatten schon Angst, er erstickt.

Bis das Foto völlig frei in der Luft schwebt! David Copperfield

Während unserer Ausstellungseröffnung von Conserving in Toronto schlagen wir zufällig eine Tageszeitung auf und sehen ein Pressebild von David Copperfield – er, umringt von zehn Händen, die ein Foto seines Auges halten. Wir sind sprachlos, das wäre doch eine schöne Idee für unser Famous-Eyes-Projekt gewesen. Vielleicht war es auch eine Art Vorsehung? Ein Jahr später jedenfalls treffen wir David während seiner Deutschlandtournee in Frankfurt. Er ist unglaublich sympathisch und von unserem Projekt begeistert. Wenn sich einem die seltene Gelegenheit bietet, den größten Magier unserer Zeit zu fotografieren, hat man natürlich schon so eine Lieblingsidee im Hinterkopf. Uns erschien es ja ein bisschen verrückt, aber dann haben wir uns doch getraut und gefragt, ob er sein Augenpolaroid nicht schweben lassen könne. Und dann passierte vor unseren Augen etwas, das wir bis heute nicht begreifen können: David hielt das Foto zwischen seinen Händen, die sich langsam auseinander bewegten, bis das Foto völlig frei in der Luft schwebte!

MAKING OF – SOME OF THE NICEST MEETINGS

"I need a gun, I need a gun!" Oliviero Toscani

We have an appointment with Oliviero Toscani in "Talk" magazine's loft. Toscani has an impressive stature and, from the first moment onwards, he emanates a winning aura of warm-heartedness. He carefully scrutinizes the *Famous Eyes* photos which we've created thus far, then suddenly jumps to his feet and shouts: "I need a gun, I need a gun!" His spontaneous idea is to use a real weapon to shoot a hole in the Polaroid of his eyes. But neither in the office nor among the security guards can we find a pistol. A little while later his assistant appears with plastic pistol. Suddenly Toscani grabs an X-acto knife and pierces his eye photo with it.

In the middle of nowhere: Jack Pierson

Jack Pierson has been living for quite some time in a house in the California desert. It takes several hours to drive from Los Angeles to Twenty-nine Palms. Pierson lives far beyond the city limits, in the middle of the desert, and despite the four-digit house number in his address, there's only one house on this road – Jack Pierson's house. It's here that he energizes himself "in the middle of nowhere." We want to take the photo outdoors, surrounded by the unique atmosphere of the desert. A strong wind is blowing. The sand blows in our faces while Jack lies on the ground, arms outstretched as though he was being crucified. His two little dogs frolic through the picture, digging holes with their paws, obviously delighted with the unexpected change of pace. We're able to create some great photos in this unreal atmosphere, but the wind nearly blows them out of our grasp.

Like in a beehive: David LaChapelle

We arrive to find David in the midst of shooting a cover for the Japanese edition of "Vogue." His studio is a beehive of activity where some twenty industrious assistants are uninterruptedly busy. It's exciting to watch David at work. Shortly before he begins shooting photos, he puts on his favorite music and turns the volume control to "extra loud." Then the shoot begins! Afterwards we rummage through his stockpile and find an old admiral's uniform jacket. He puts it on and everyone agrees that he looks very much like Michael Jackson. We photograph him from a tall ladder, beneath the American flag. His entire team helps. Ultimately, everyone involved is enthusiastic about the results.

"In art, I'm still a young talent." Udo Lindenberg

Udo Lindenberg, the German rock legend, his hat pulled down low over his eyes, a bodyguard in tow, awaits us in the twilit bar at Hamburg's Atlantik Hotel. It proves to be rather difficult to photograph his eyes. The diffuse light in the bar may flatter the physiognomy of elderly ladies, but it's not the best illumination for shooting good photos. The boxlike Polaroid camera repeatedly snags on the brim of the hat which seems to have become inextricably attached to Udo Lindenberg's head. Finally, we begin to philosophize about art. Udo says, "In art, I'm still a young talent." The time passes quickly and Udo must depart for another appointment, perhaps in another hotel bar…

Finally, the veils fall! Uschi Obermaier

The landscape is very impressive as we drive from Santa Monica to Topanga Canyon, where Uschi Obermeier lives. She greets us cordially with her pleasantly deep voice and lovable Bavarian accent. Her dog watches us shyly from a safe distance. We think things over for a long time, try a few experiments, and then, finally, the veils fall: Uschi is in her element! Nudity is so natural for her that she almost seems fully dressed. Who would imagine, when viewing this charming, almost youthful image, that countless men had already dreamt of her in the 1970s? We drink another fresh peppermint tea in her backyard and enjoy a wonderful California sunset.

A truly eerie atmosphere. Rammstein

The musicians in the German band called "Rammstein" are enthusiastic about our *Conserving* project. No wonder: they're both famous and infamous for their peculiar tastes. For example, they want us to create the artwork for their new album "Mutter" in the same style as *Conserving*. All of the musicians must immerse themselves into a huge, water-filled basin and float there as though they were dead bodies preserved in formaldehyde. It's a truly eerie atmosphere. That, of course, is also a marvelous situation for *Famous Eyes* because Rammstein under water is a once-in-a-lifetime sight that surely won't be repeated again anytime soon.

"Rani, Rani!" Fredy and Franco Knie

We meet Fredy and Franco, two members of the world-famous Knie circus family, during their tour of Switzerland. Franco poses for the photo shoot together with an elegant elephant. He calls out "Rani, Rani" and the elephant curls her trunk obediently. The other elephants form two opposing lines just one meter behind Daniel. Princess Stéphanie is in a good mood. She jokes with the artistes. Even though elephants are reputed to be tame and kind creatures, when one turns one's back on so many of them, one can't help but feel a bit anxious. Fortunately, everything goes according to plan. Then Fredy's proud palomino stallion is led in. The horse is very nervous at first, but gradually he calms down. Inside, the circus family is rehearsing. Applause and peals of laughter are heard on this gorgeous, warm afternoon in Bern.

His steel-blue eyes sparkle in the morning light. Jürgen Prochnow

Diffuse morning light, clouds of fog stream over the ocean, and the first seagulls come flying and screeching above the pier. The air is cold. Everything is perfectly planned. We gaze expectantly towards the dark sea and, just as we'd agreed, "THE BOAT" appears at a safe distance. A black dinghy approaches the pier. We become progressively more nervous with each subsequent pull on the oars. Suddenly the alarm clock wakes us from our dreams. Was it only a dream? No, not entirely, because at 11 a.m. we're scheduled to meet Jürgen Prochnow on Santa Monica pier. His steel-blue eyes sparkle in the morning light...

"Colorful little lamps and golden altars" Pierre & Gilles

Together with our publisher Matthias Reuss, we immerse ourselves with astonishment in the marvelous realm of Pierre and Gilles. Rooms, filled full with kitsch artworks, colorful little lamps, and golden altars: all this transports us into another world. The two artists radiate an unbelievable harmony. With tremendous pleasure, they rummage through the dwelling, searching for the most ingenious motif. Finally they find what they've been looking for in the atelier: a huge, neo-baroque, gold frame provides the perfect contrast to the wild tattoos on the bodies of Pierre and Gilles.

We were afraid he'd suffocate. Ben Becker

We meet Ben Becker in Berlin. We try everything conceivable before the kamikaze actor and musician has the perfect idea. He grabs a roll of duct tape and disappears with it into the bathroom, only to return a few minutes later with the tape wrapped around his entire head, leaving only his mouth and nose exposed. Now we have to work fast because he's gradually having trouble breathing. He pulls his T-shirt over his head and we shoot nine Polaroids for the tableau while he tears the tape off his head, nearly in a panic. Ben gasps, breathes deeply. We were afraid he'd suffocate.

Until the photo floats freely in space! David Copperfield

During the opening of our "Conserving" exhibition in Toronto, we chance to open a newspaper and find a press photo of David Copperfield. It shows him surrounded by ten hands which hold a photo of his eye. We're speechless: this would have been a beautiful idea for our *Famous Eyes* project. Was it a kind of clairvoyance? Be that as it may, one year later we meet David in Frankfurt during his tour of Germany. He's unbelievably simpatico and is very enthusiastic about our project. When one has the rare opportunity to photograph the greatest magician of our era, one naturally has a favorite idea in the back of one's mind. It seemed a little crazy, but then we dared to ask him: Could he make our Polaroid of his eye float in midair? And then something happened, right in front of our eyes, which we couldn't explain – and still can't explain today. David holds the photo between his hands, then gradually separates his hands – and leaves the photo floating freely in midair!

MAKING OF – QUELQUES-UNES DES RENCONTRES LES PLUS SYMPAS

« I need a gun, I need a gun ! » Oliviero Toscani

Nous avons rendez-vous avec Oliviero Toscani dans le loft du magazine Talk. De corpulence imposante, celui-ci s'avère d'emblée incroyablement chaleureux. Très concentré, il regarde les photos réalisées pour Famous Eyes, avant de sauter brusquement en criant : « I need a gun, I need a gun! » Il lui vient spontanément l'idée de tirer sur le polaroid de ses yeux avec une vraie arme. Mais impossible de mettre la main sur un pistolet au bureau. Les gardiens n'étant pas non plus armés, son assistant revient peu après muni d'un simple pistolet en plastique. Il prend alors son cutter et lacère la photo de ses yeux.

Au milieu d'un no man's land – Jack Pierson

Jack Pierson habite depuis un certain temps une maison dans le désert de Californie. De Los Angeles, il ne faut que quelques heures de voiture pour atteindre Twentyne Palms. Excentrée, au milieu du désert, sa maison, la maison de Jack Pierson est, malgré son numéro à quatre chiffres, la seule en bordure de route. C'est ici qu'il puise son énergie « in the middle of nowhere ». Nous voulons prendre notre photo à l'extérieur dans le cadre unique du désert. Un vent fort souffle et le sable nous fouette le visage tandis que Jack est étendu sur le sol, tel le christ sur la croix. Ses deux chiens sautillent sur la photo et creusent des trous dans la terre tout à la joie du changement. C'est dans ce climat presque irréel que naissent des images superbes prêtes à s'envoler.

Comme dans une ruche – David LaChapelle

Lorsque nous arrivons chez David, il est encore en plein travail pour Le Vogue japonais. Son studio rappelle une ruche avec ses 20 assistants en permanente effervescence. Il est très intéressant de regarder David travailler. Juste avant de photographier, il augmente le son de sa musique préférée, puis se lance. Nous tombons sur une veste d'amiral qu'il revêt pour lui donner l'apparence de Michael Jackson. Nous le photographions du haut d'une échelle en dessous du drapeau américain. Toute son équipe nous aide. Pour finir, la photo soulève l'enthousiasme général.

« En tant qu'artiste, je suis encore jeune ». Udo Lindenberg

Le feutre bien enfoncé sur la tête, Udo Lindenberg, légende du rock allemand, nous attend en compagnie de son garde du corps dans la pénombre du bar de l'hôtel Hamburger Atlantik. Il s'avère difficile de photographier son œil. Si la lumière diffuse du bar est plutôt favorable aux dames d'un certain âge, elle n'est guère appropriée pour réaliser de bonnes photos. Sans compter notre appareil polaroid qui sans cesse se prend dans son chapeau, cet accessoire semblant ne faire qu'un avec le personnage. Nous terminons en philosophant sur l'art, Udo déclare être encore jeune en tant qu'artiste. Le temps passe et notre homme doit se rendre à un autre rendez-vous, peut-être dans le bar d'un autre hôtel…

Quand tombent les vêtements – Uschi Obermaier

Le voyage de Santa Monica aux canyons Topanga, où habite Uschi Obermaier, nous offre des paysages d'une beauté impressionnante. Celle-ci nous accueille chaleureusement de son agréable voix grave aux charmantes tonalités bavaroises. Son chien nous observe craintivement à distance. Nous réfléchissons longuement, faisons beaucoup d'essais avant qu'elle ne se dévêtisse. Uschi est dans son élément ! La nudité est pour elle si naturelle qu'elle paraît presque habillée. Qui pourrait croire en regardant cette photo hardie, presque juvénile, que tant d'hommes rêvaient déjà d'elle dans les années 70. Dans son jardin, nous buvons alors un thé à la menthe bien frais et profitons d'un splendide coucher de soleil californien.

Une atmosphère vraiment morbide – Rammstein

Il n'est pas étonnant que notre projet Conserving ait soulevé l'enthousiasme des musiciens du groupe allemand Rammstein, bien connus pour leur goût des plus insolites. Ainsi sommes-nous chargés de photographier la pochette de leur nouvel album Mutter dans le style de l'ouvrage. Tous les musiciens seront immergés dans un immense bassin rempli d'eau où ils flotteront tels des corps conservés dans le formol. Une atmosphère vraiment morbide, un moment formidable à saisir dans Famous Eyes, car on n'est pas prêt de revoir le groupe Rammstein sous l'eau d'ici tôt.

« Rani, Rani ! » Fredy et Franco Knie

C'est au cours de leur tournée à travers la Suisse et la France que nous avons rencontré Fredy et Franco Knie de la famille du cirque connue dans le monde entier. Franco pose pour la photo en compagnie d'un élégant éléphant et lui lance des « Rani, Rani » pour l'inviter à tire-bouchonner sa trompe. Le reste des éléphants font la haie à un mètre seulement derrière Daniel. La princesse Stéphanie est de bonne humeur et blague avec les artistes. Même si ces animaux sont de nature particulièrement complaisante, s'en approcher de trop près en leur tournant le dos s'avère une expérience un peu angoissante. Mais tout se passe bien. Survient alors le fier étalon Palomino de Fredy. D'abord très nerveux, il finit par se calmer. Sous le chapiteau, l'orchestre du cirque répète, l'on entend applaudissements et rires en cet après-midi merveilleusement chaud à Berne.

Ses yeux bleu acier scintillent dans l'aurore. Jürgen Prochnow

Lumière diffuse de l'aurore et nappes de brouillard se répandent au-dessus de la mer, tandis que les premières mouettes forment des cercles au-dessus de la jetée. L'air est froid. Tout est parfaitement organisé. Pleins d'espoir, nous regardons l'eau sombre et finalement, à l'heure dite, LE BATEAU se profile à bonne distance. Un canot de bord noir s'approche de la jetée. Notre nervosité croît à chaque coup de rame quand soudain nous sortons de notre rêve. Mais était-ce seulement un rêve ? Non pas tout à fait. A 11 heures, nous rencontrons Jürgen Prochnow sur la jetée de Santa Monica. Ses yeux bleu acier scintillent dans l'aurore…

« Petites lampes de couleur et autels dorés« Pierre & Gilles

C'est en compagnie de notre éditeur Matthias Reuss, au comble de l'étonnement, que nous pénétrons dans l'univers merveilleux de Pierre et Gilles. Ainsi des pièces remplies d'objets d'art kitch, de petites lampes de couleur et d'autels dorés nous mènent-elles dans un autre monde. Il émane des deux artistes une incroyable harmonie et nous trouvons beaucoup de plaisir à chercher dans leur maison le motif génial que nous finissons par trouver dans leur atelier : un cadre en or massif néo-baroque contraste avec les tatouages sauvages de Pierre et Gilles.

Nous avons craint qu'il n'étouffe. Ben Becker

Nous rencontrons Ben Becker à Berlin. Nous faisons toutes sortes d'essais quand soudain notre artiste connu pour son goût de l'extrême et les musiciens ont une idée de génie. Celui-ci prend un rouleau de gros scotch et disparaît dans la salle de bain. Il revient quelques minutes plus tard le scotch enroulé autour de sa tête recouvrant sa bouche et son nez. Alors il faut faire vite, car il commence à respirer difficilement. Il retire son tee-shirt et alors qu'il arrache le scotch dans un état presque de panique, nous réalisons les nouveaux polaroids destinés au tableau. Ben respire profondément. Nous avons craint qu'il n'étouffe.

Quand la photo plane dans les airs ! David Copperfield

Alors que notre exposition Conserving commence à Toronto, nous achetons un quotidien où David Copperfiled fait la une avec une photo le faisant apparaître entouré de dix mains tenant un cliché de son œil. Nous restons bouche bée : ne voilà pas une idée superbe pour notre projet Famous Eyes ? Un coup de la providence ? Quoi qu'il en soit, un an plus tard, nous rencontrons David Copperfield à Francfort au cours de sa tournée en Allemagne. Il nous apparaît incroyablement sympathique et enthousiaste quant à notre projet. Quand on a la chance de photographier le plus grand magicien de notre époque, on a bien sûr une idée de génie derrière la tête. Si la nôtre nous semble au départ un peu folle, nous osons tout de même la lui soumettre et lui demandons s'il ne pourrait pas faire planer son polaroid. Il se passe alors quelque chose que nous ne nous expliquons pas jusqu'à aujourd'hui : David tient sa photo entre ses mains qui s'écartent lentement pour laisser la photo planer librement dans les airs !

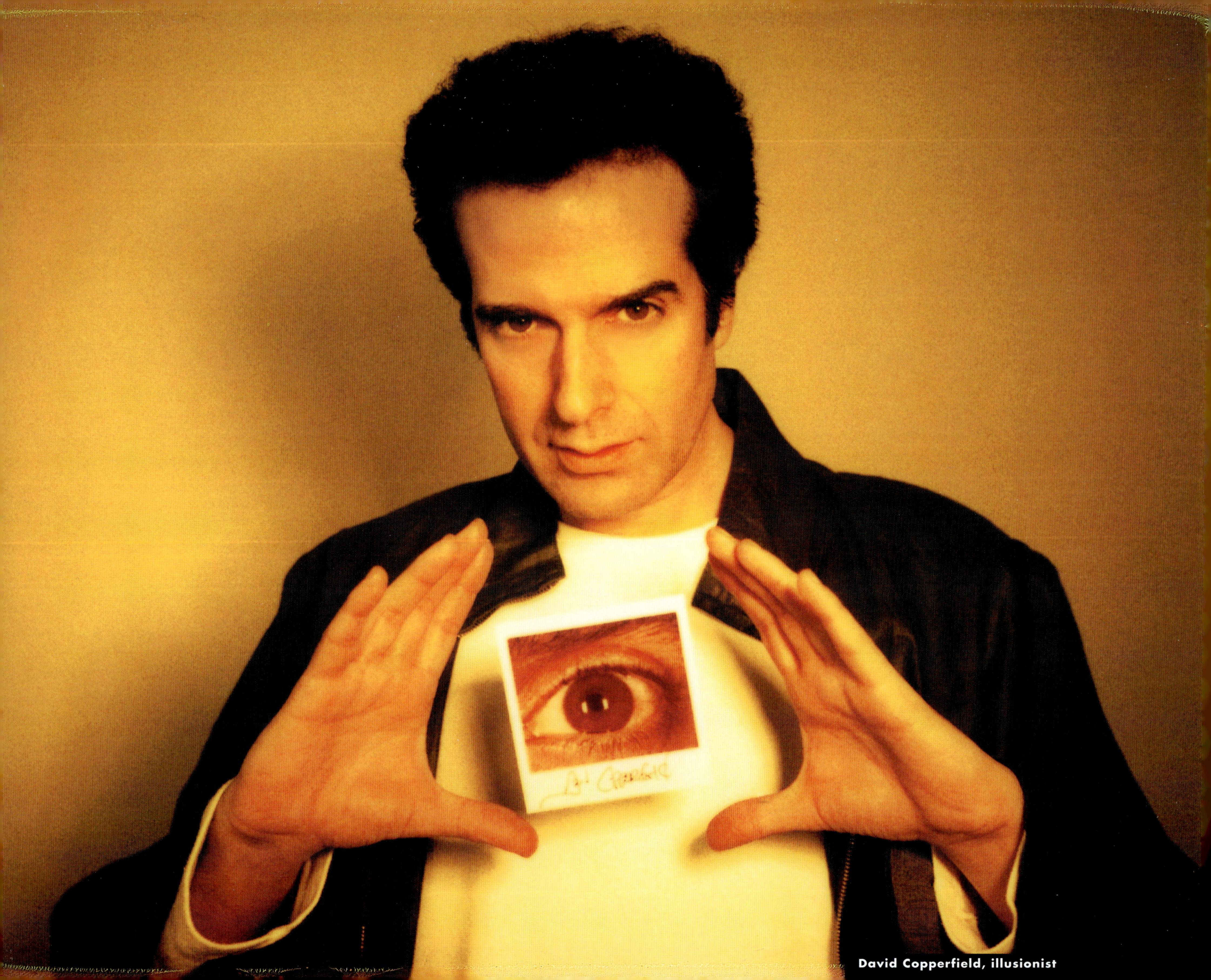

David Copperfield, illusionist

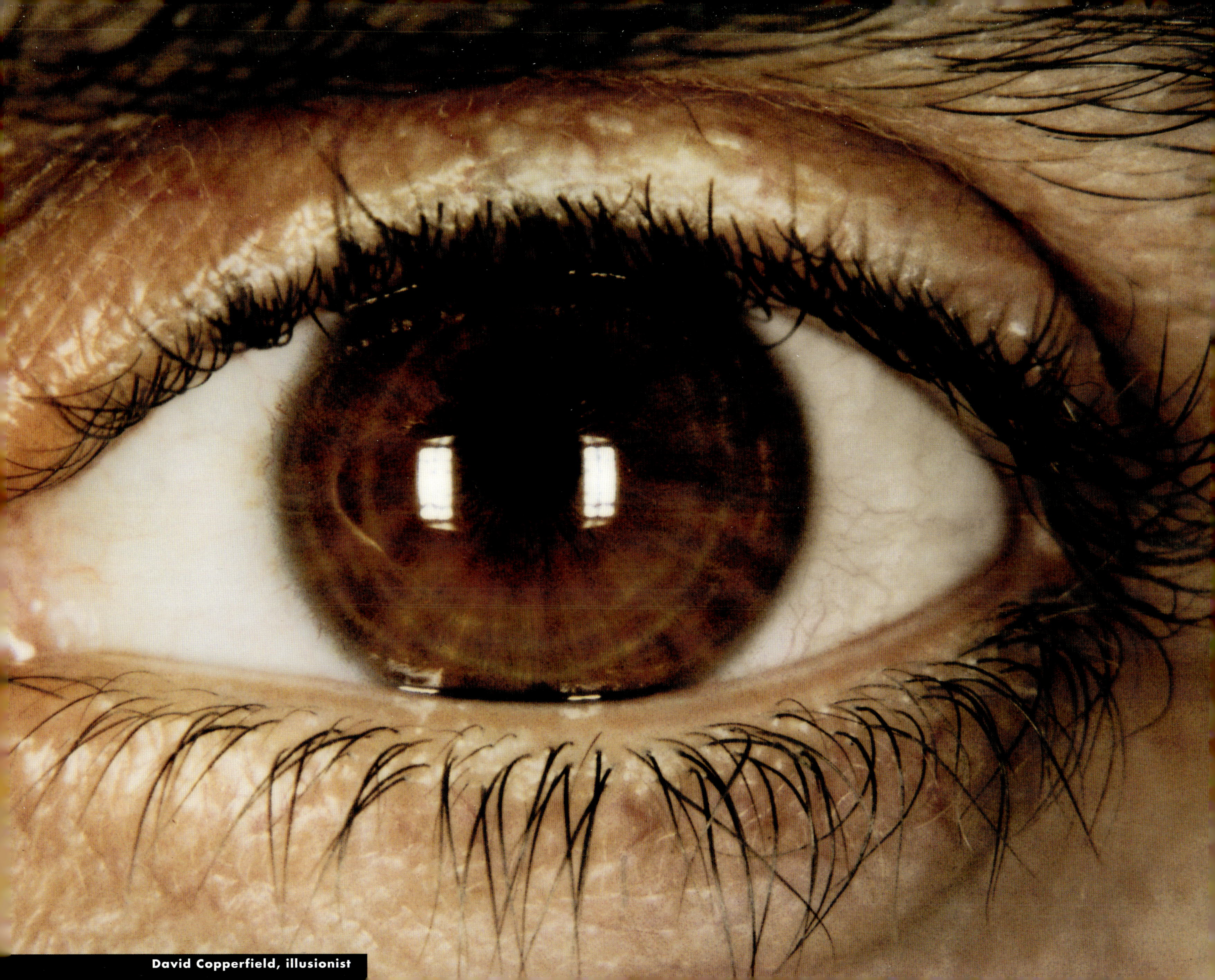

David Copperfield, illusionist

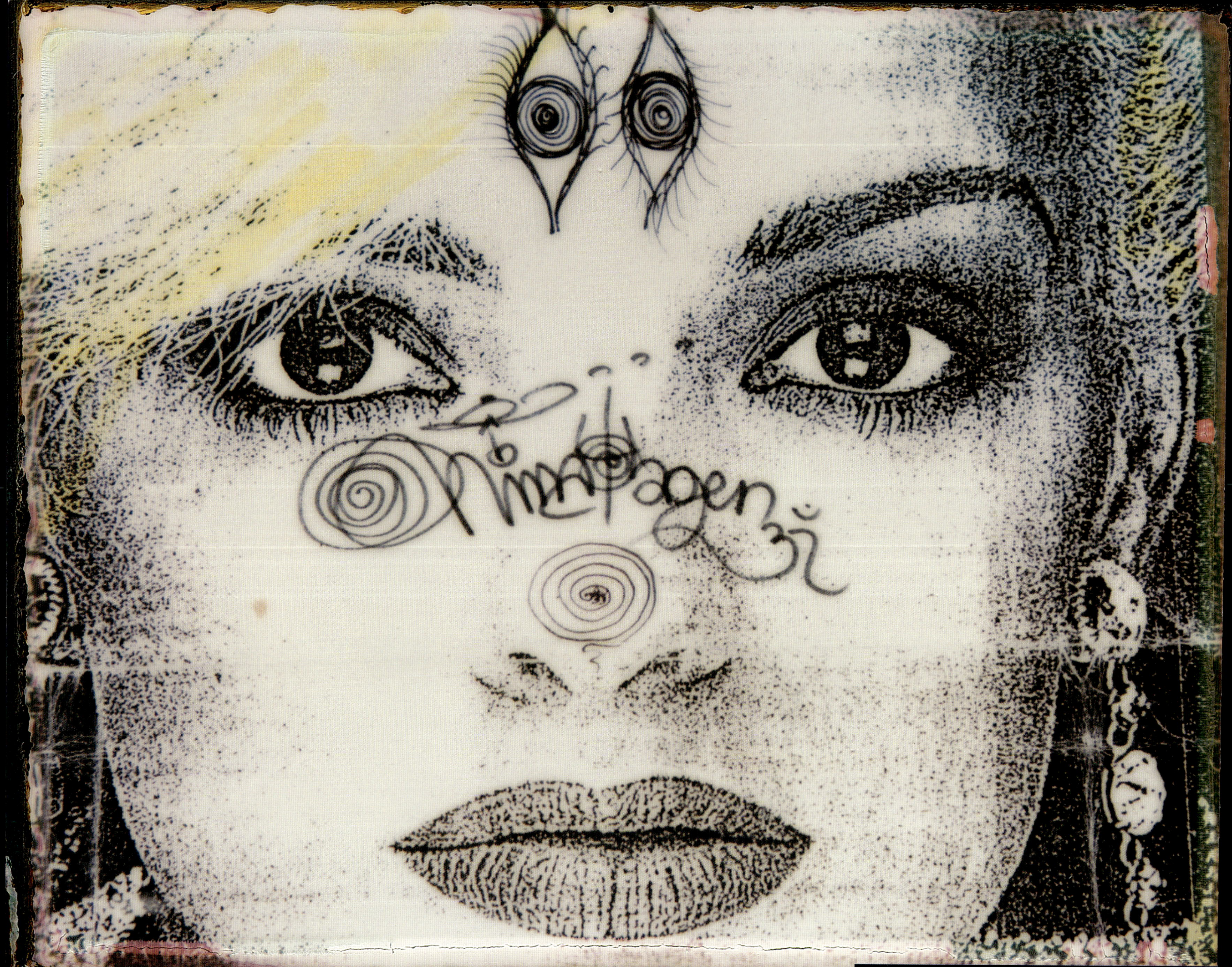

Nina Hagen, musician

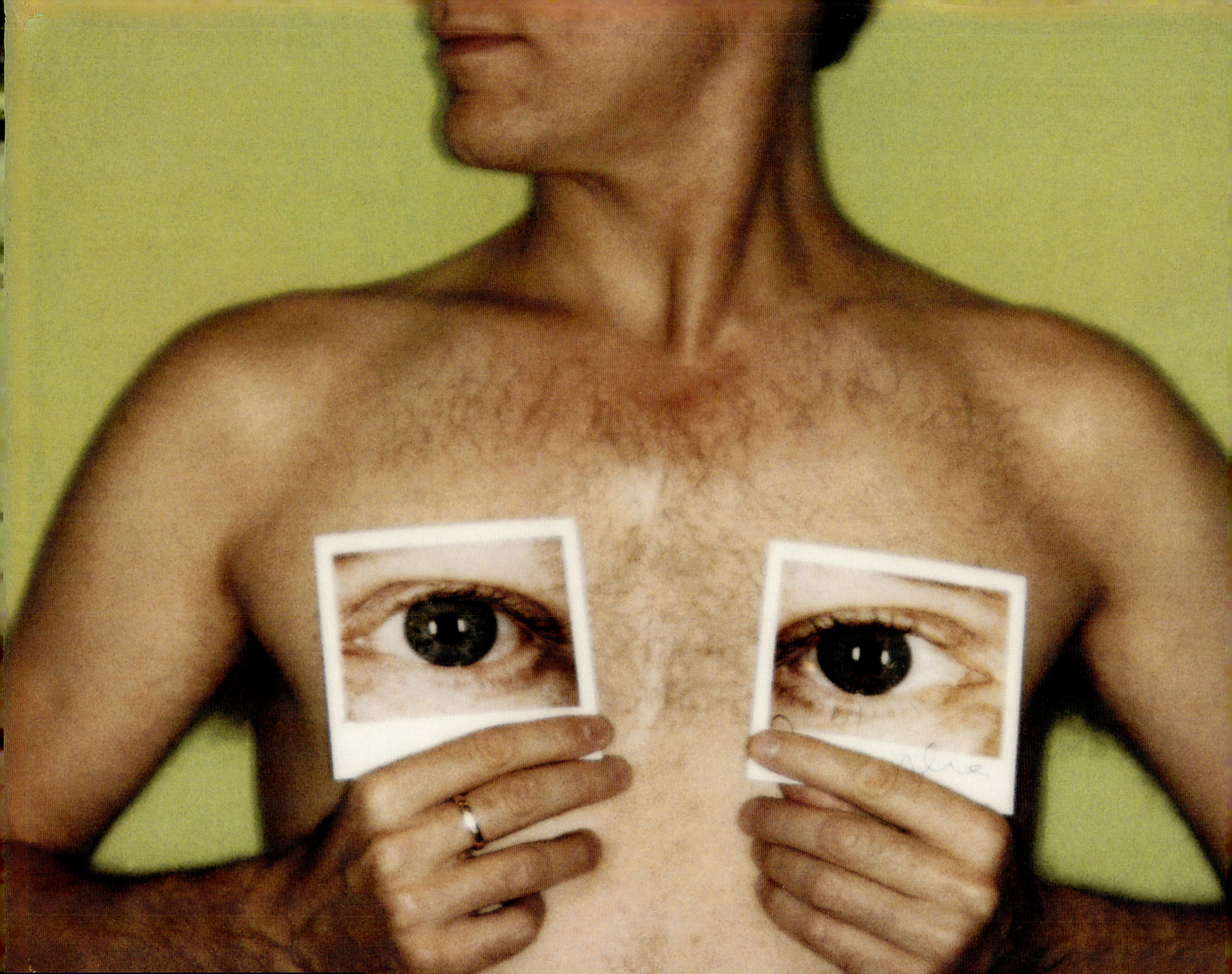

Bryan Adams, musician

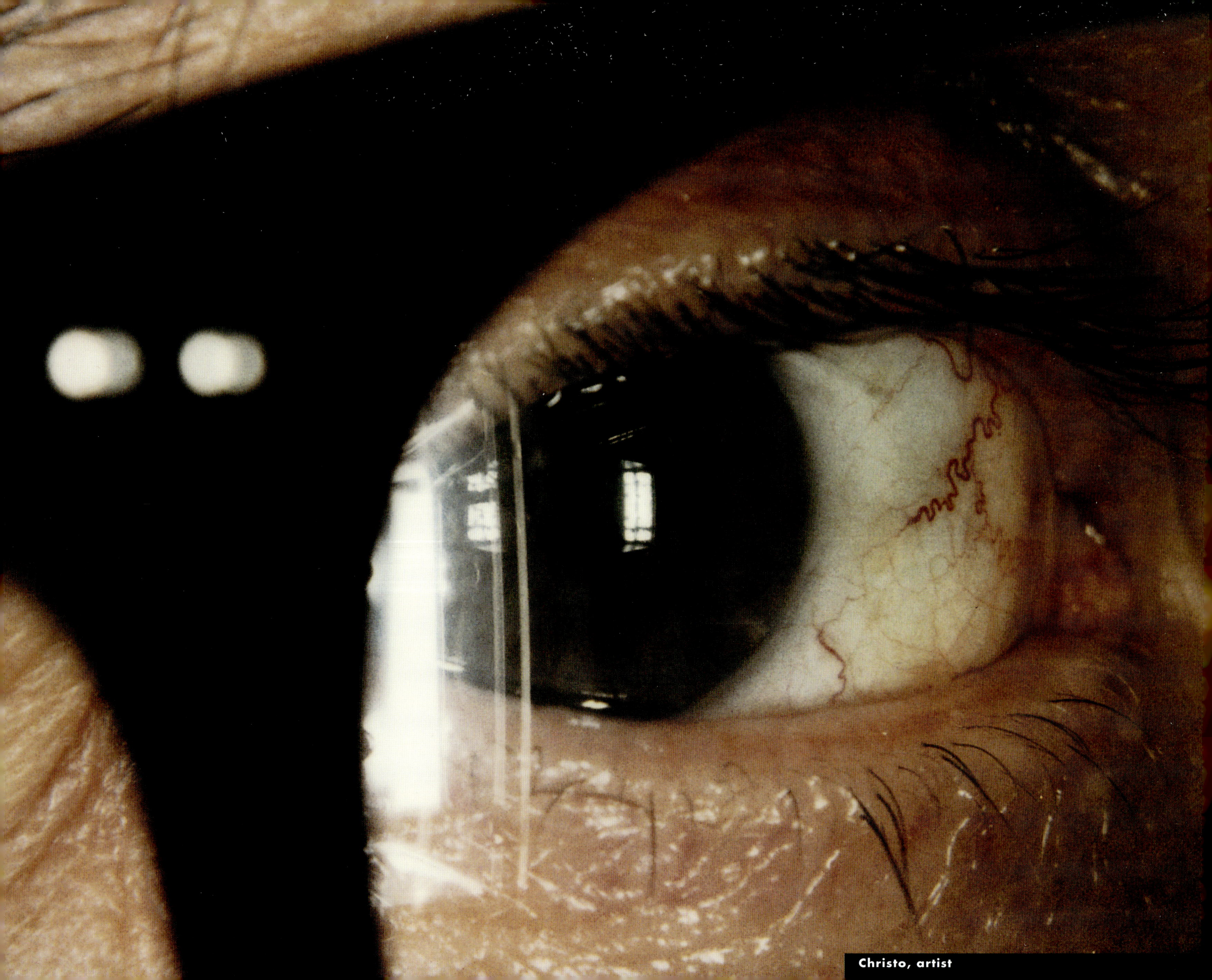

Christo, artist

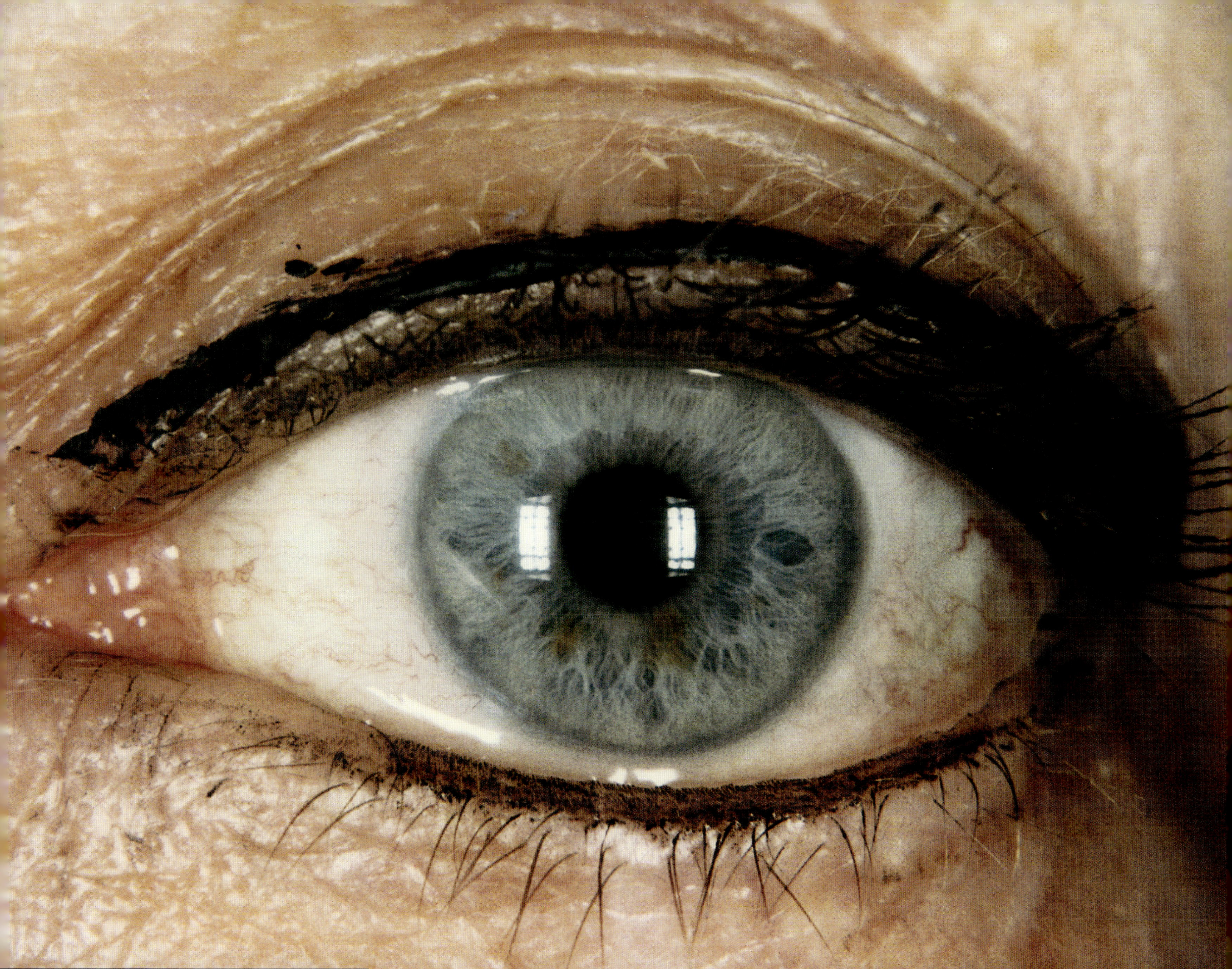

Jeanne-Claude, artist

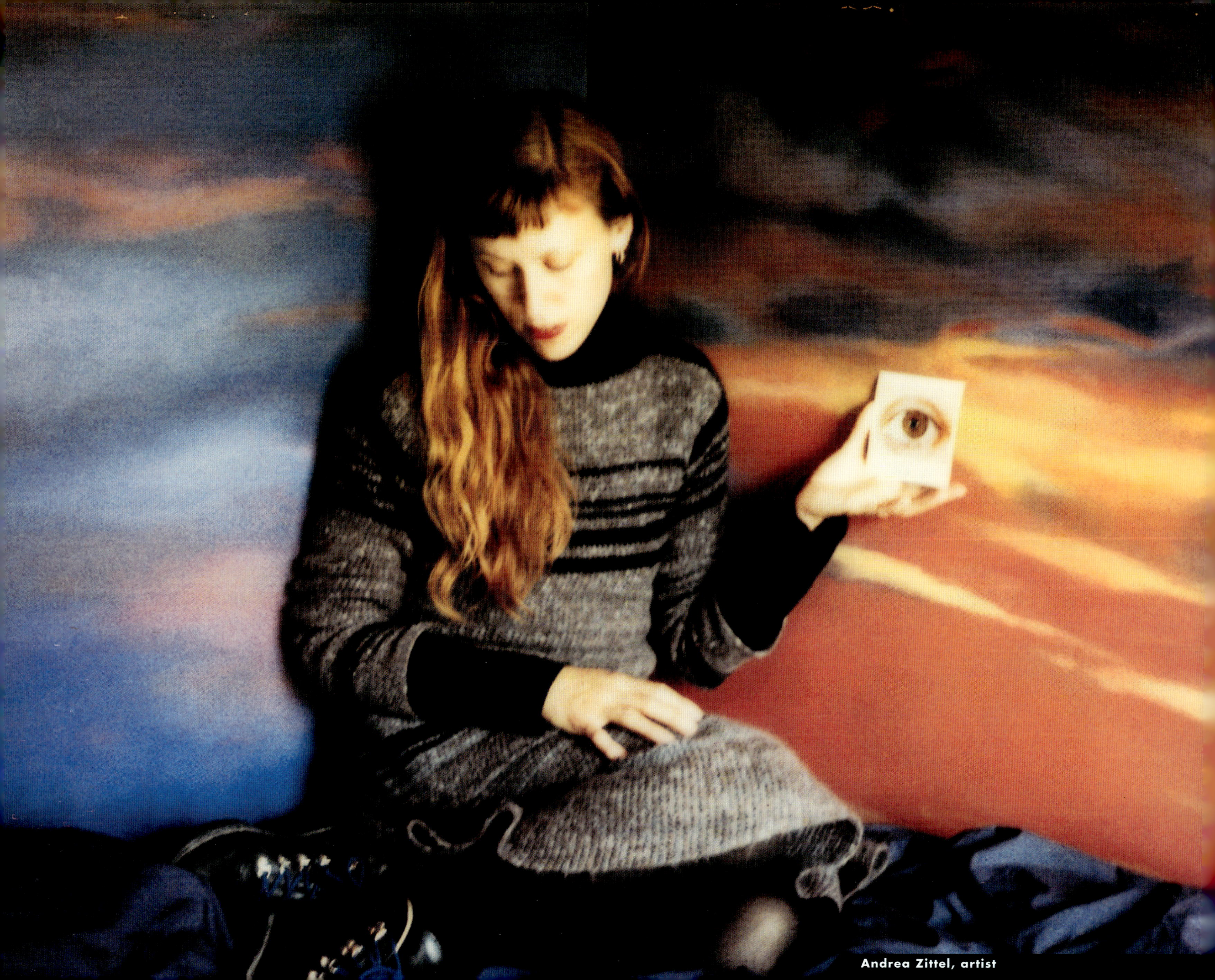

Andrea Zittel, artist

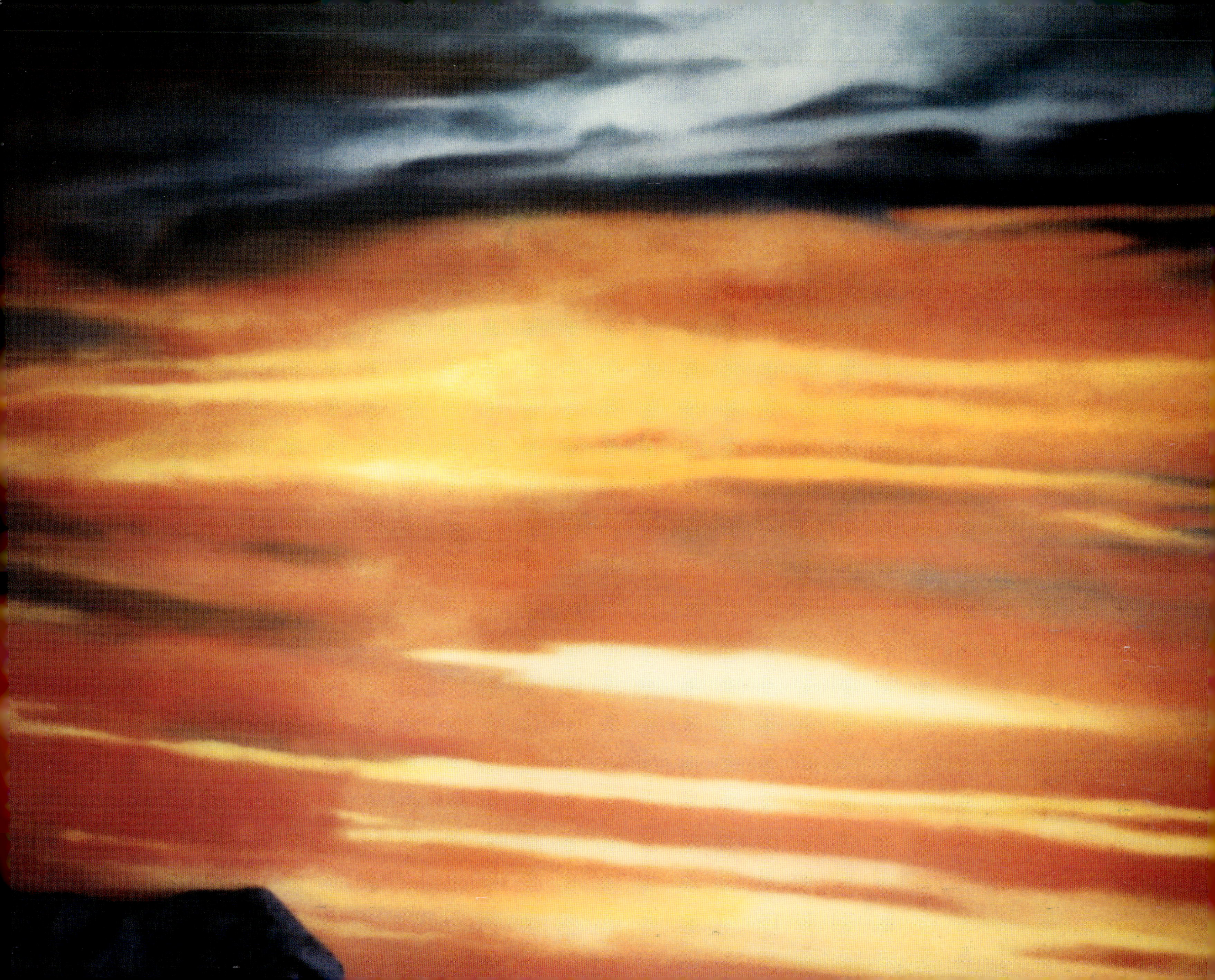

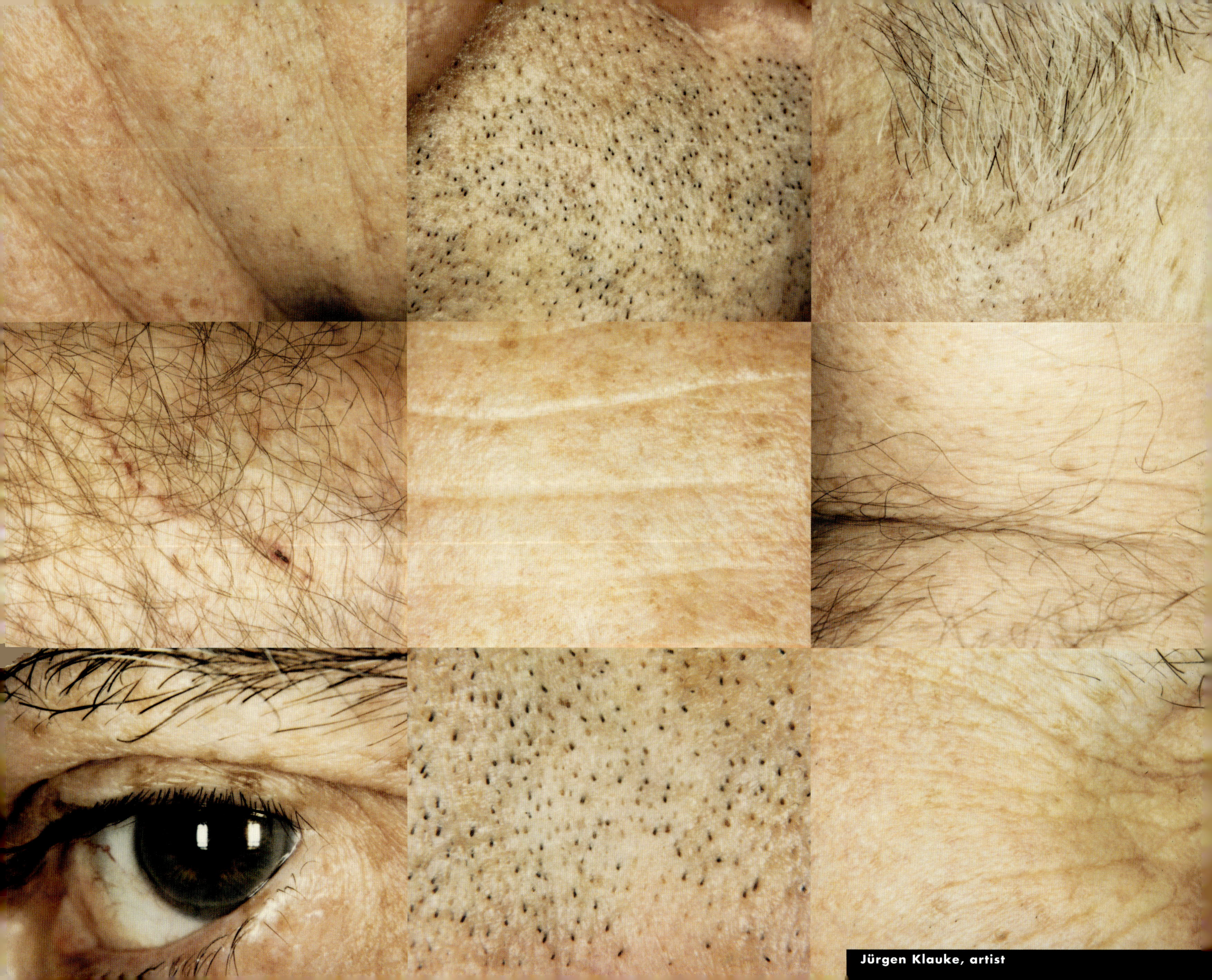

Jürgen Klauke, artist

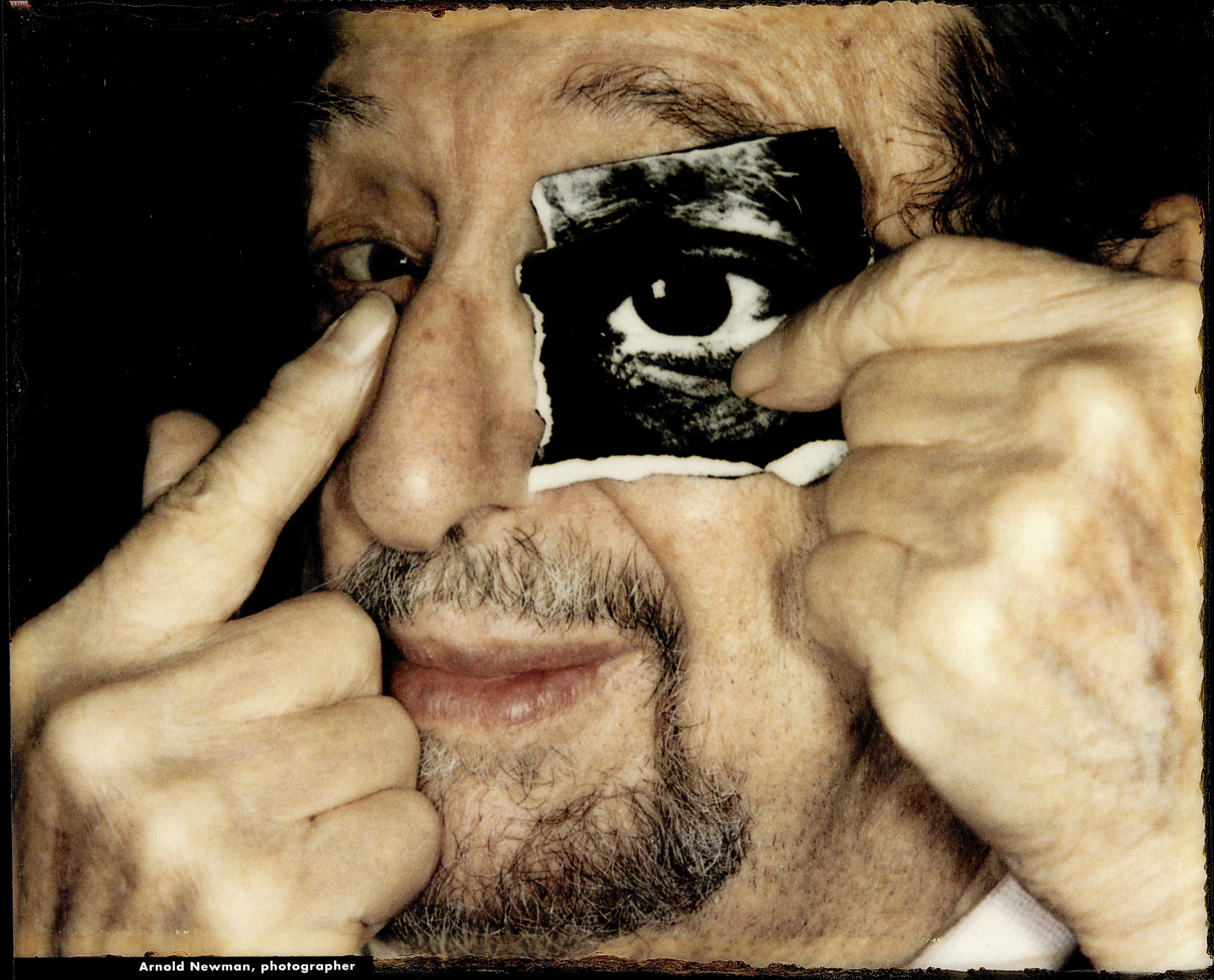

Arnold Newman, photographer

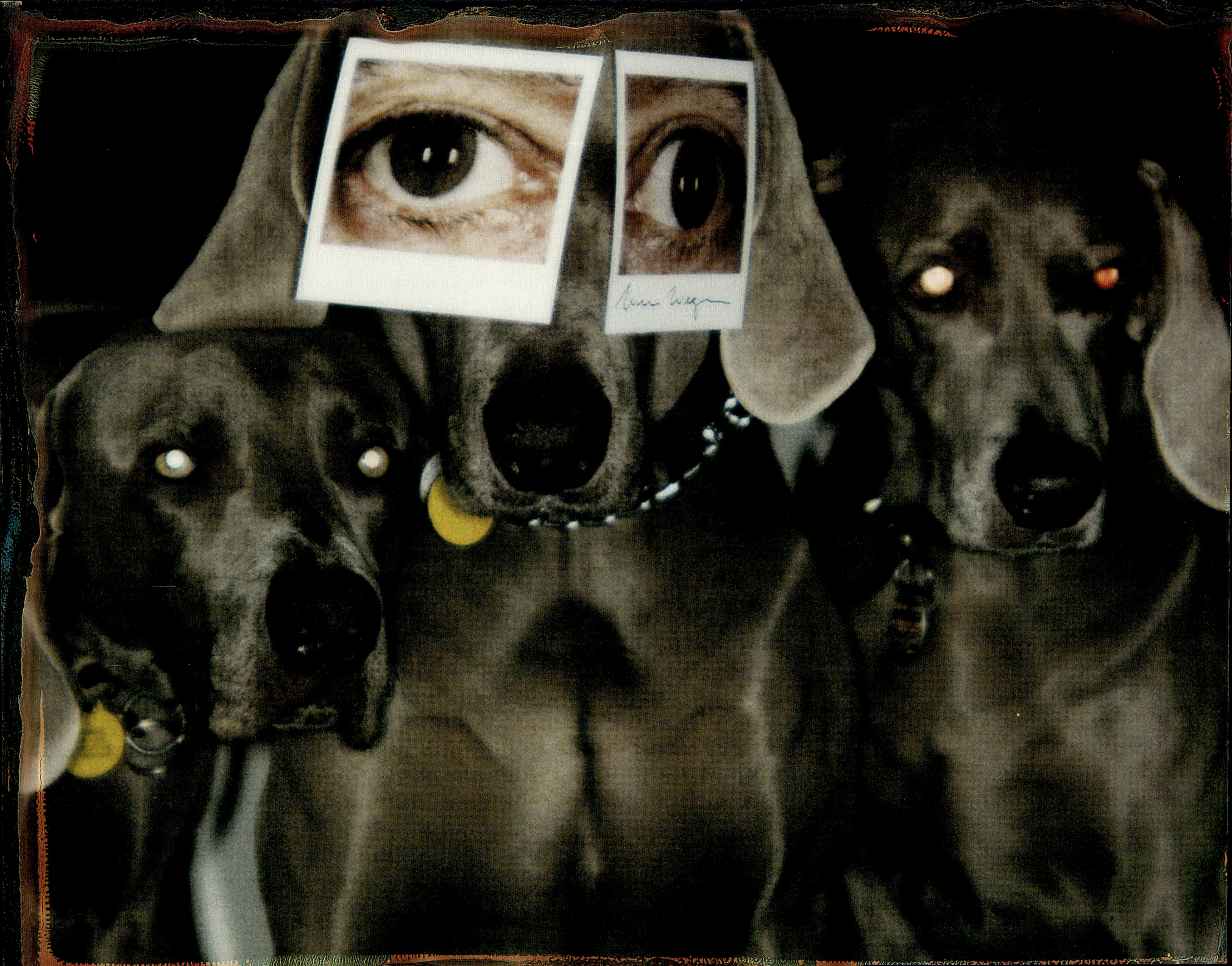

William Wegman, artist

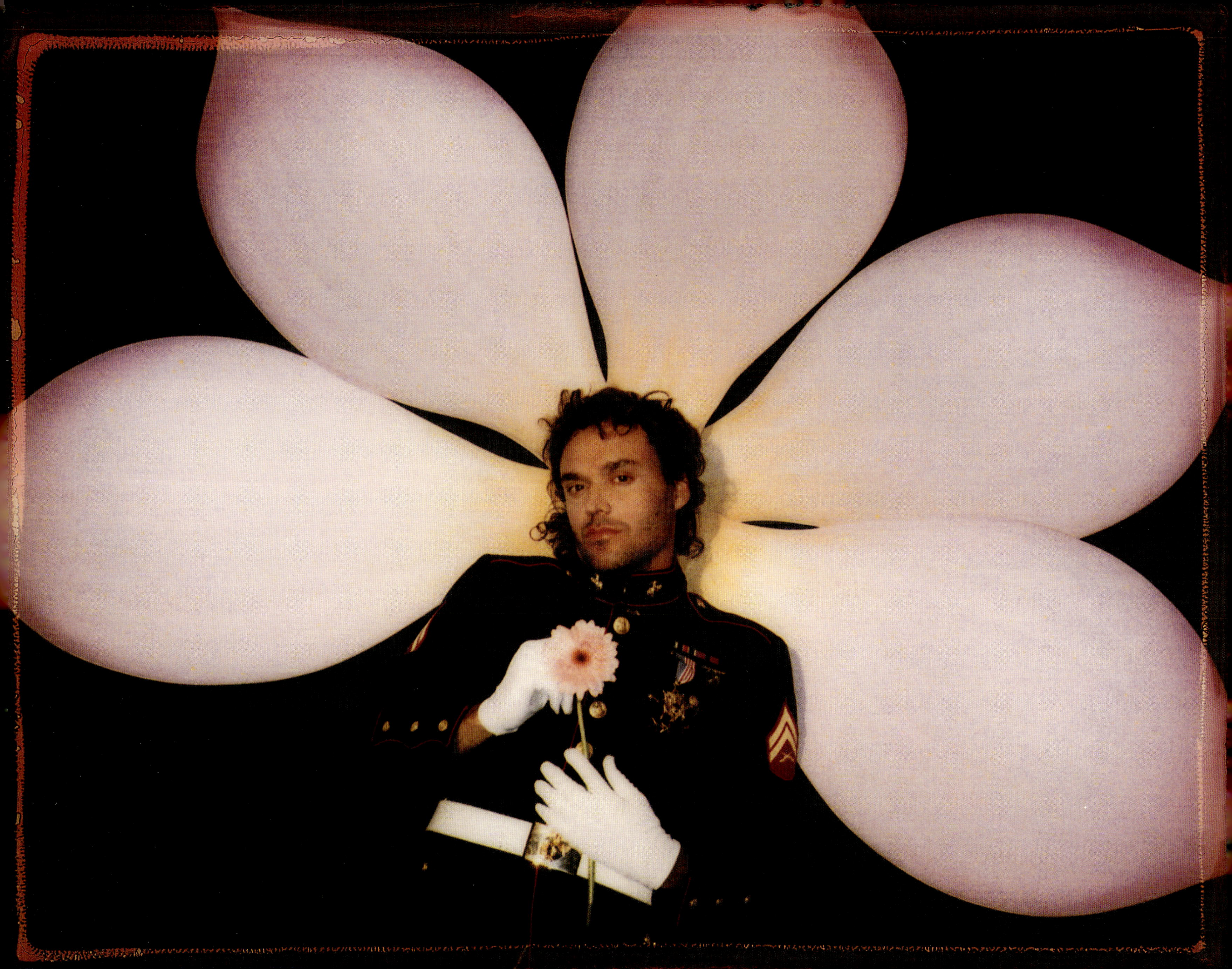

David LaChapelle, photographer

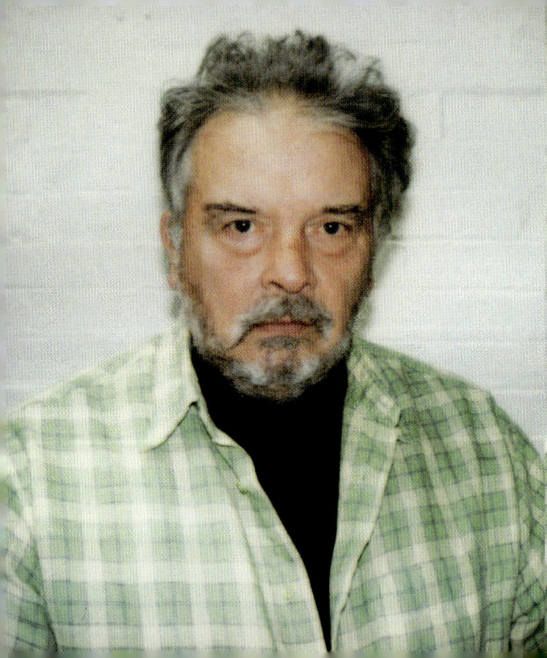
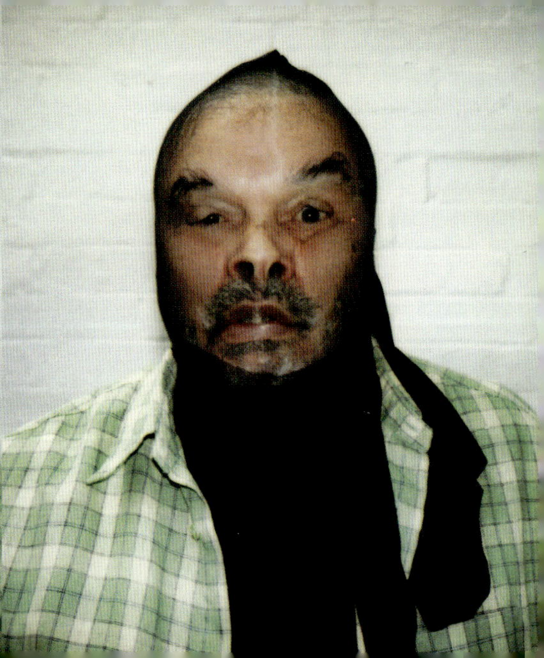

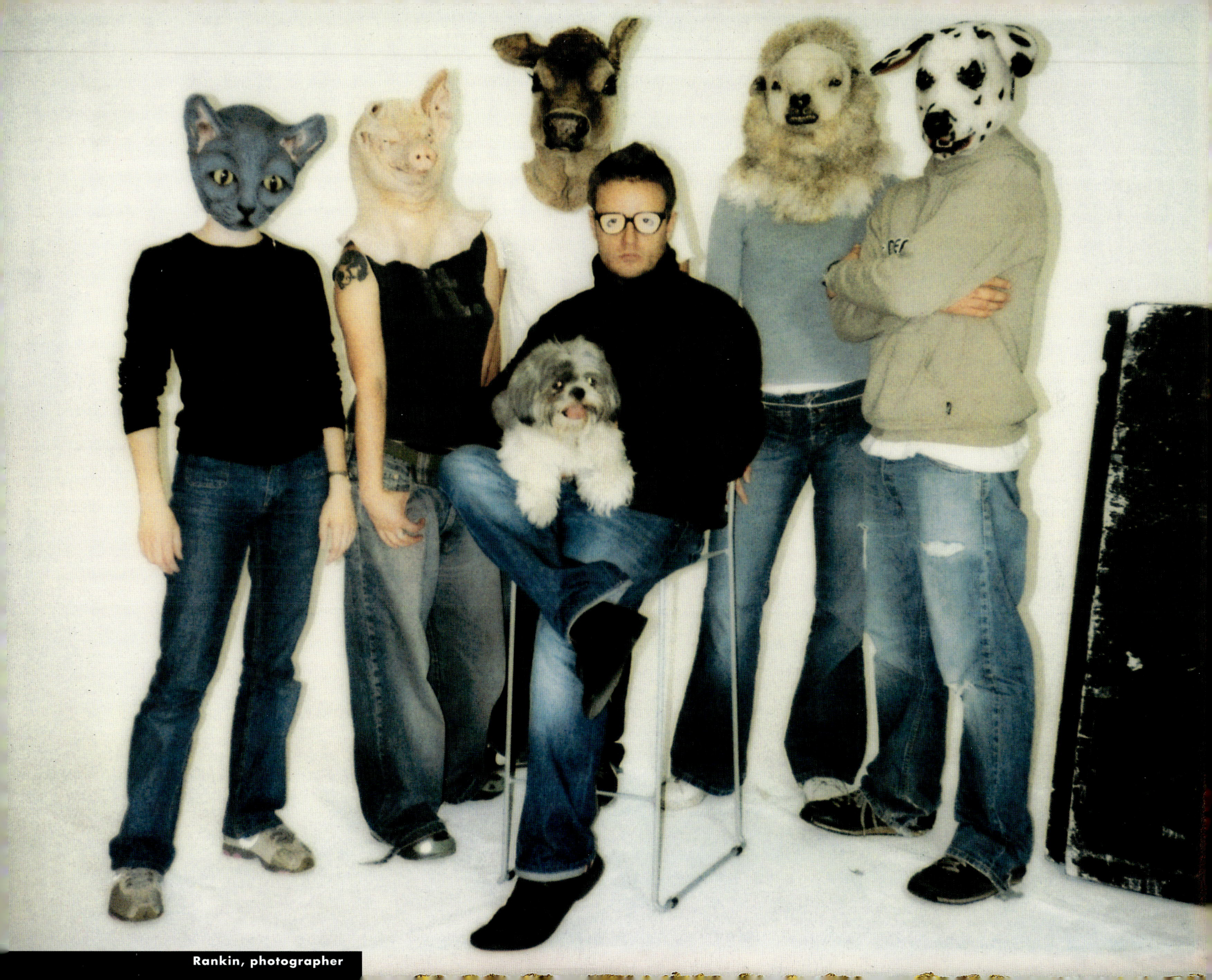

Rankin, photographer

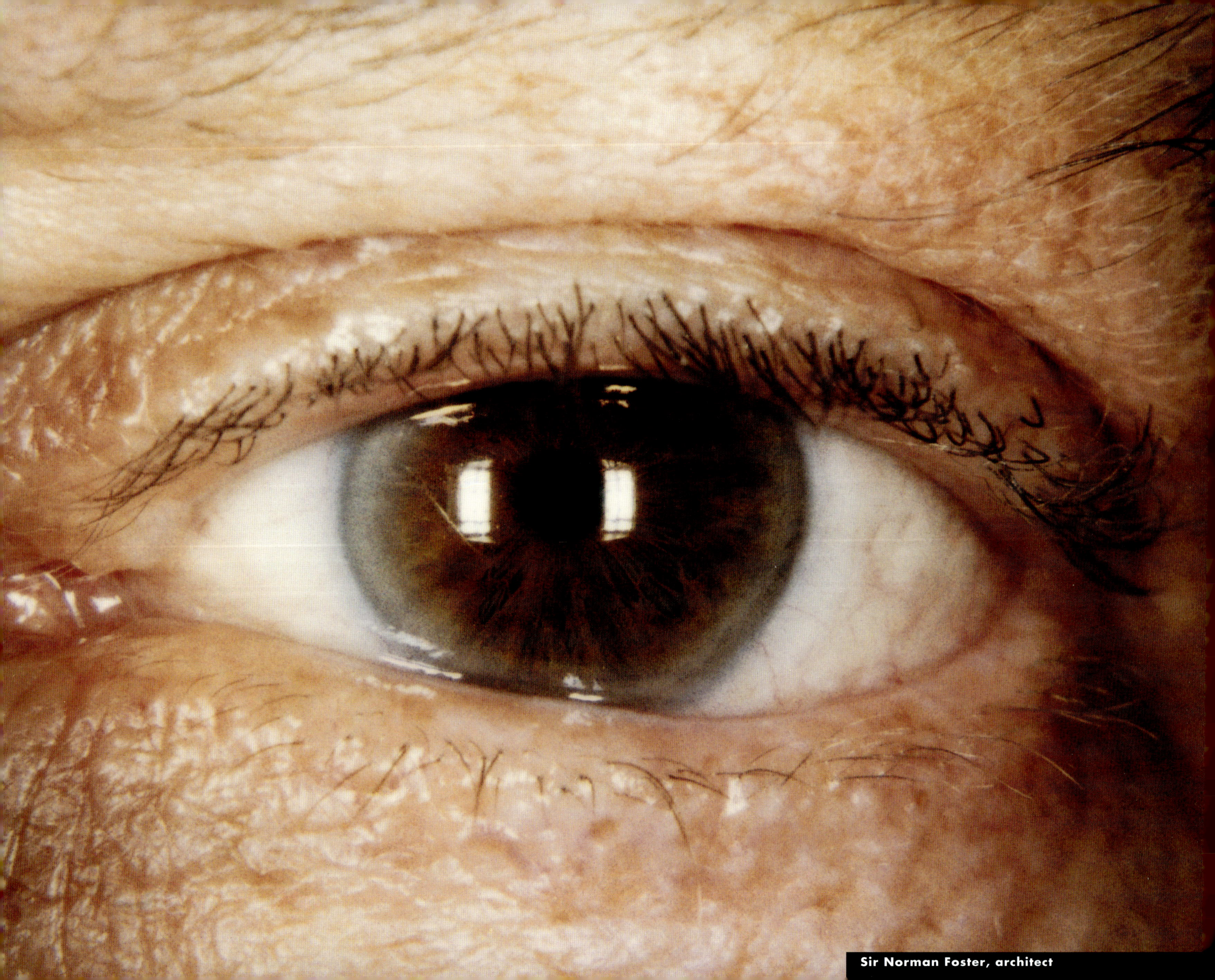

Sir Norman Foster, architect

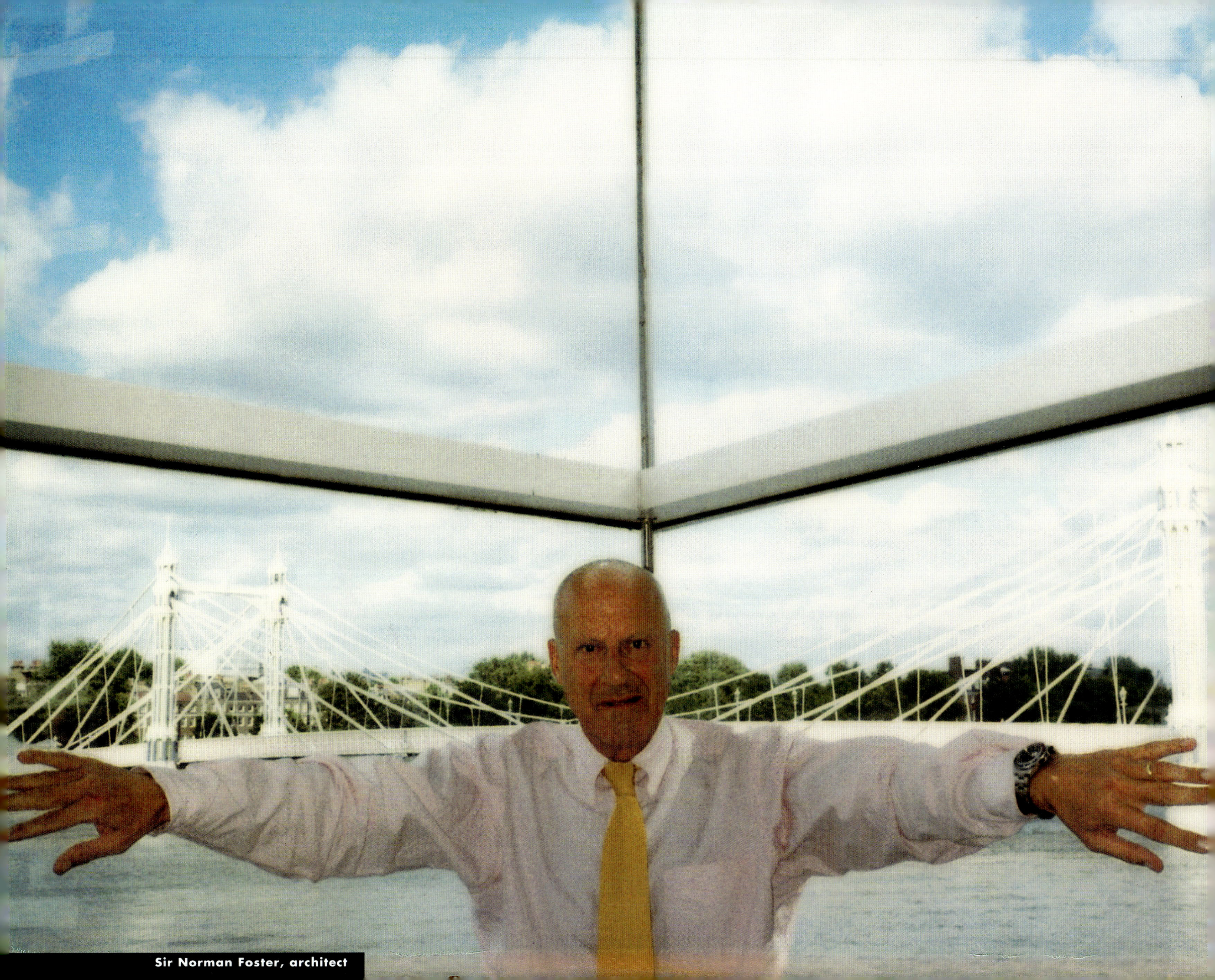

Sir Norman Foster, architect

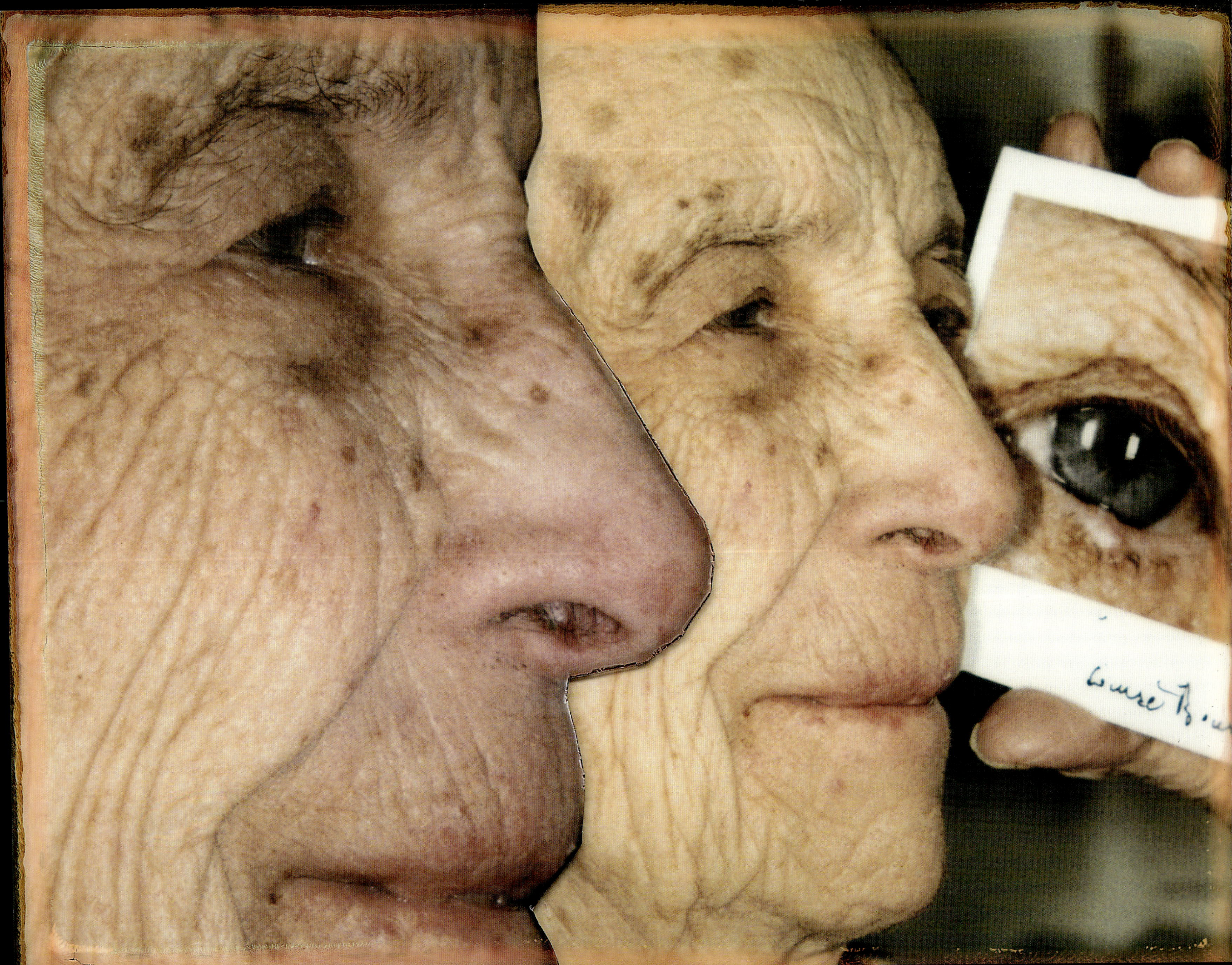

Louise Bourgeois, artist

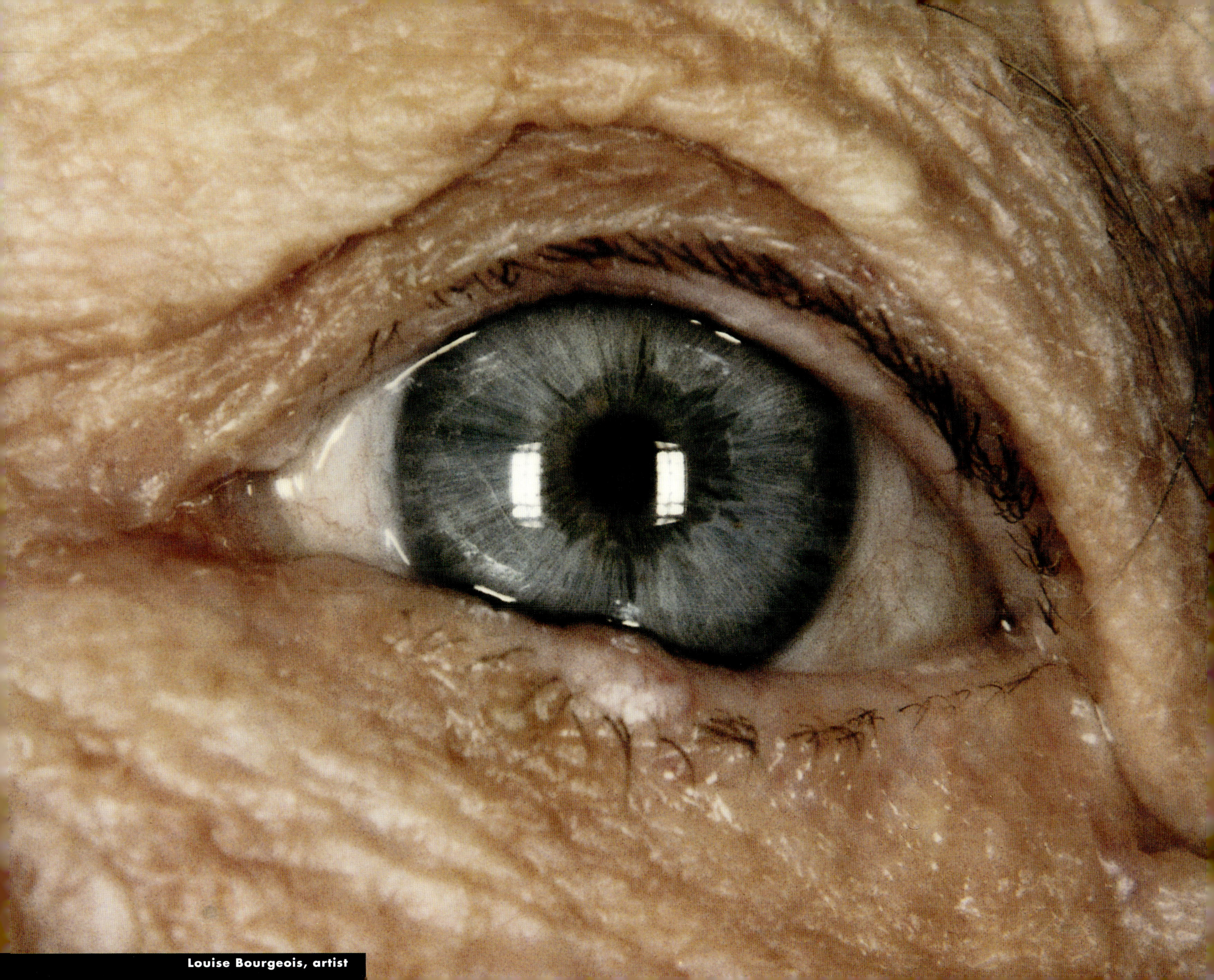

Louise Bourgeois, artist

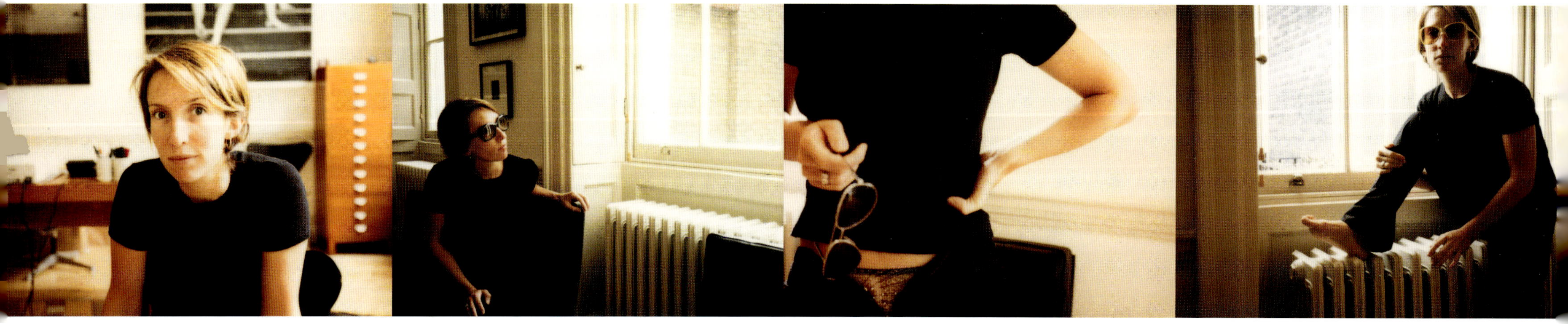

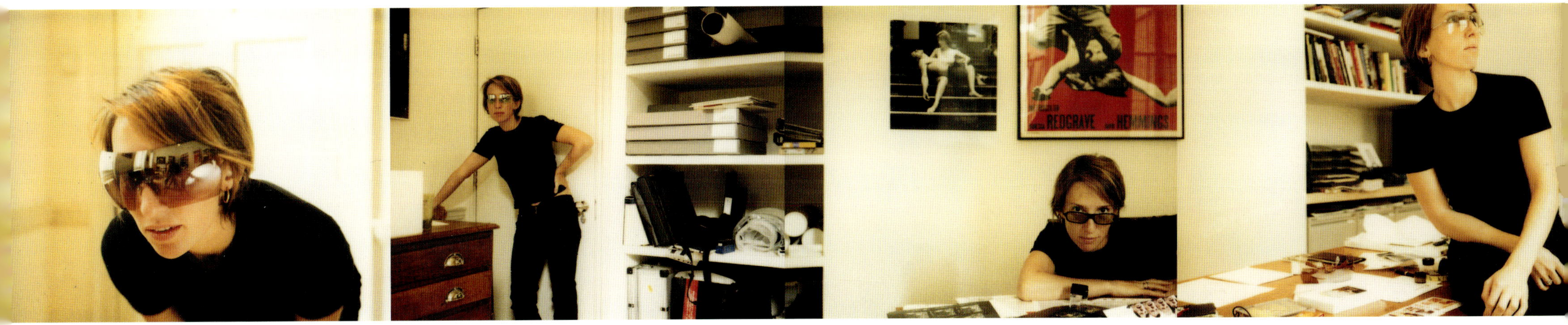

Sam Taylor-Wood, artist

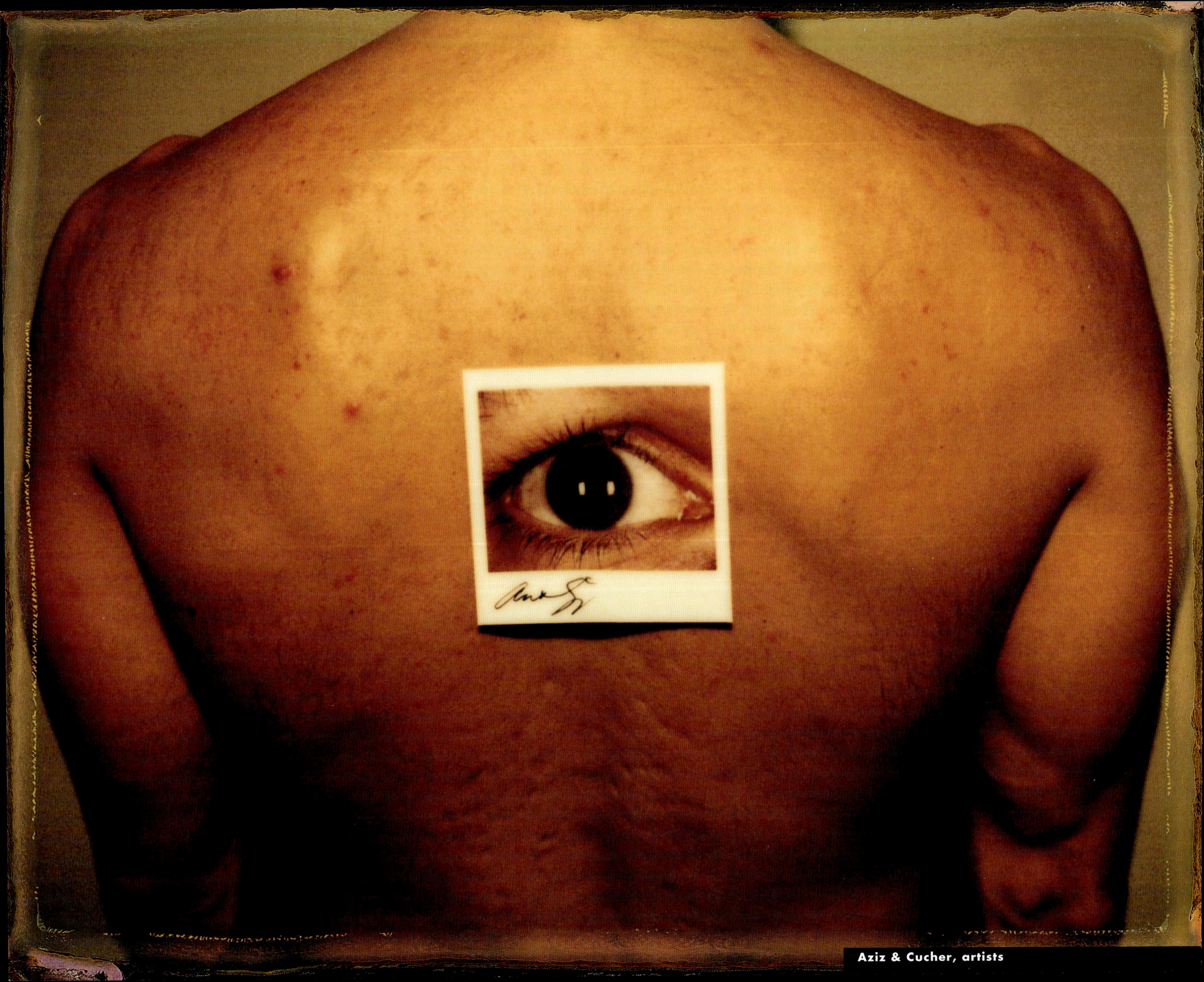

Aziz & Cucher, artists

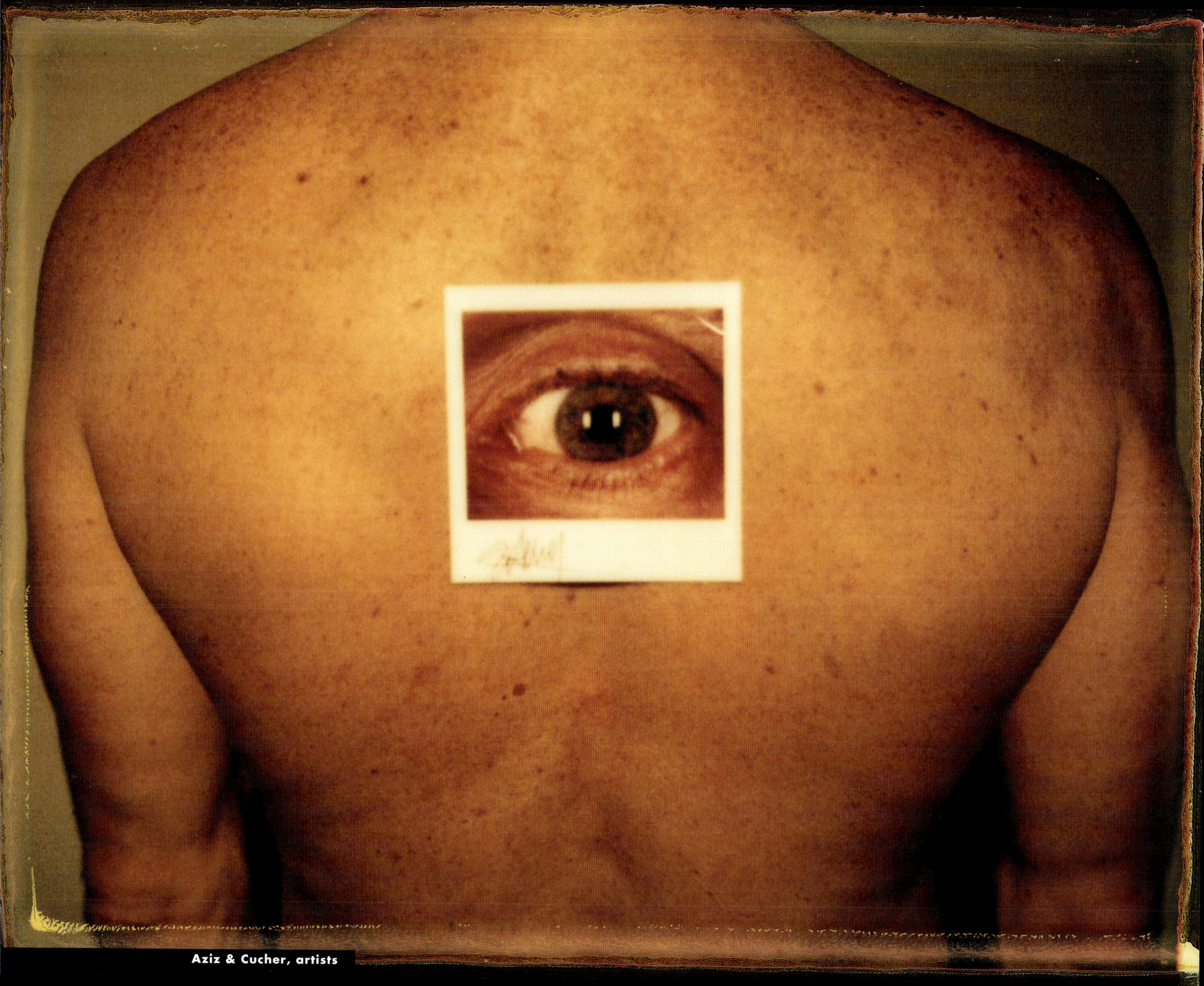

Aziz & Cucher, artists

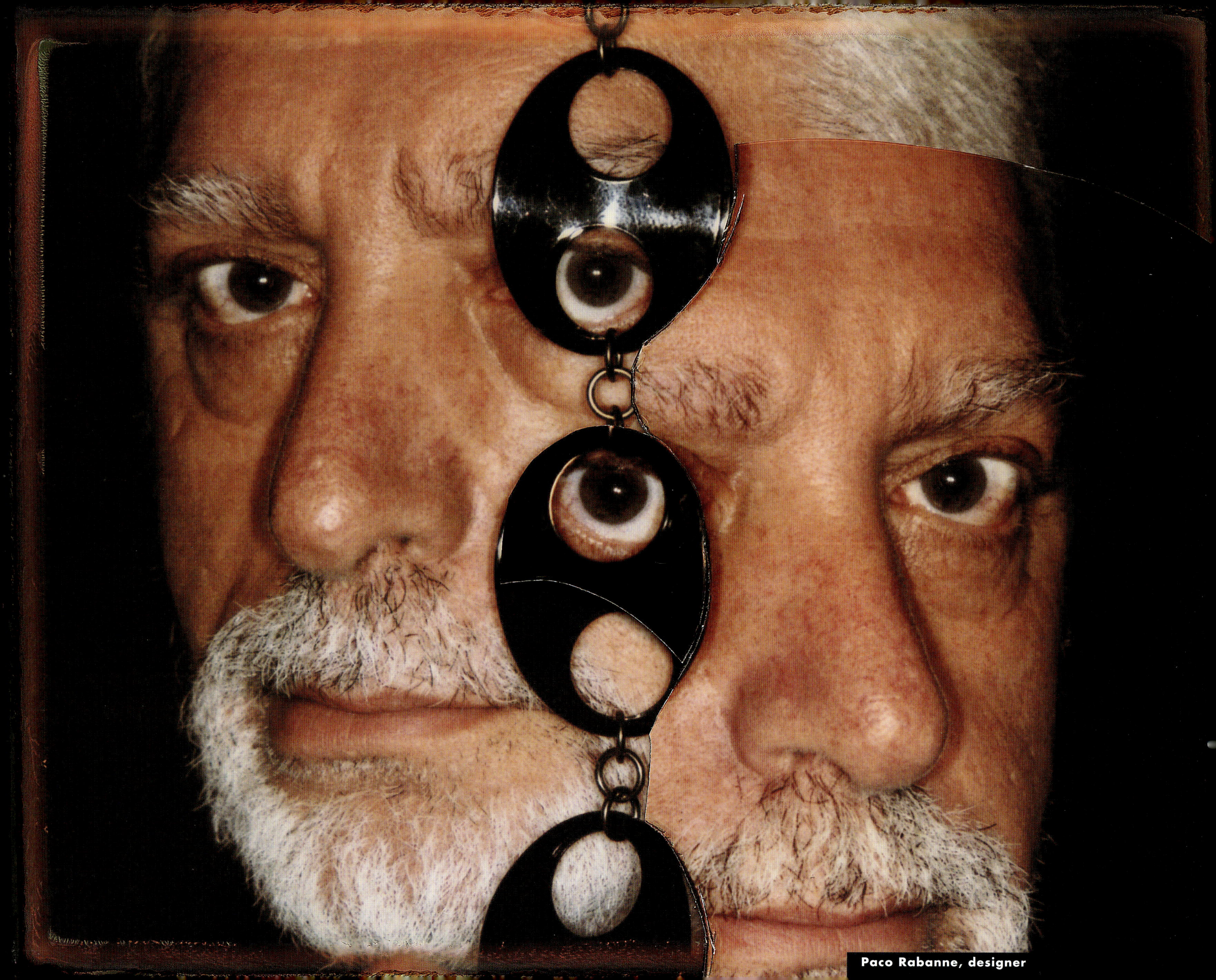

Paco Rabanne, designer

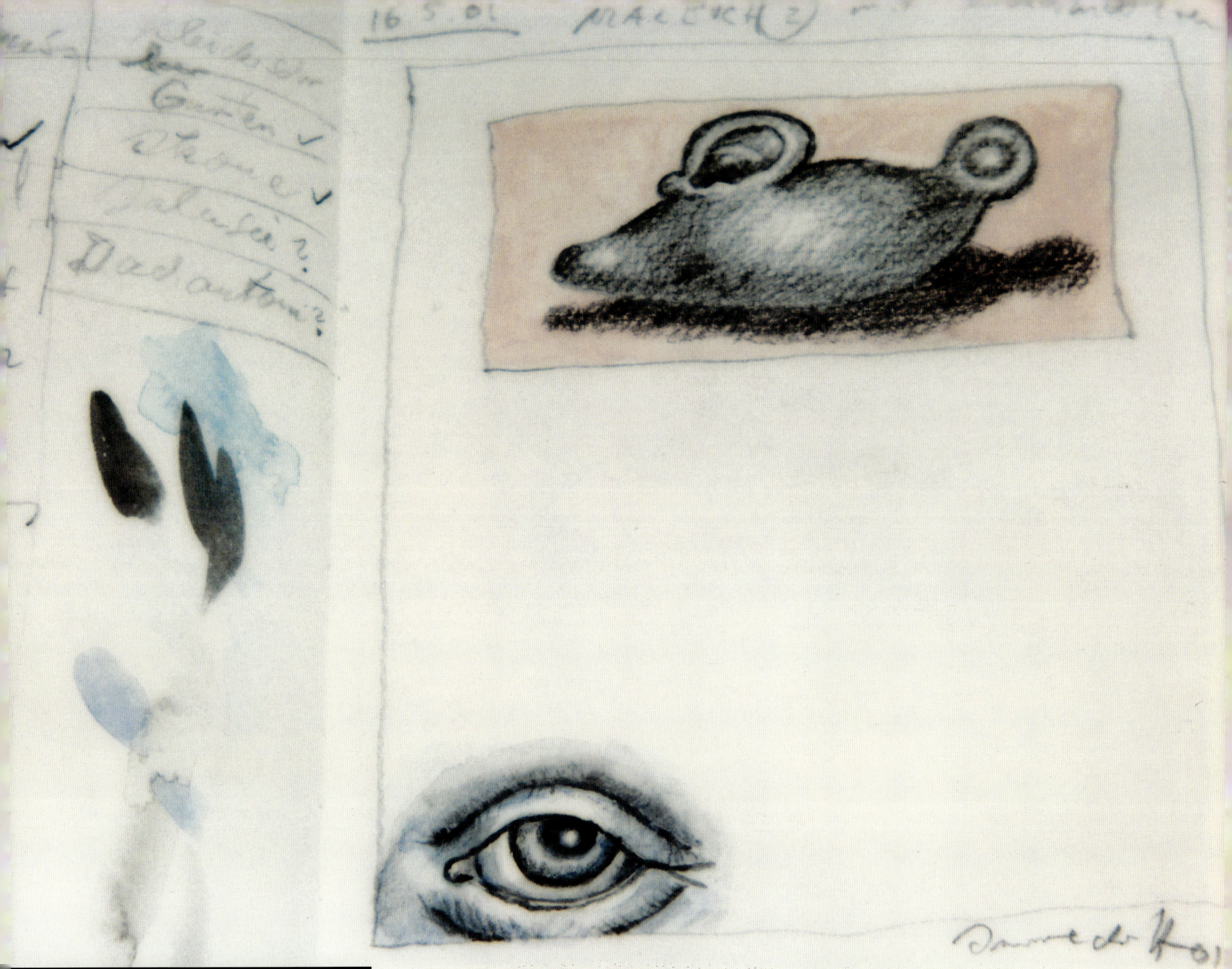

Jörg Immendorff, artist

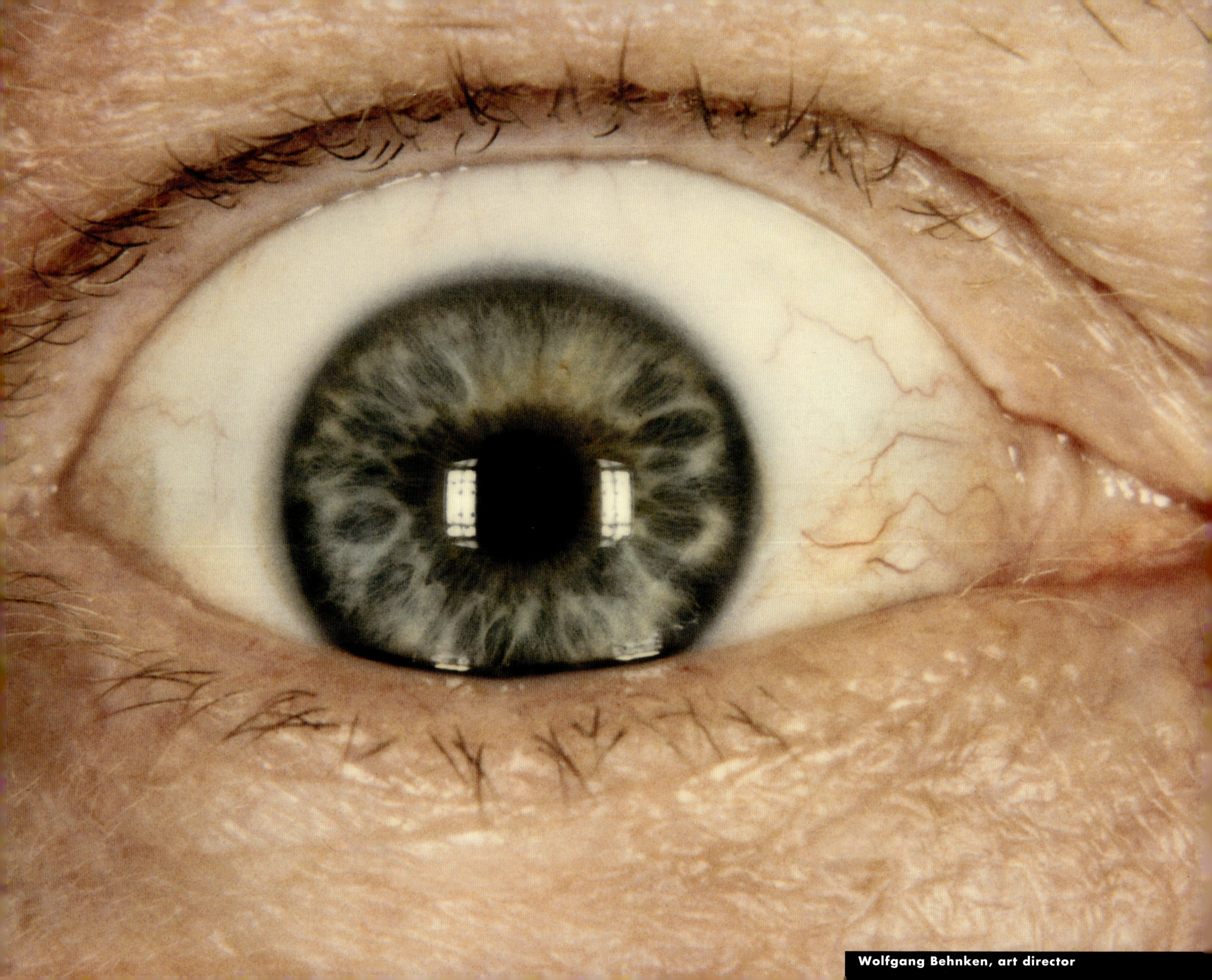

Wolfgang Behnken, art director

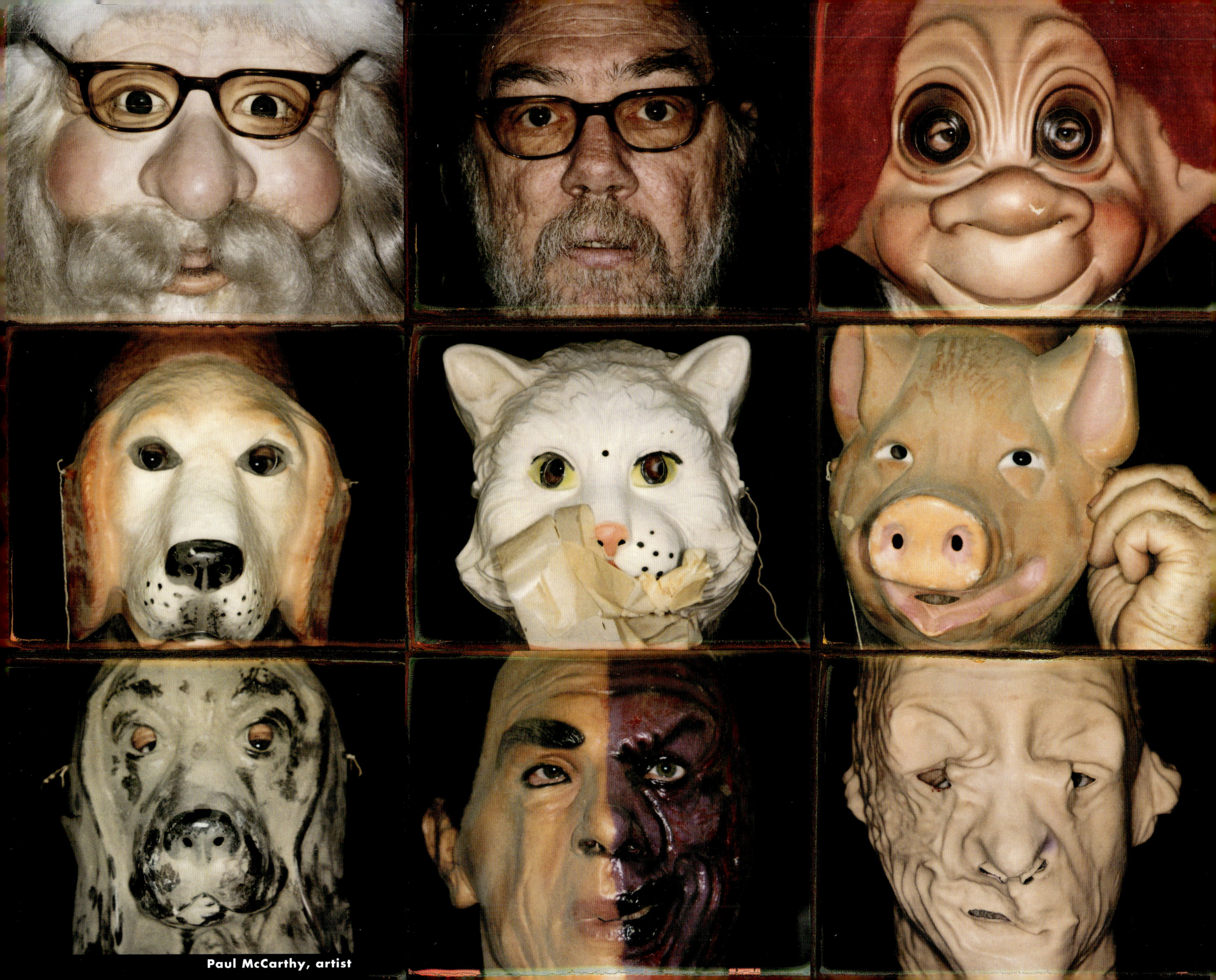

Paul McCarthy, artist

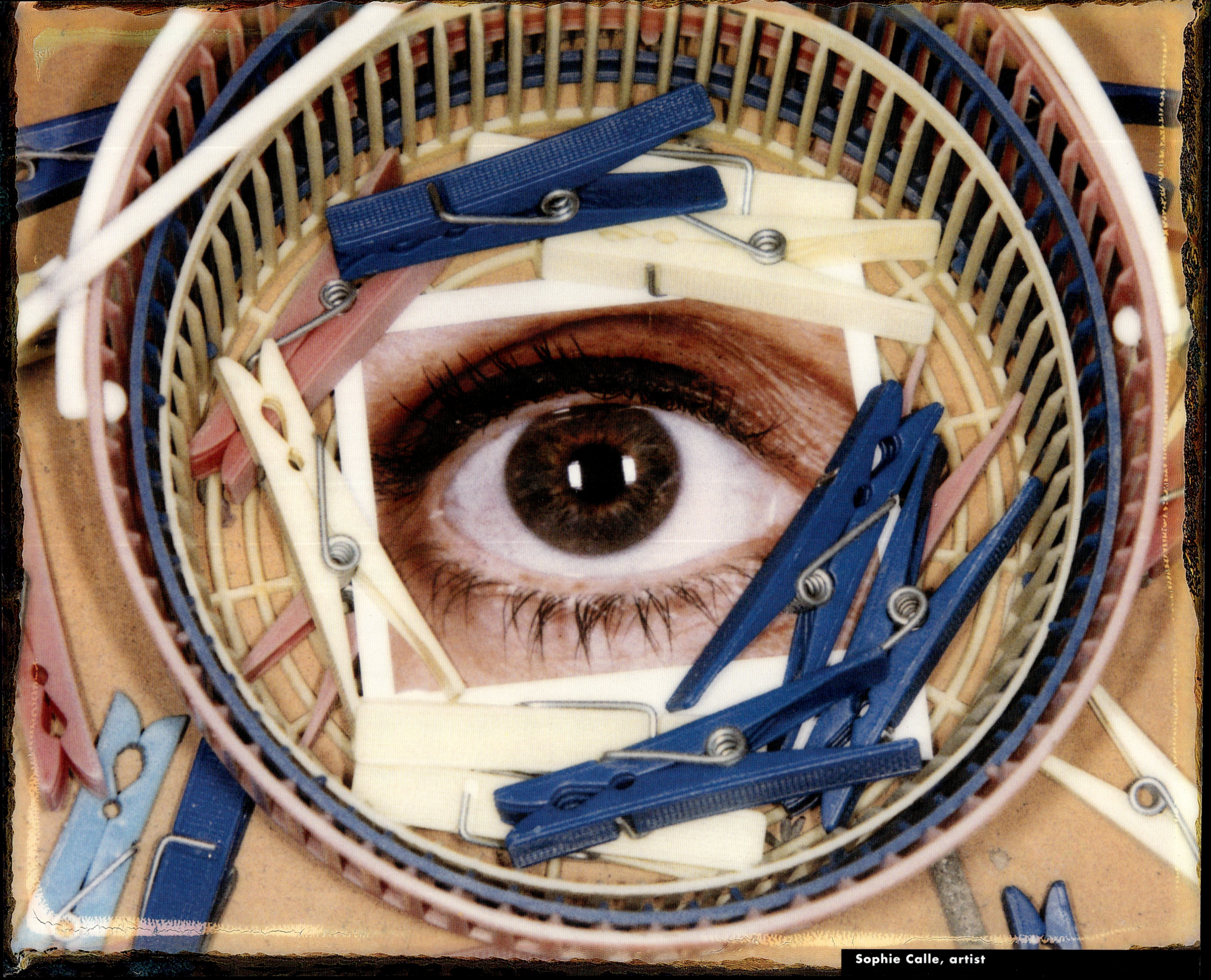

Sophie Calle, artist

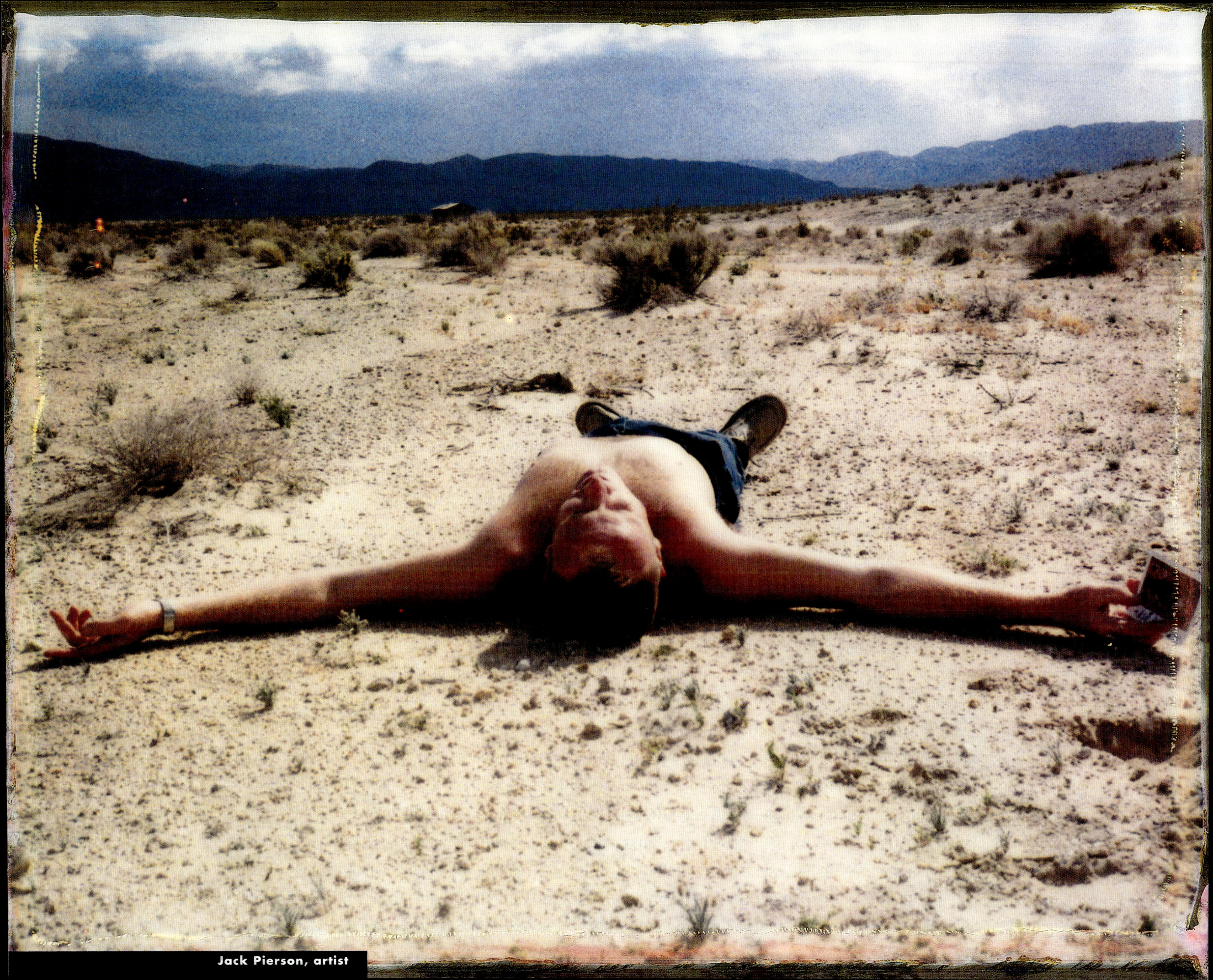
Jack Pierson, artist

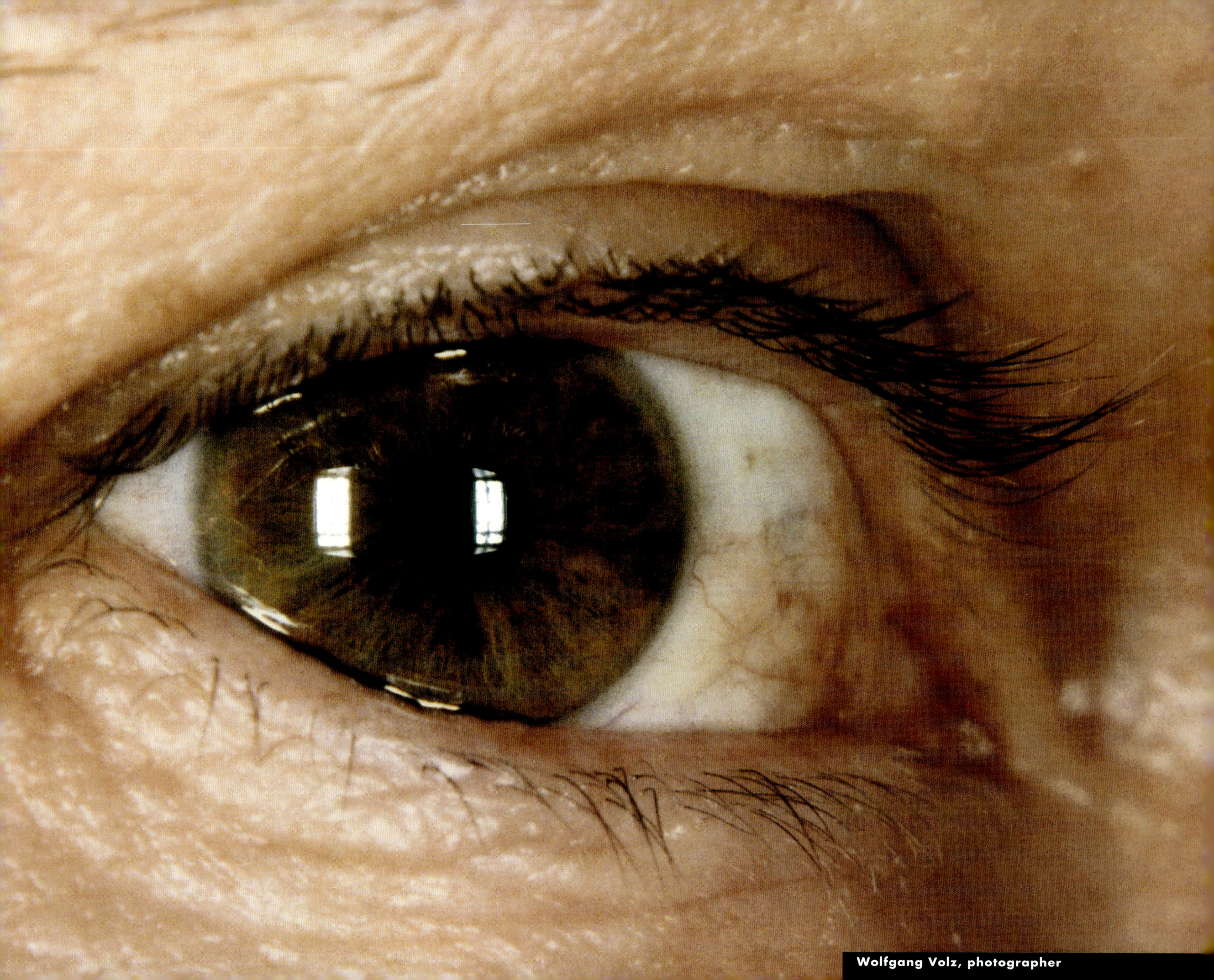

Wolfgang Volz, photographer

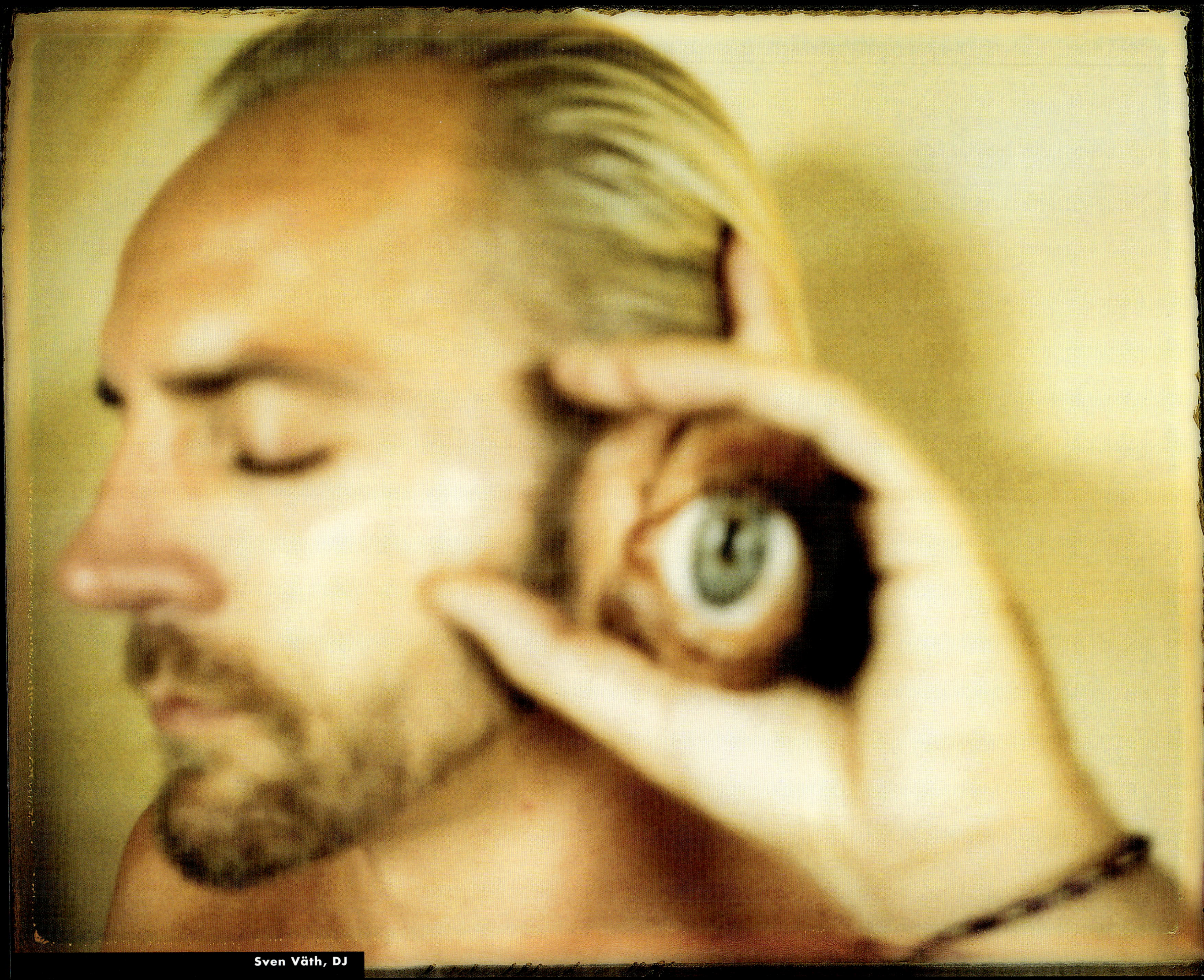

Sven Väth, DJ

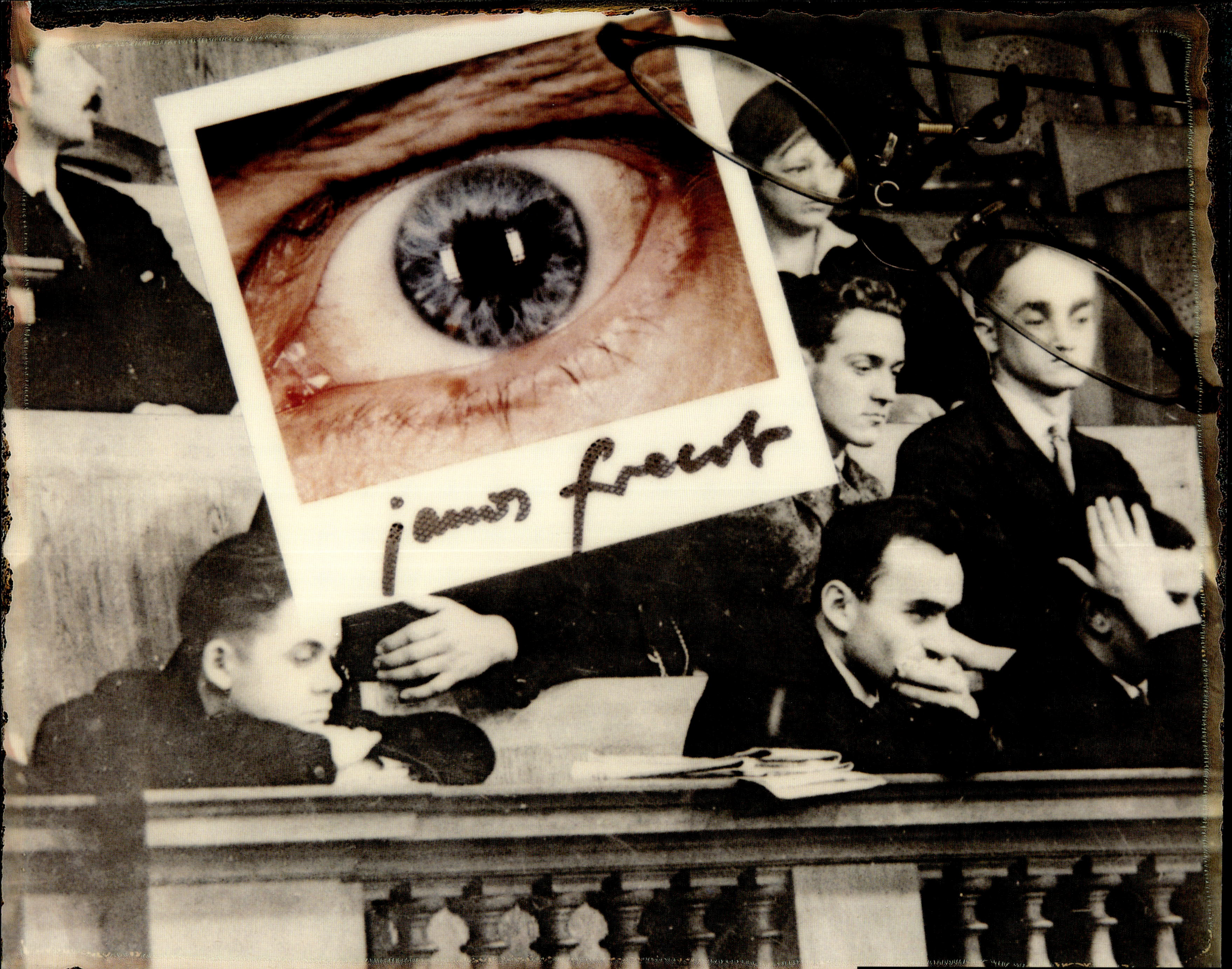

Janos Frecot, curator

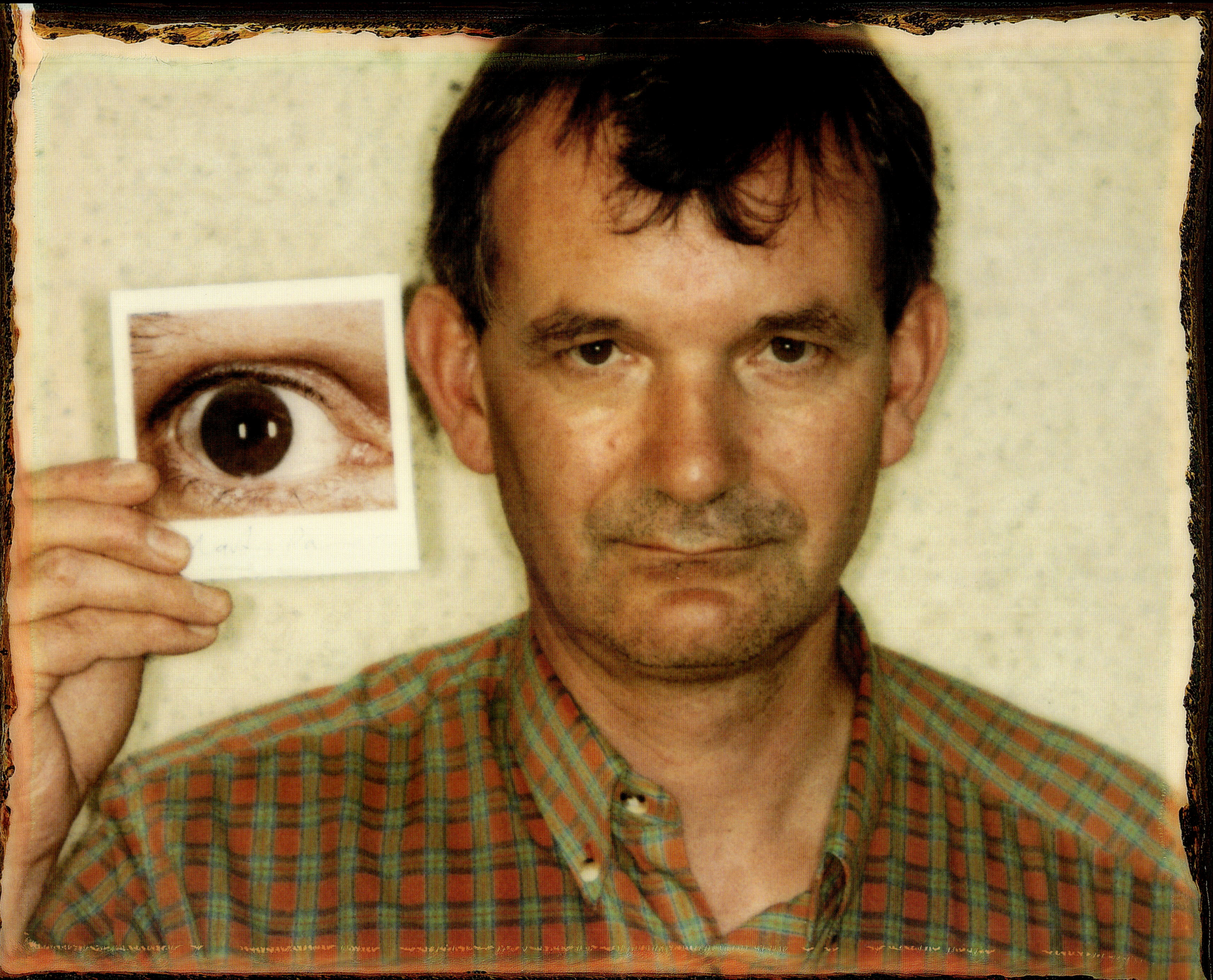

Martin Parr, photographer

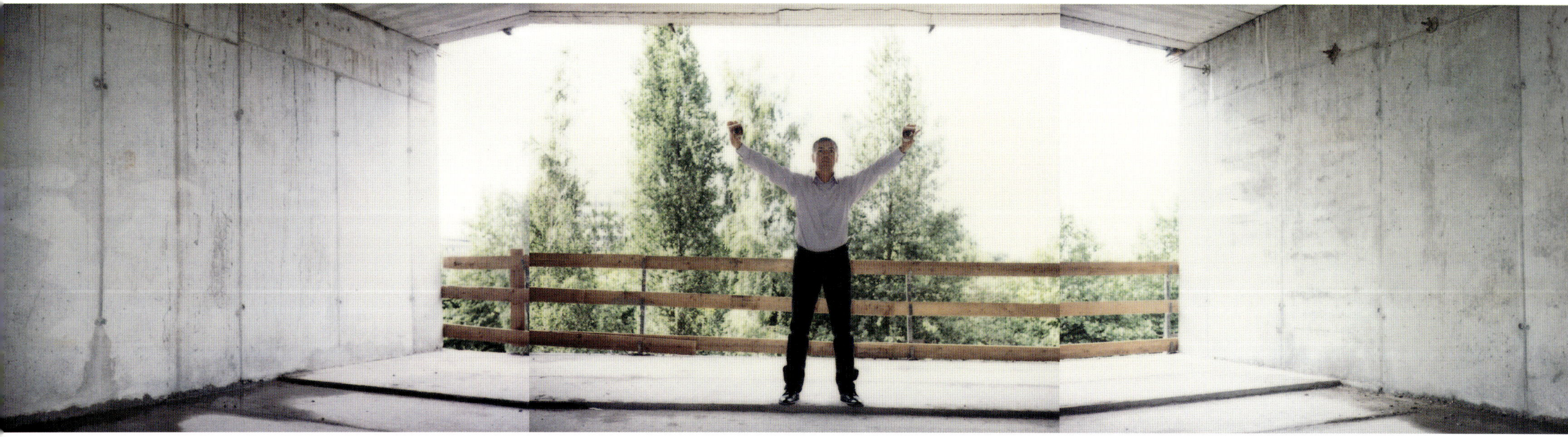

Andreas Gursky, artist

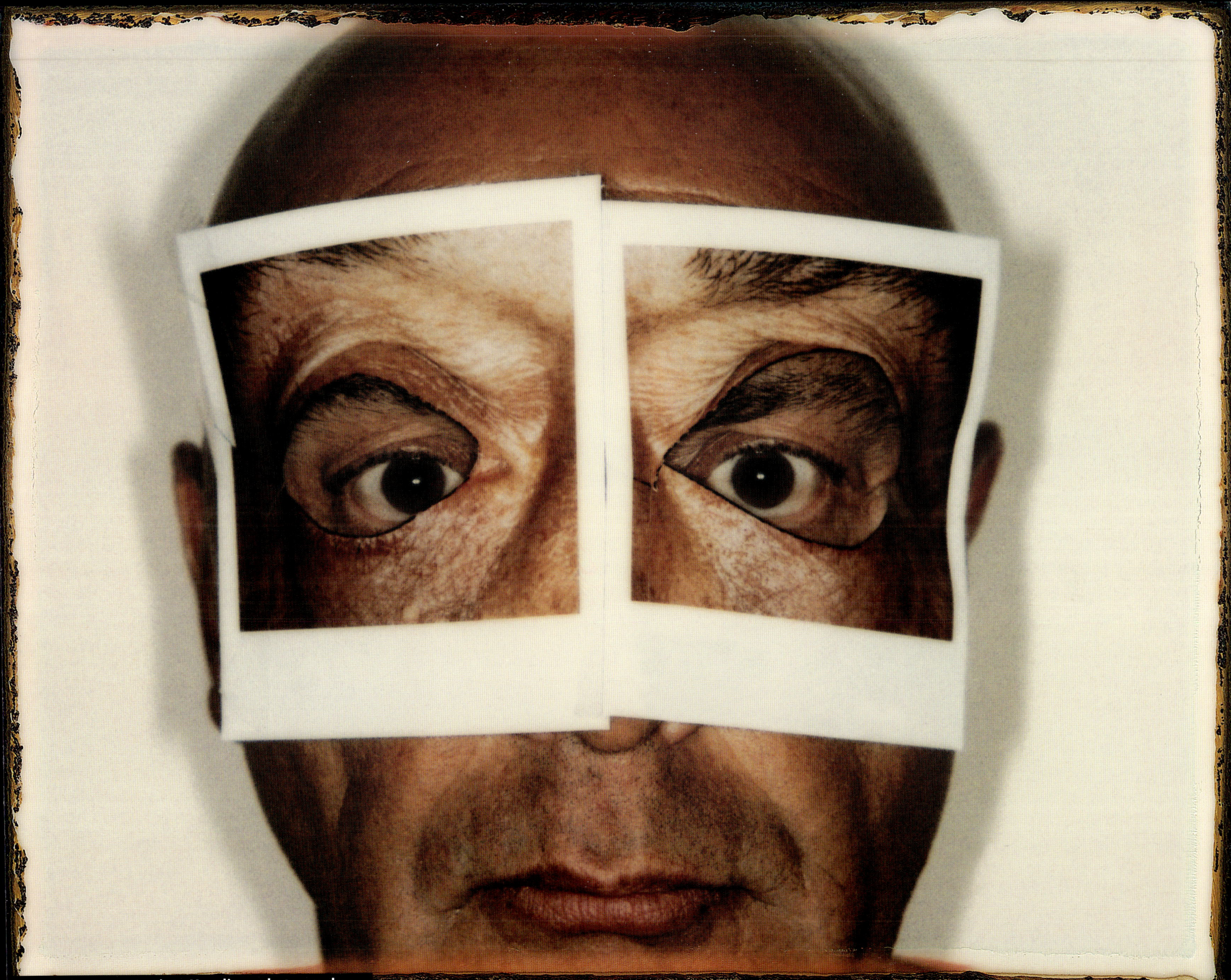

Jean-Baptiste Mondino, photographer

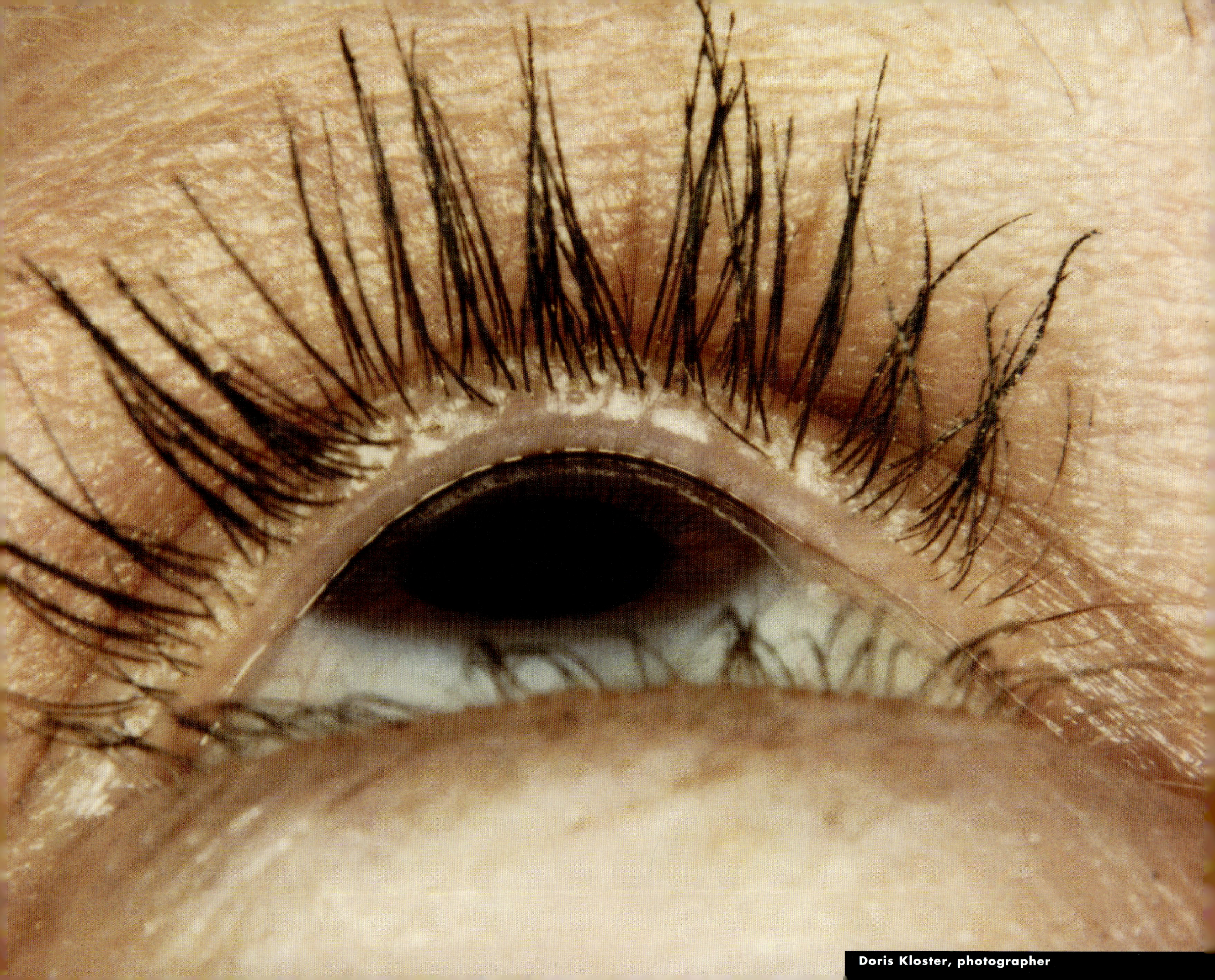

Doris Kloster, photographer

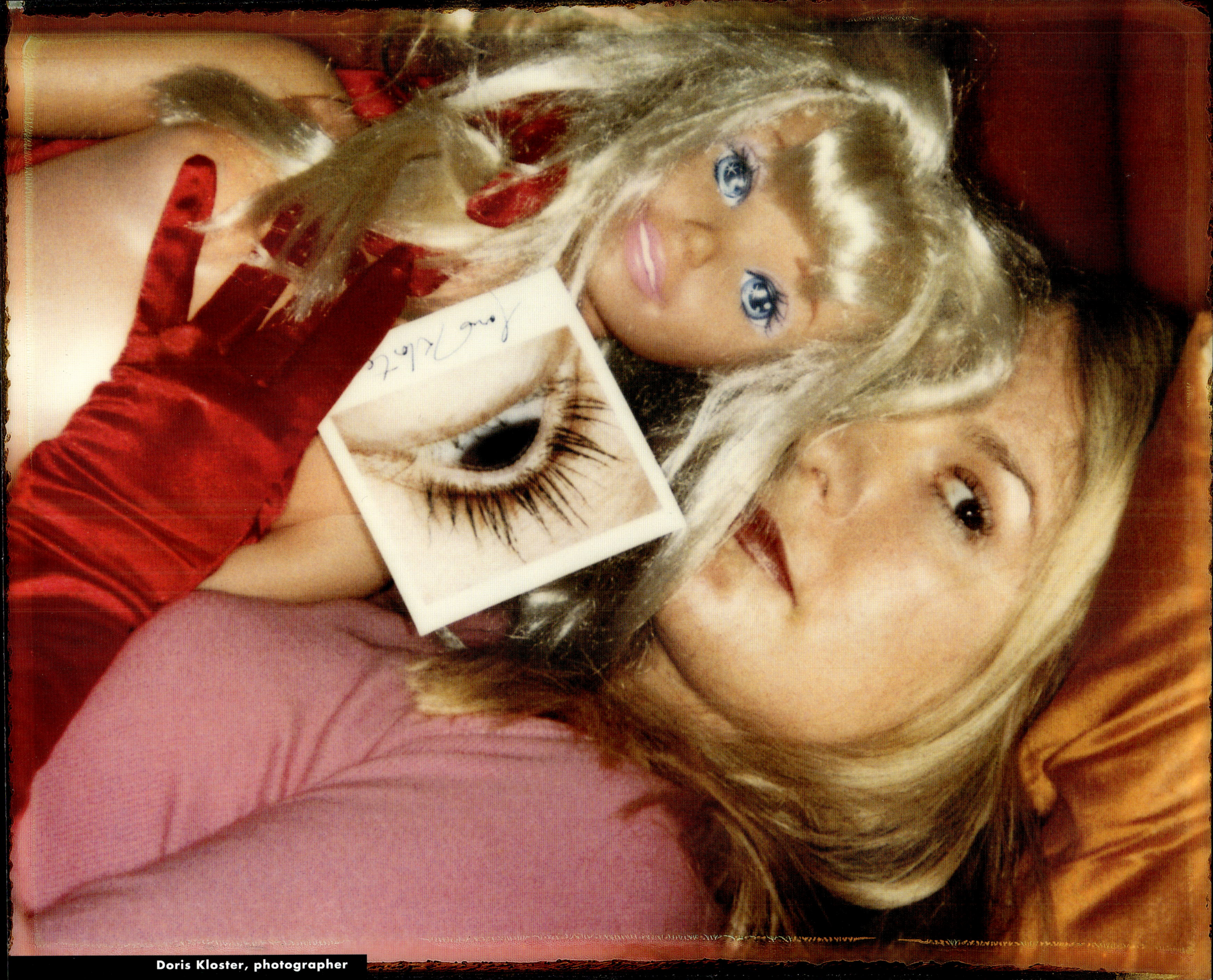

Doris Kloster, photographer

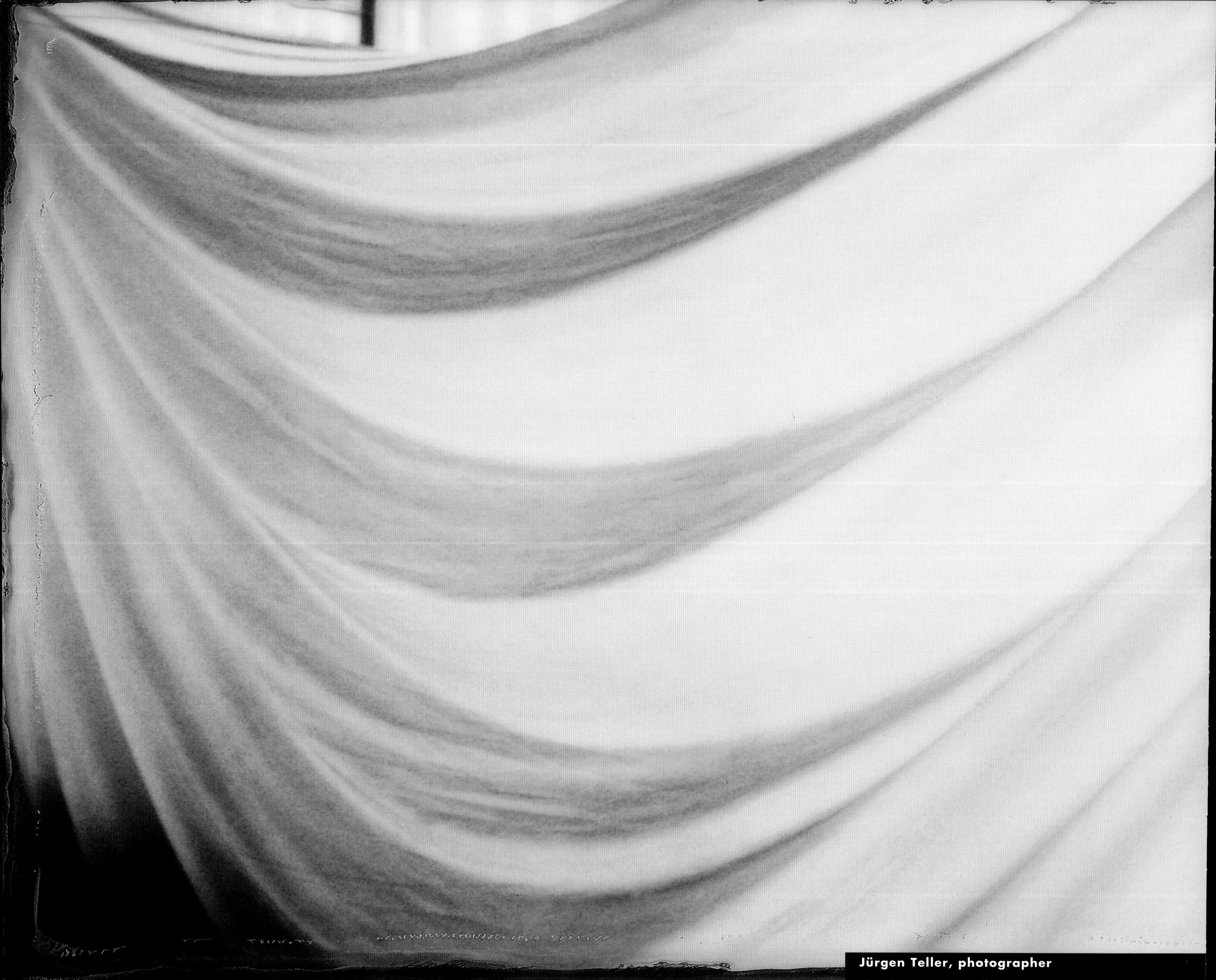

Jürgen Teller, photographer

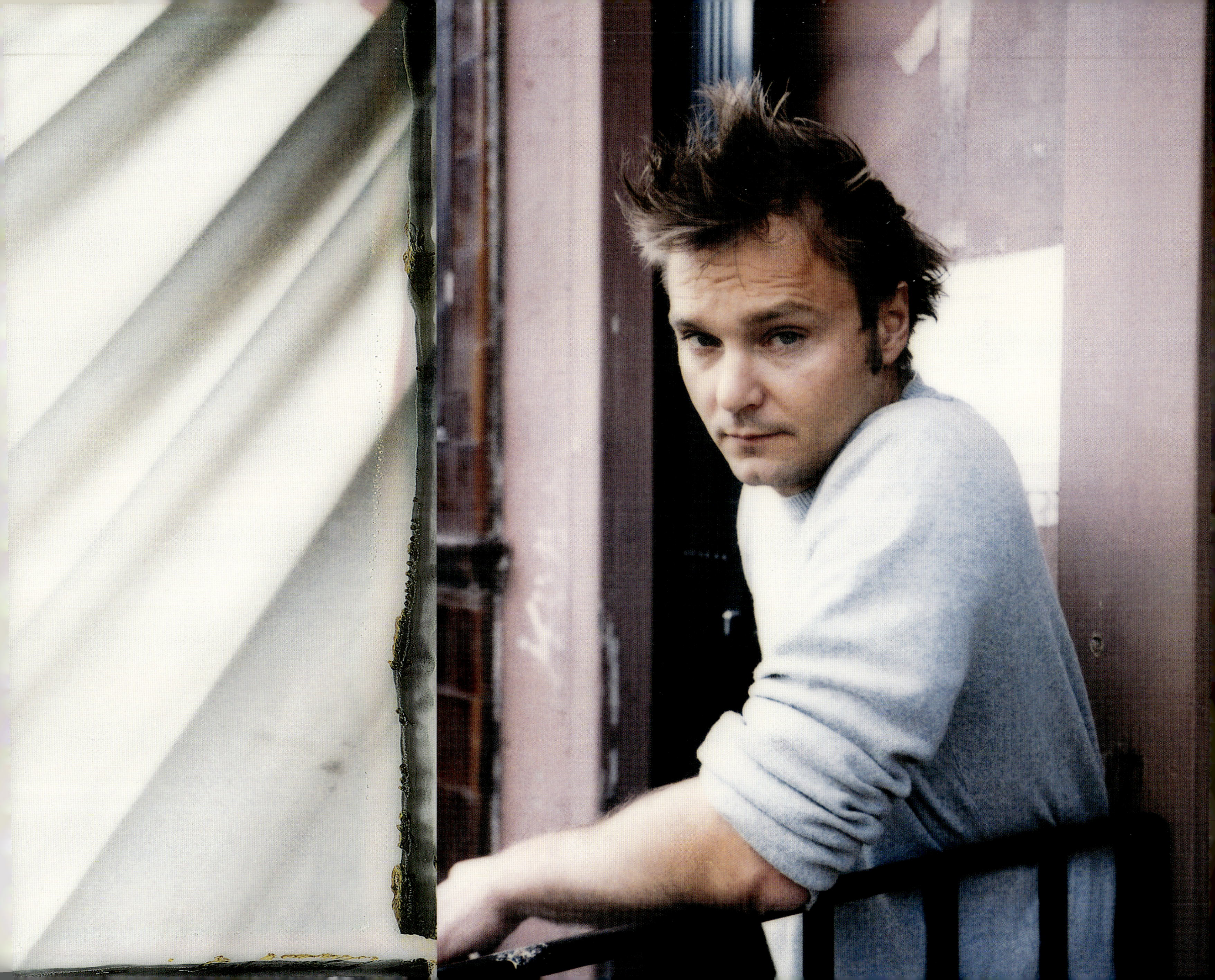

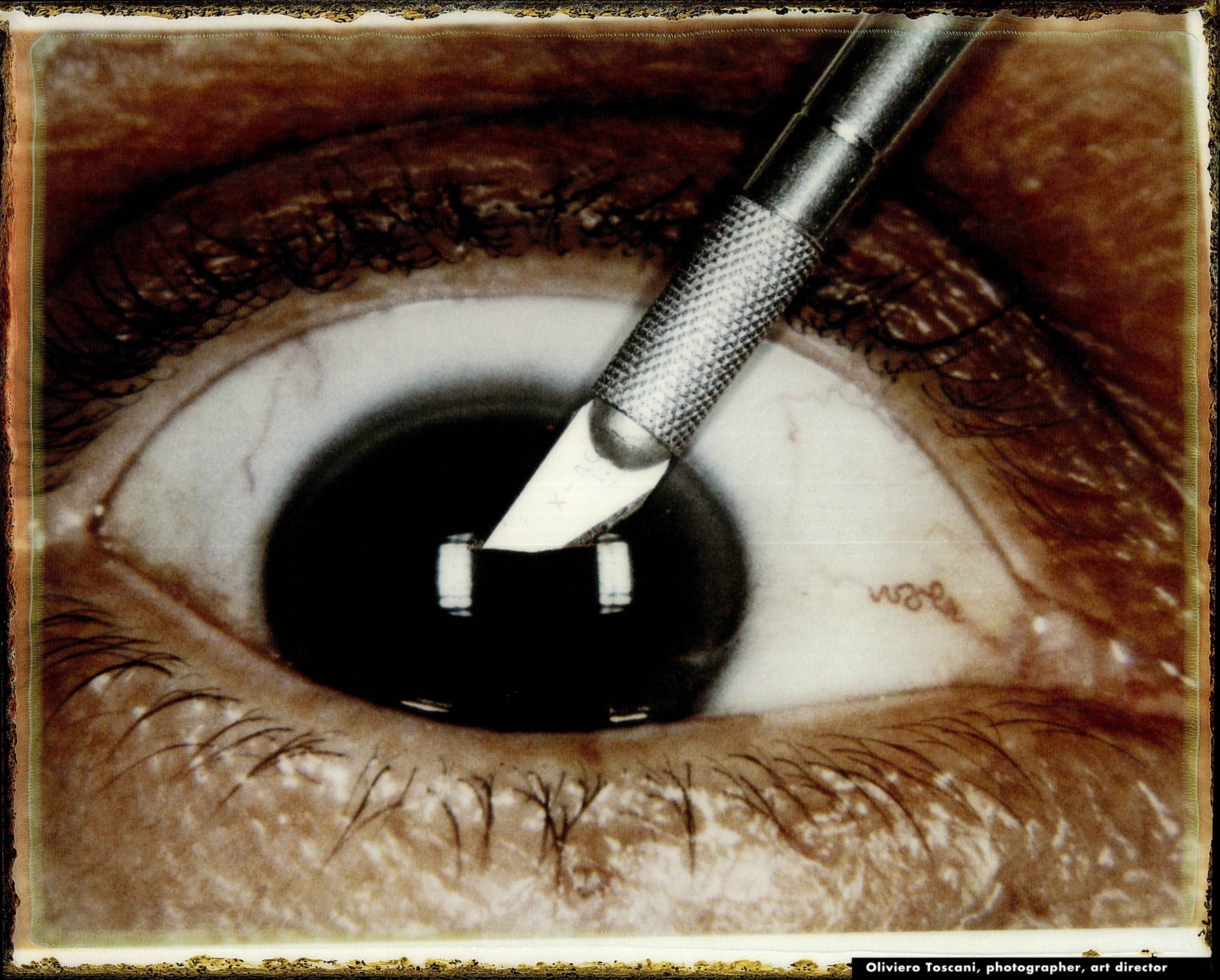

Oliviero Toscani, photographer, art director

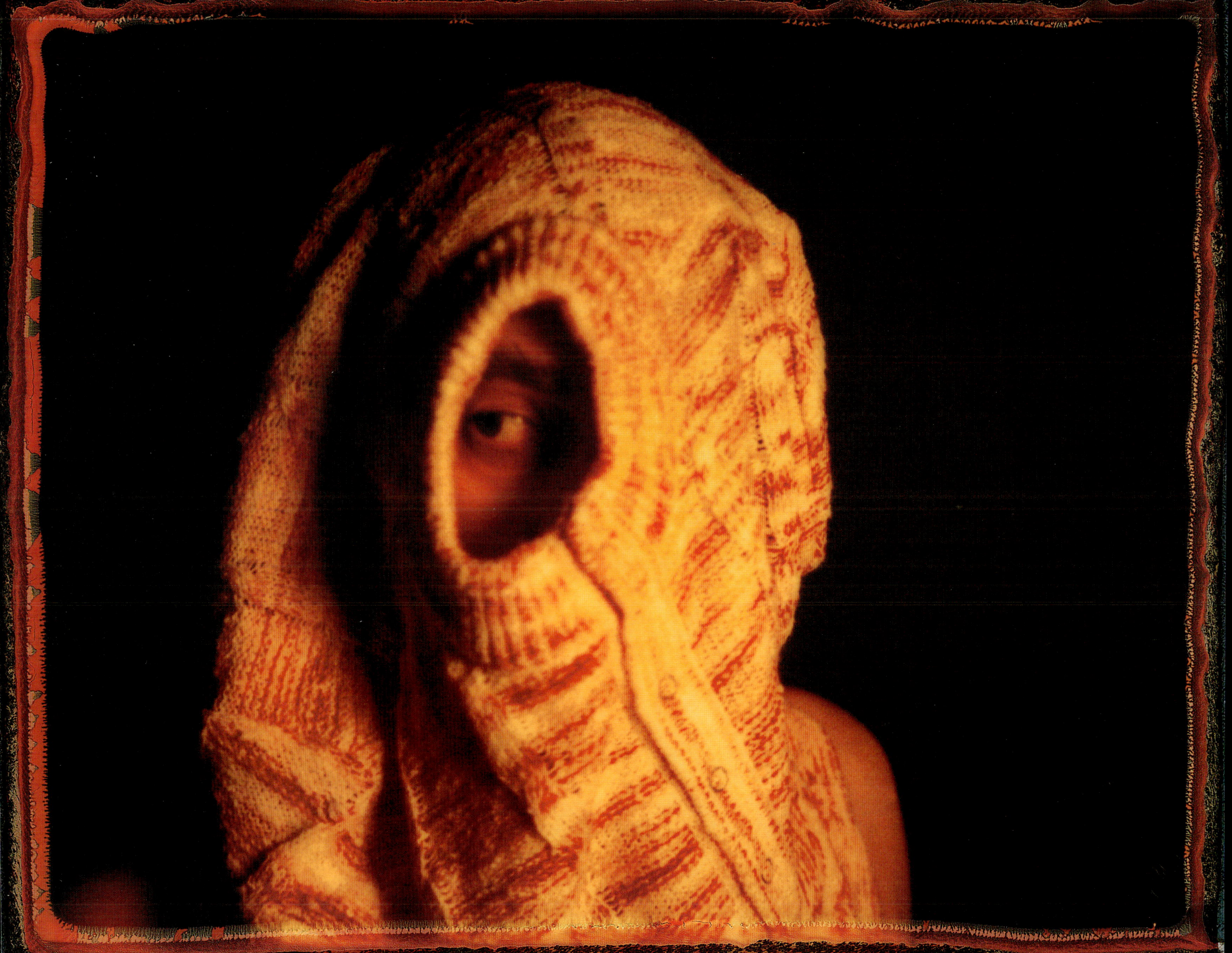

Bernhard Prinz, artist

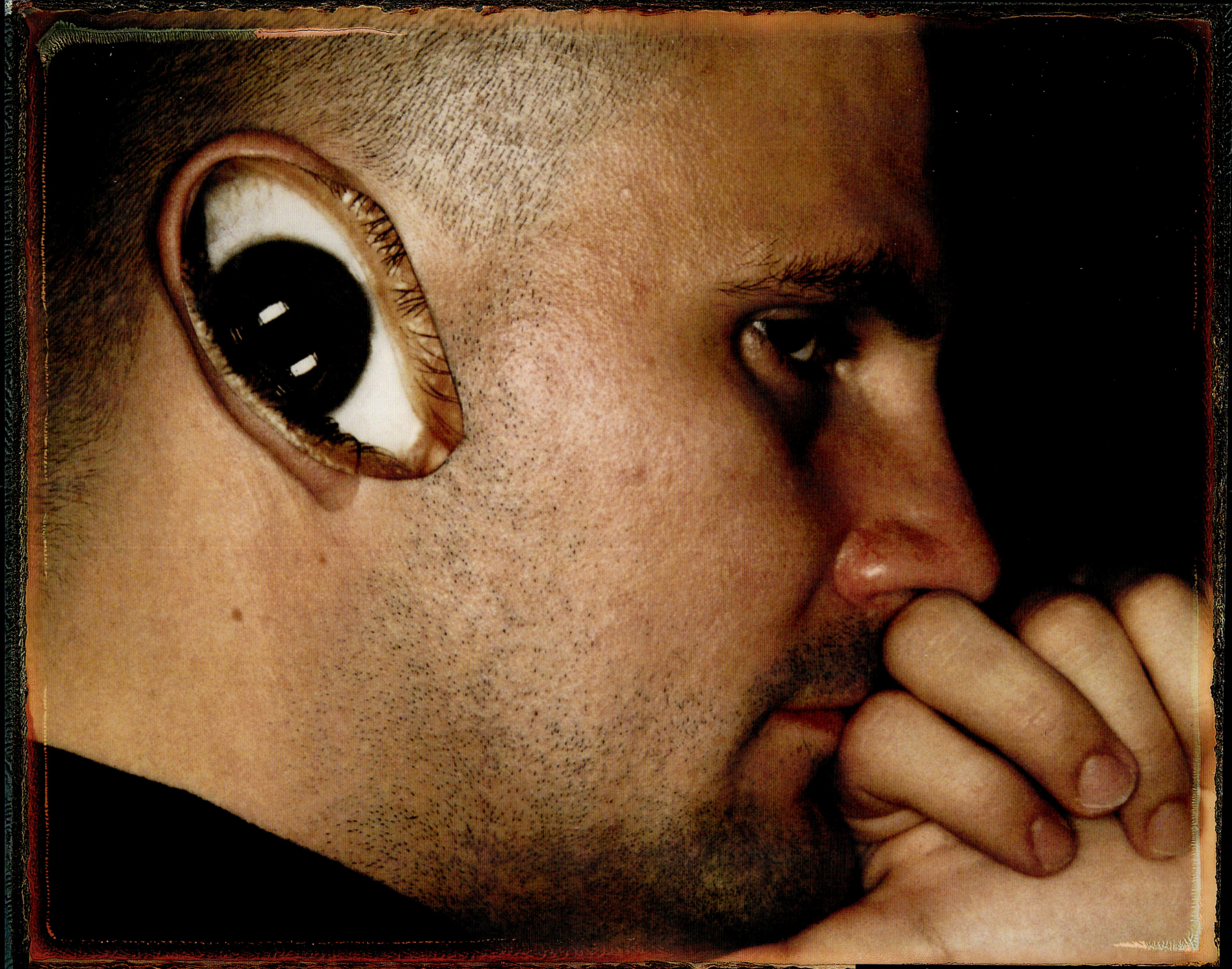
DJ Westbam, DJ

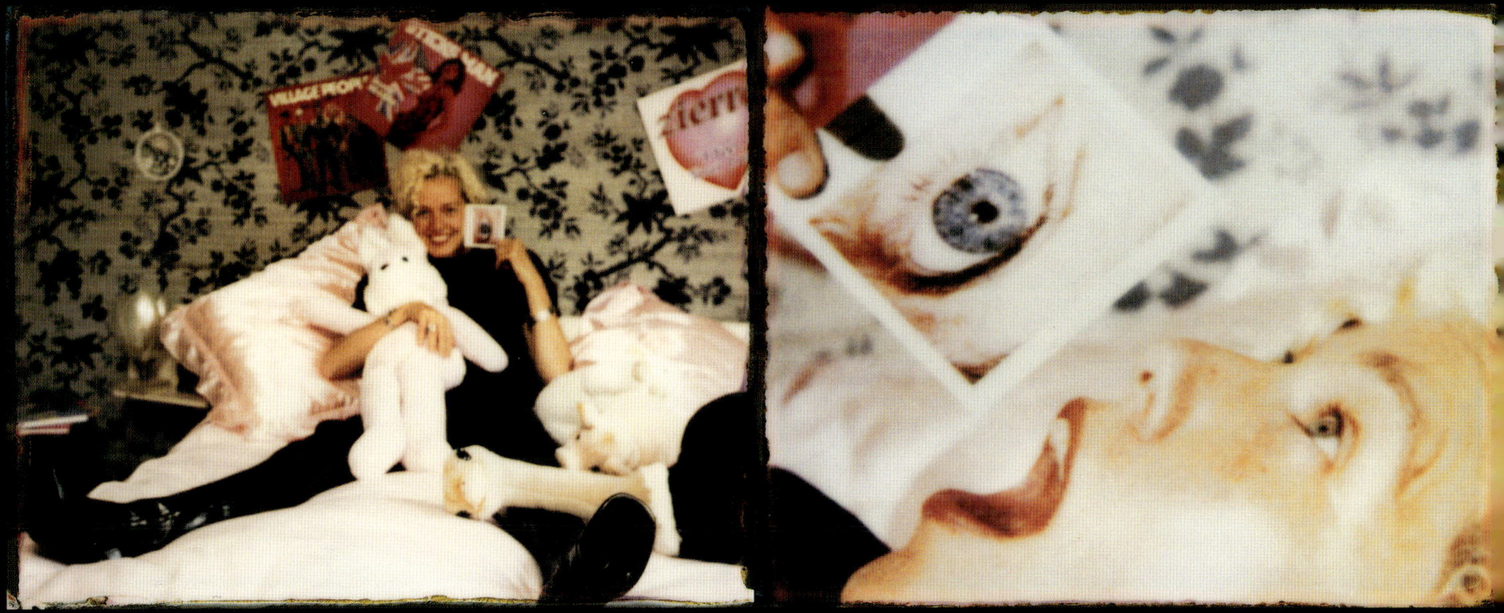

Ellen von Unwerth, photographer

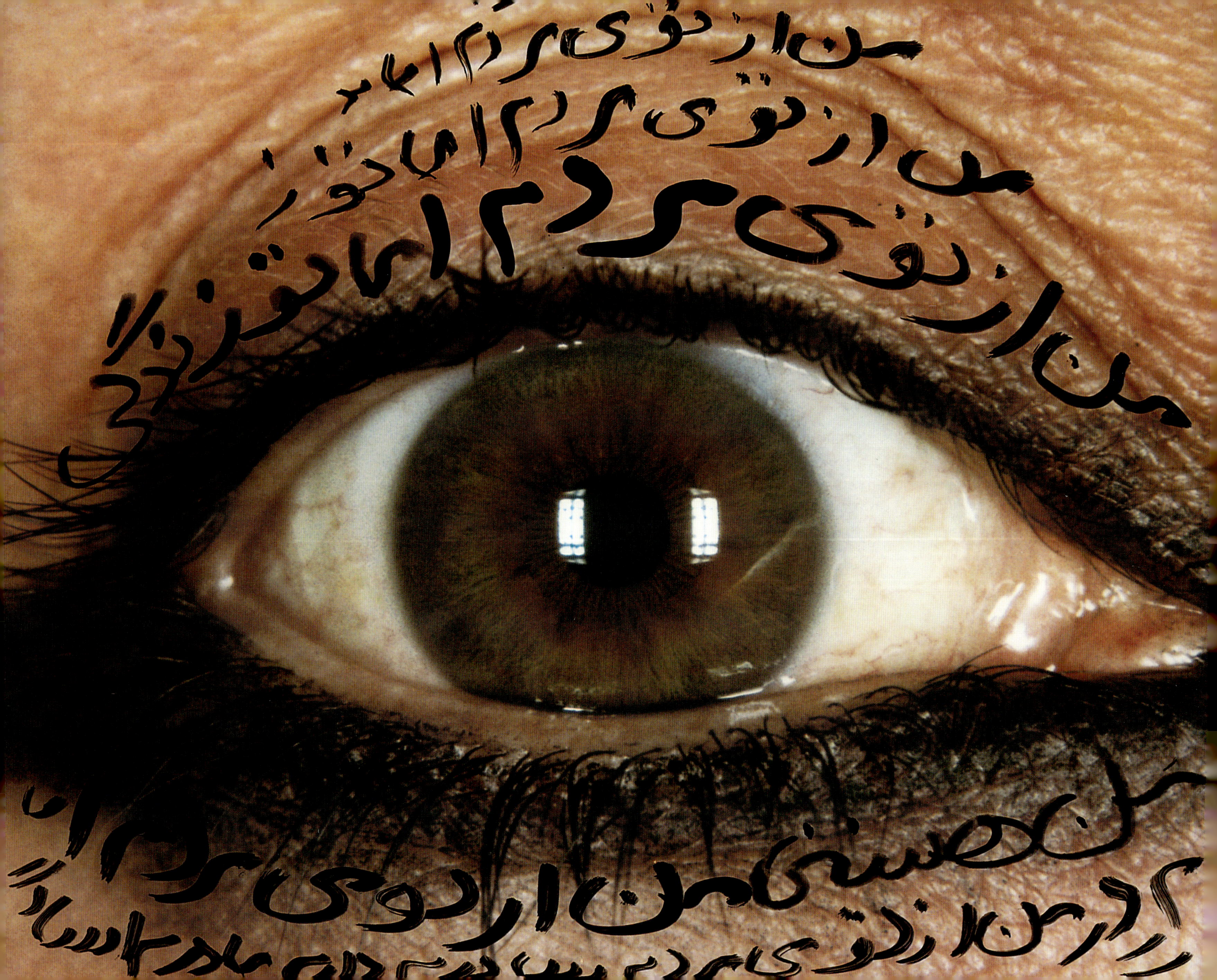

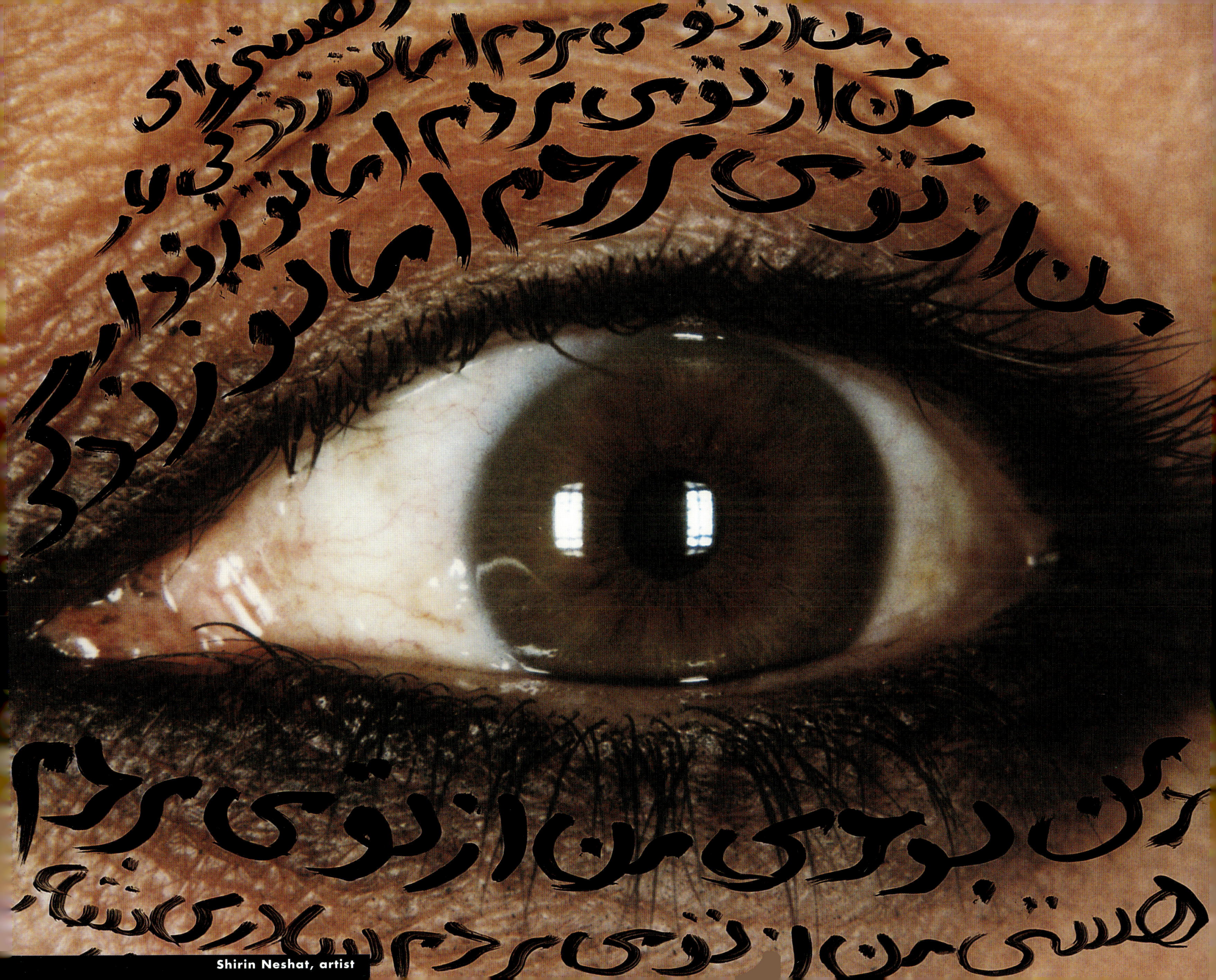

Shirin Neshat, artist

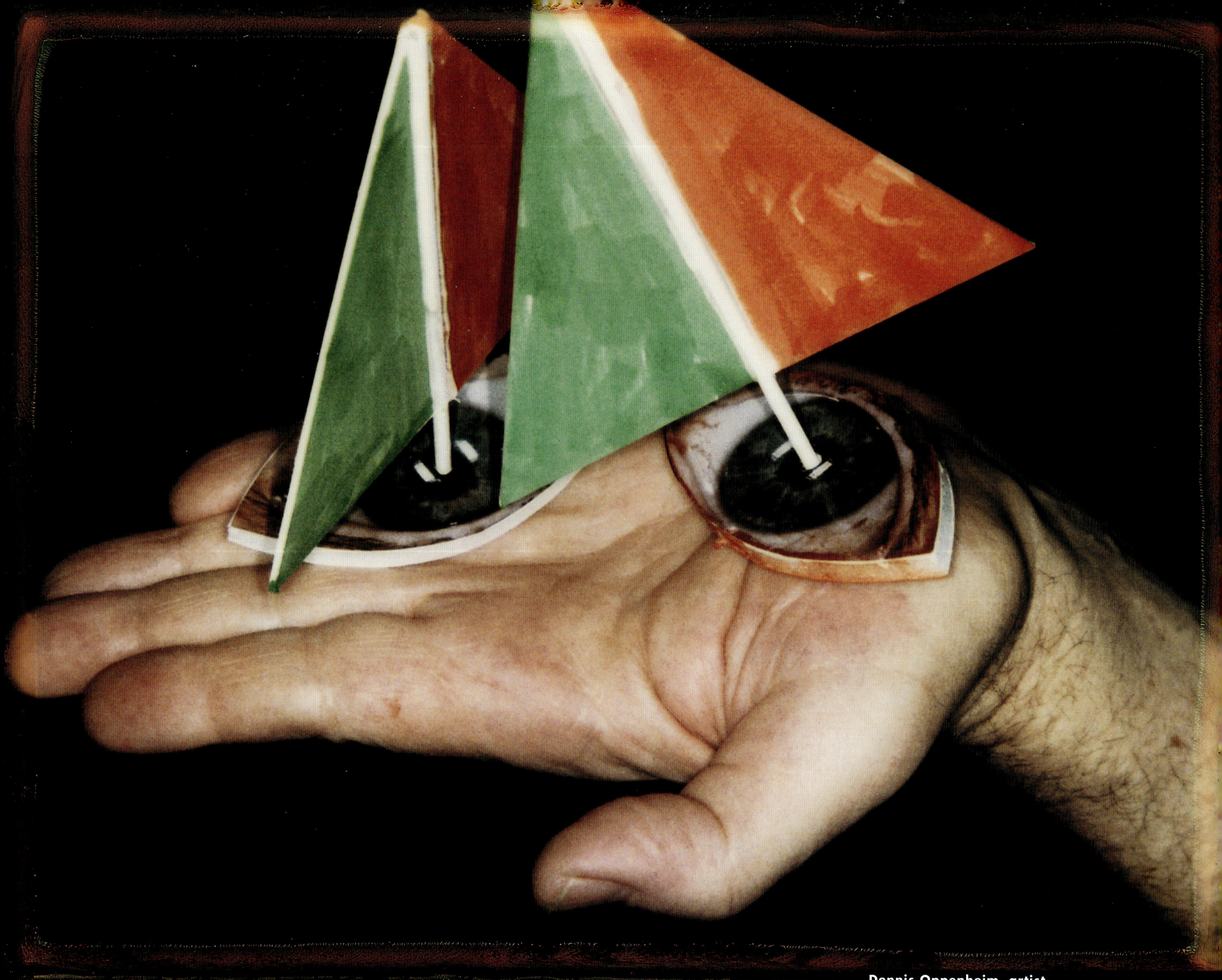

Dennis Oppenheim, artist

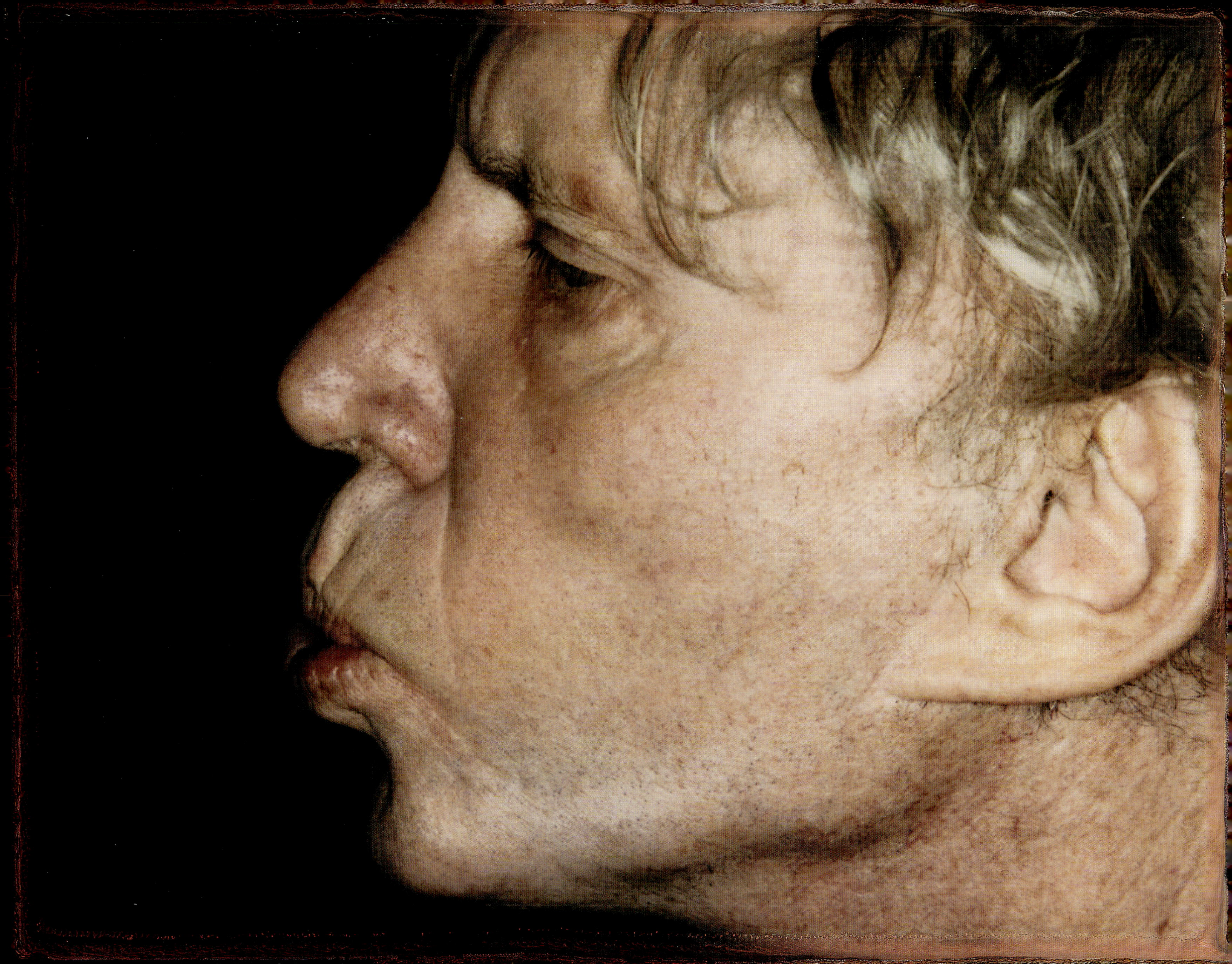

Dennis Oppenheim, artist

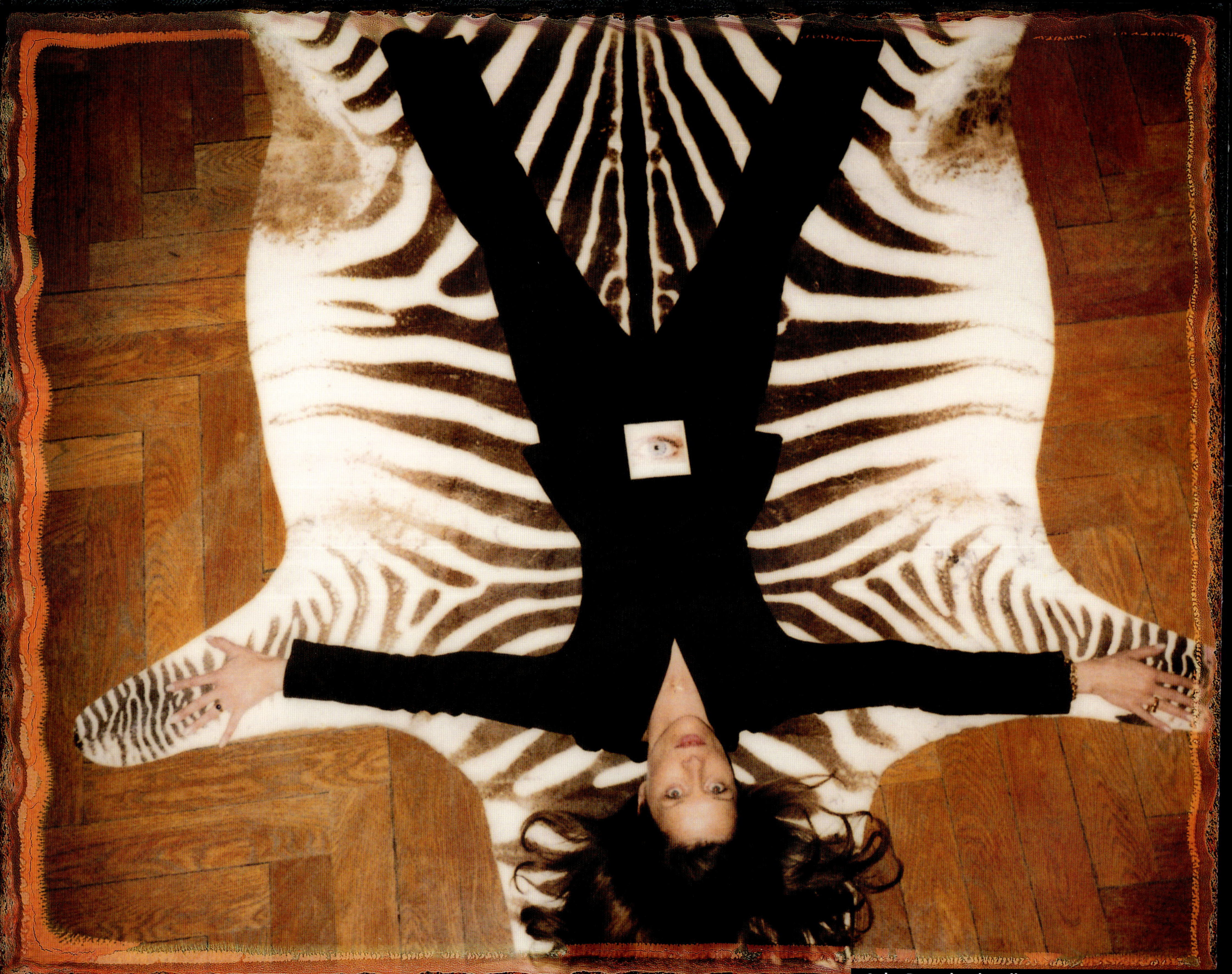

Anke Degenhard, gallery owner

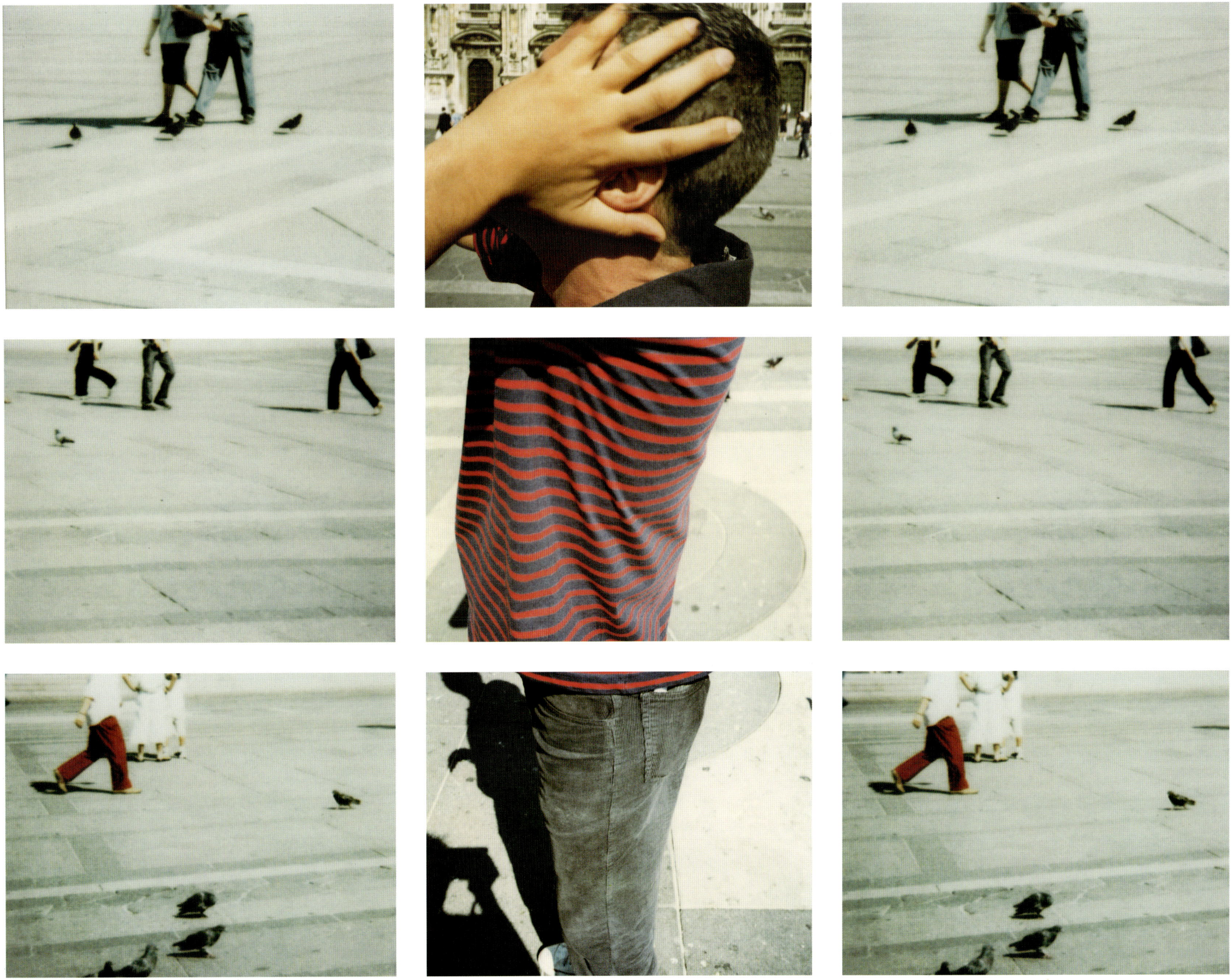

Maurizio Cattelan, artist

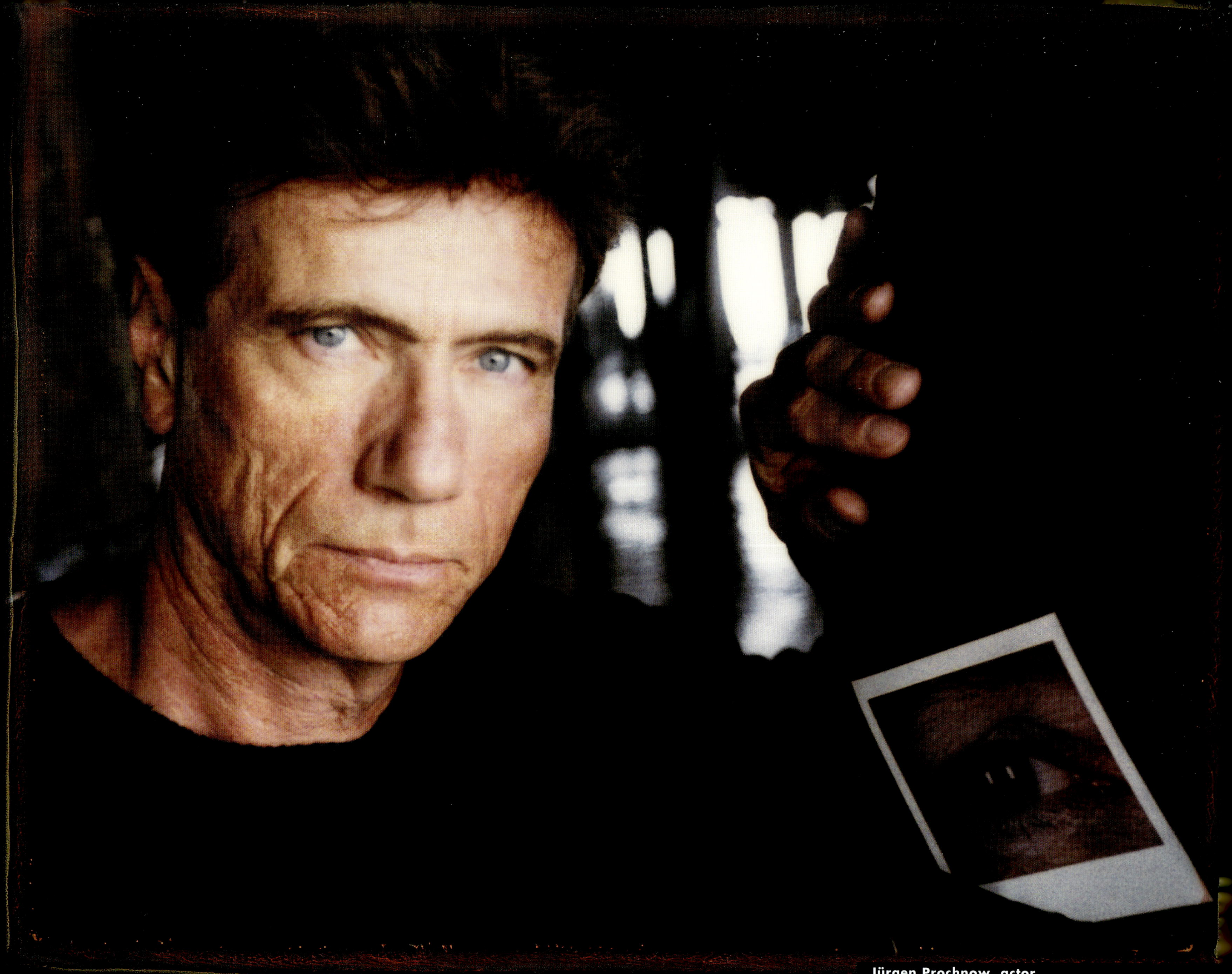

Jürgen Prochnow, actor

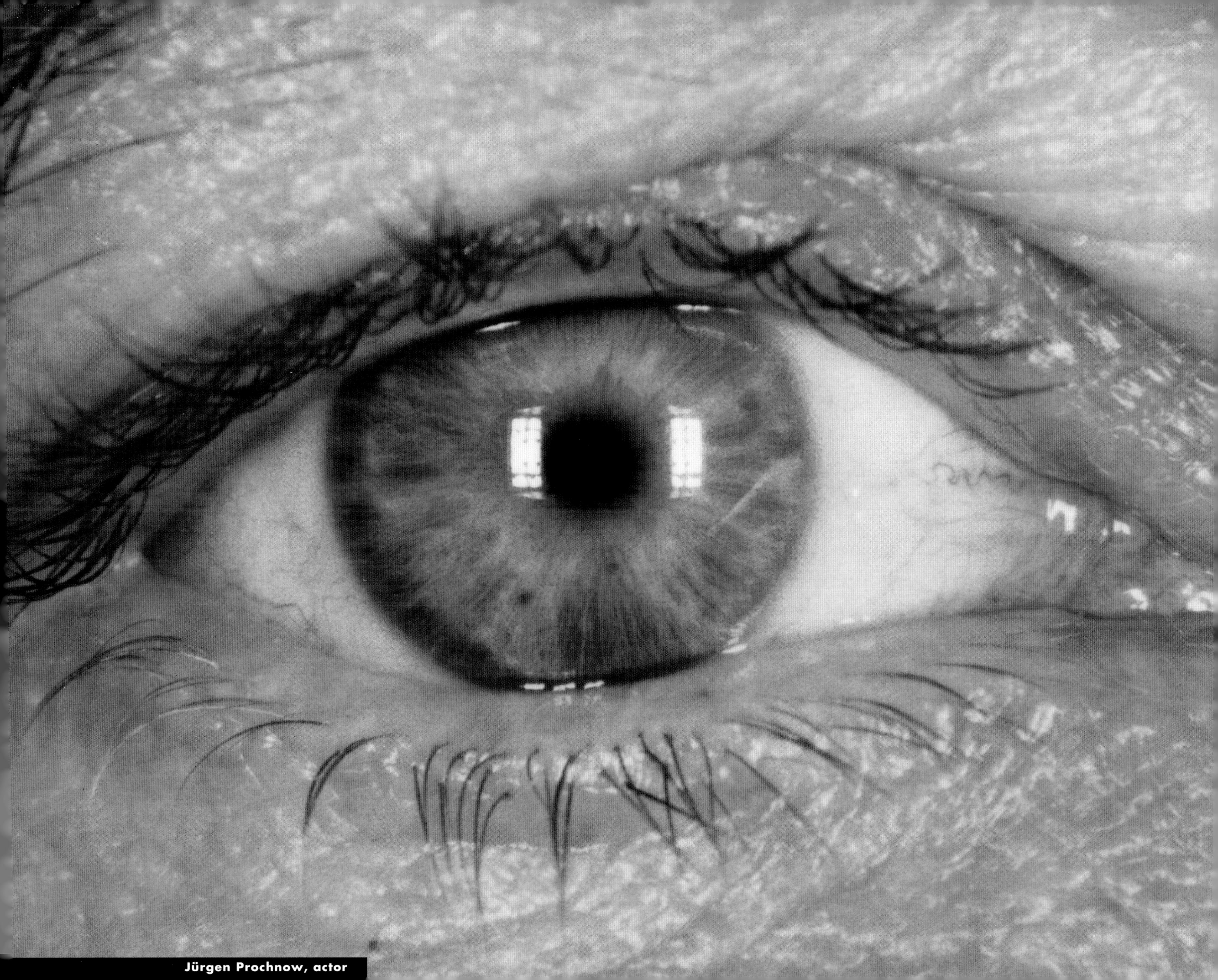

Jürgen Prochnow, actor

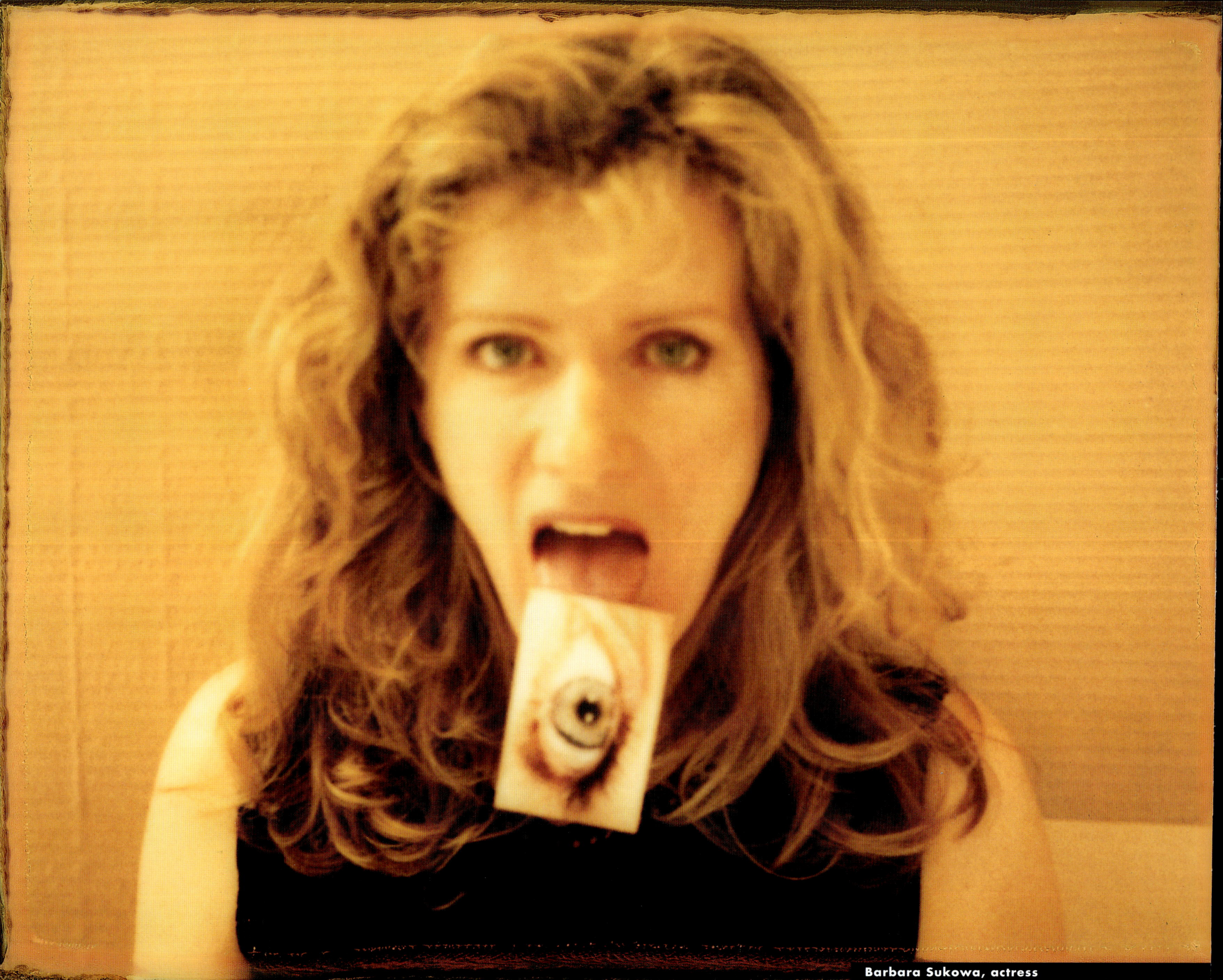

Barbara Sukowa, actress

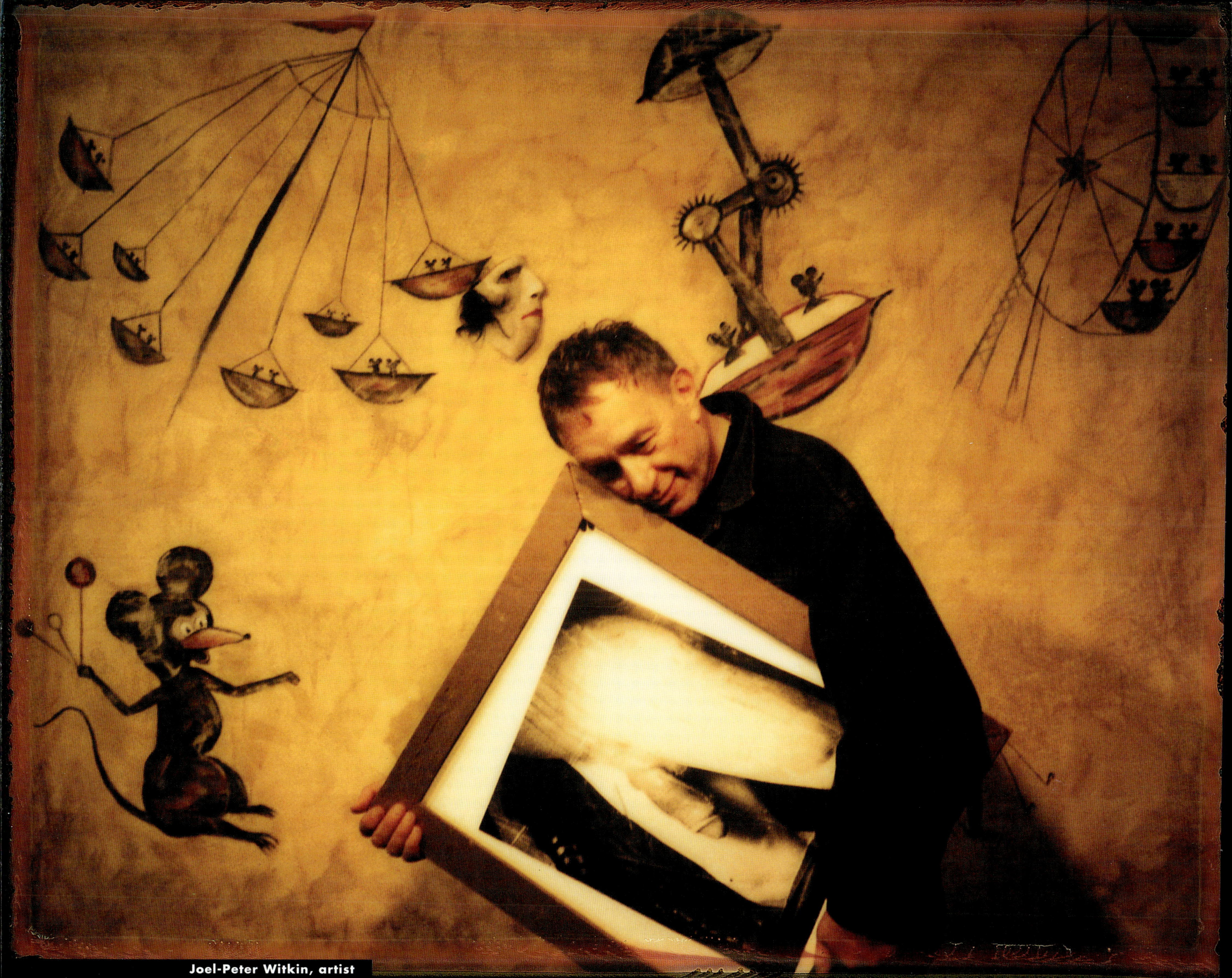

Joel-Peter Witkin, artist

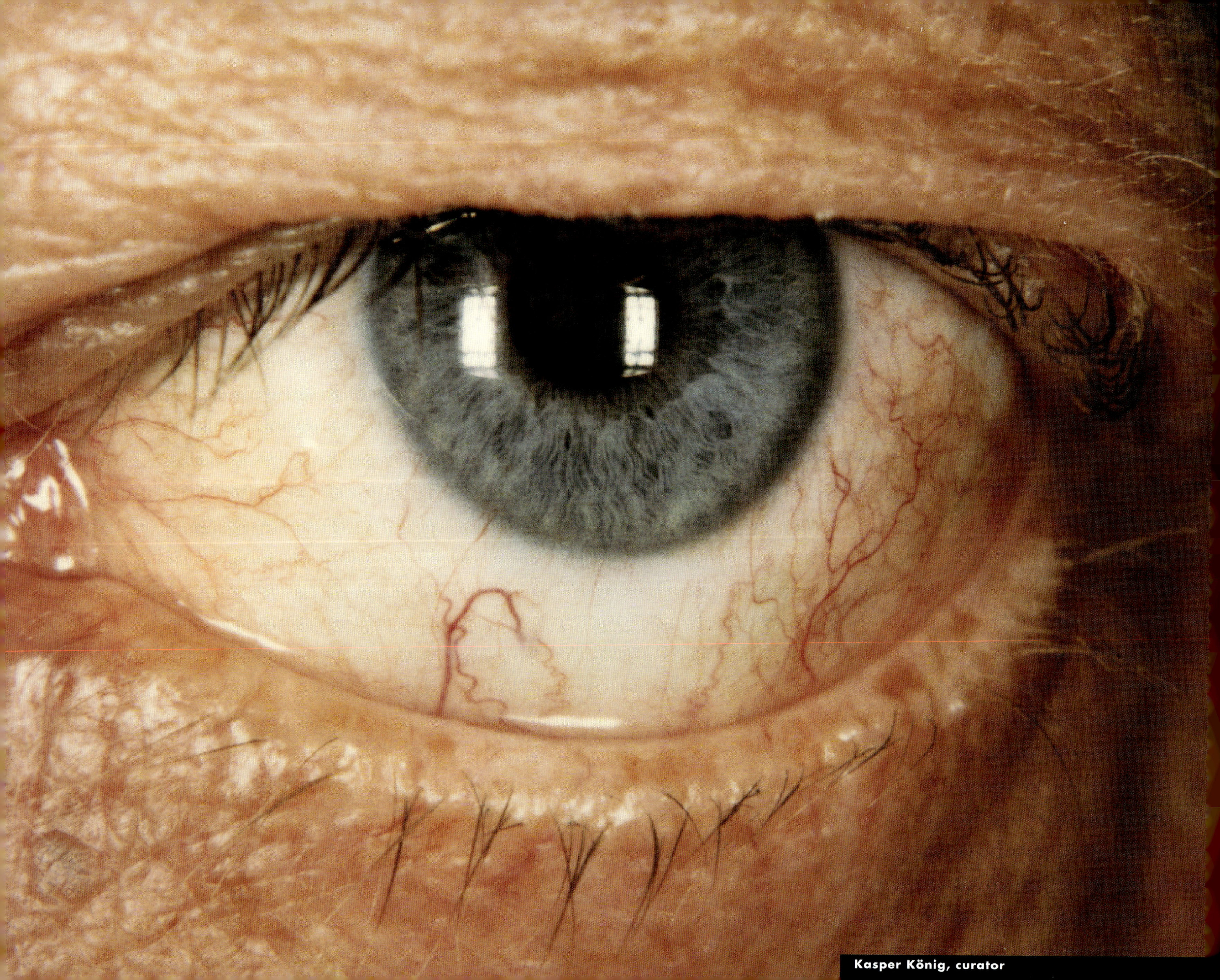

Kasper König, curator

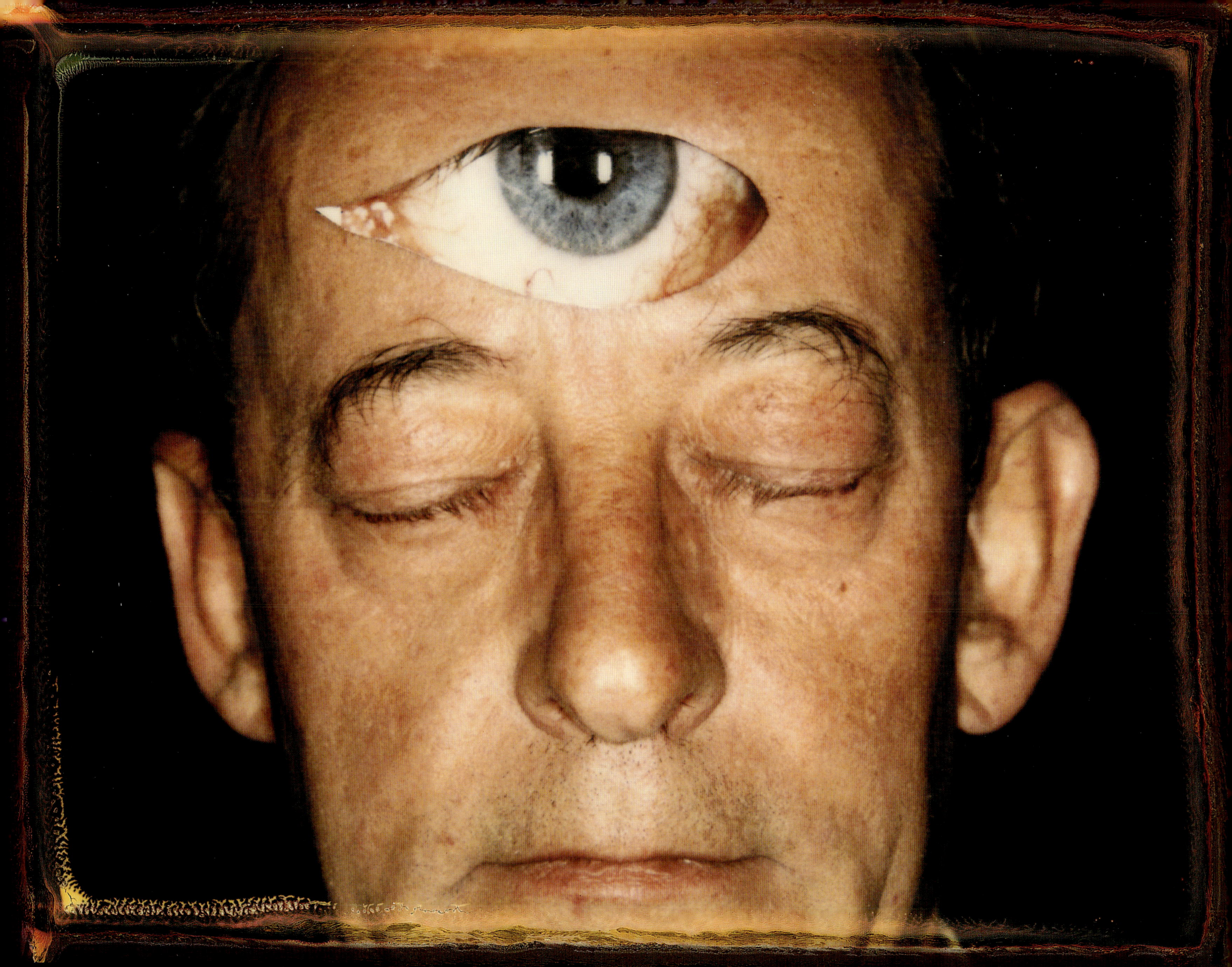

Kasper König, curator

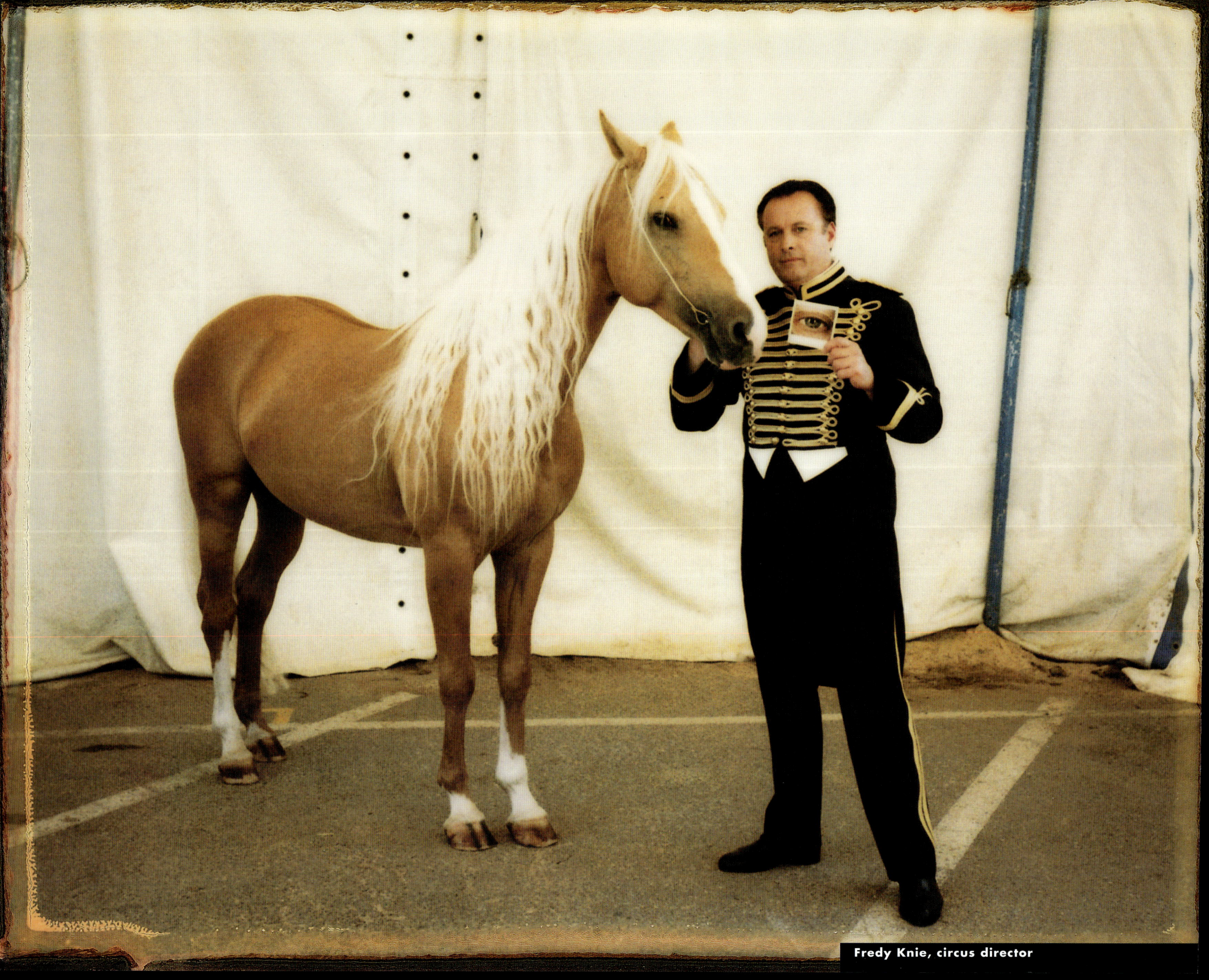

Fredy Knie, circus director

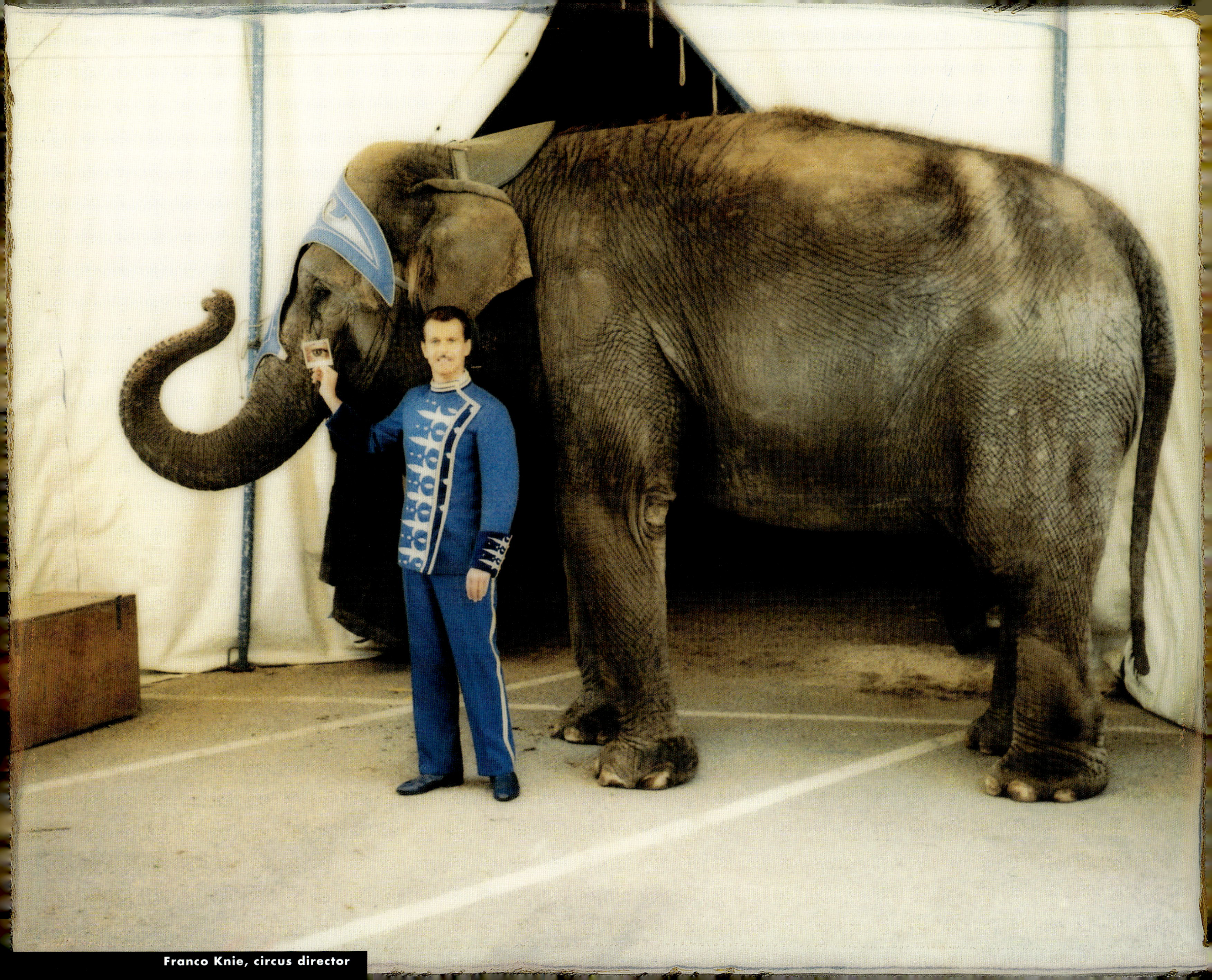

Franco Knie, circus director

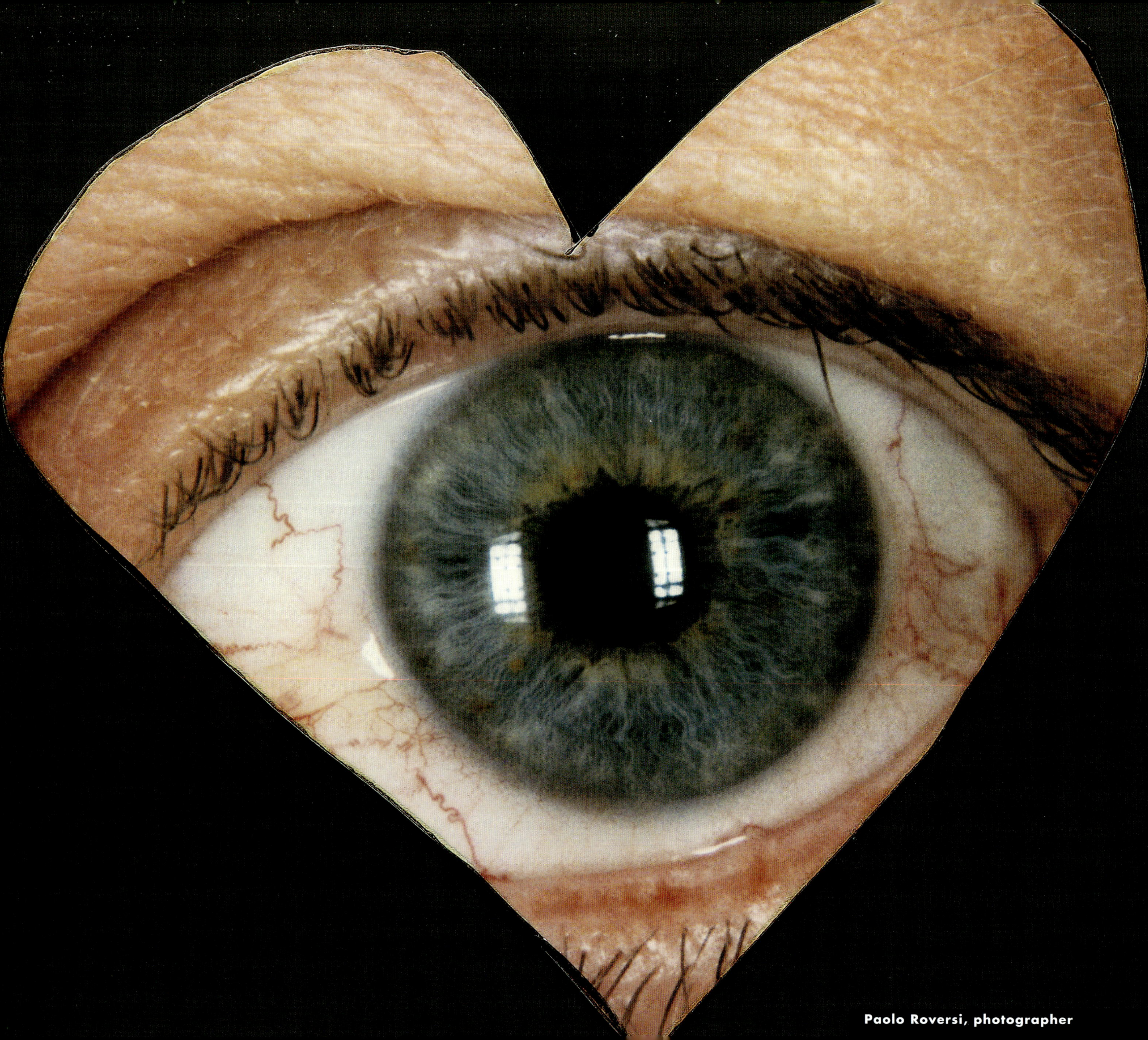

Paolo Roversi, photographer

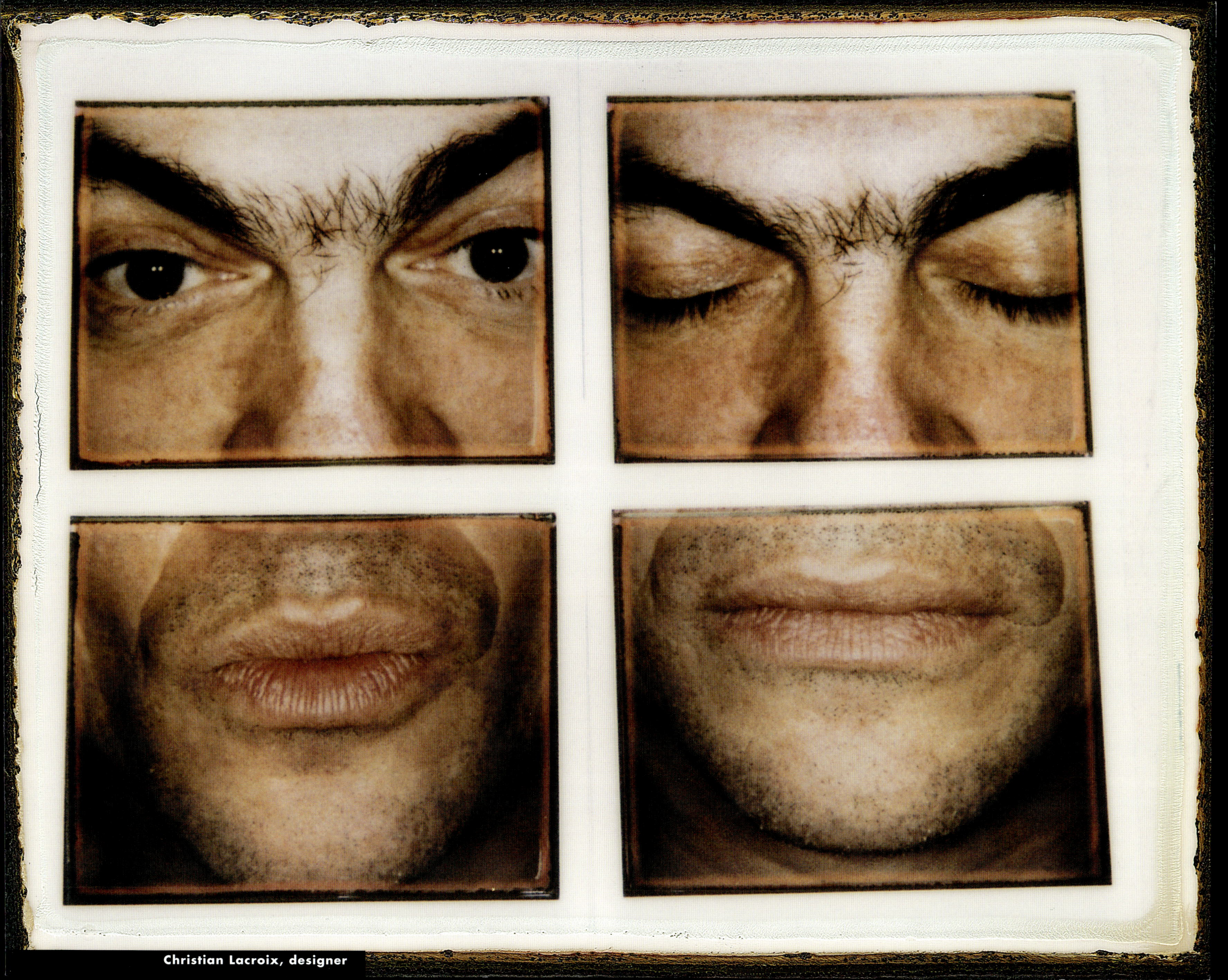

Christian Lacroix, designer

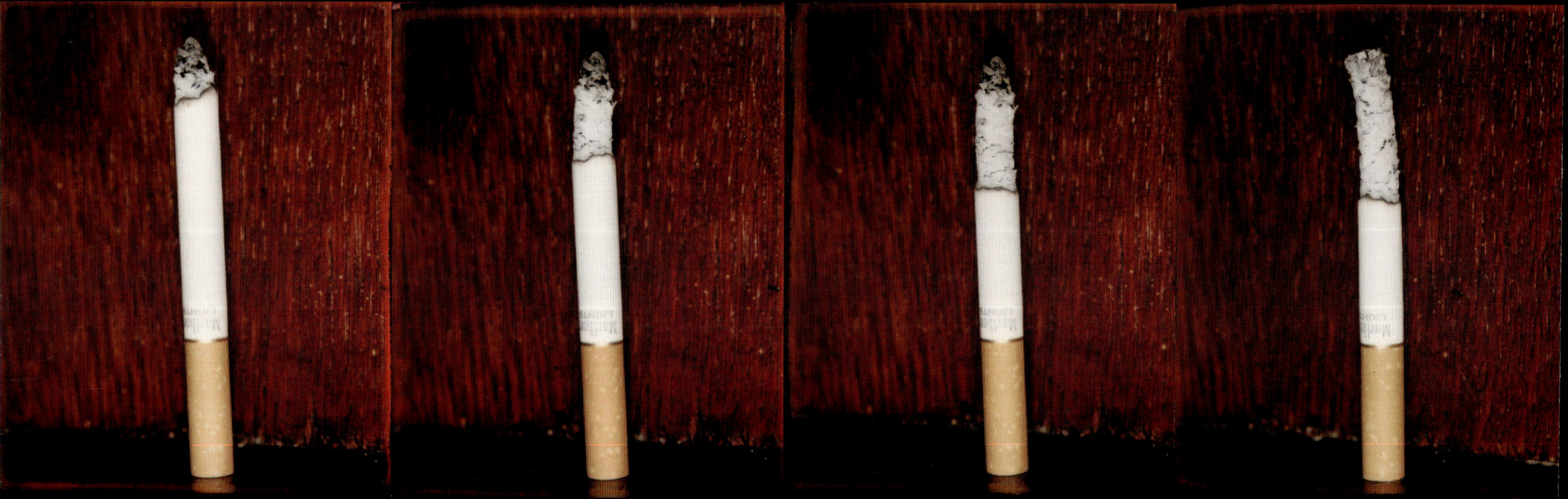

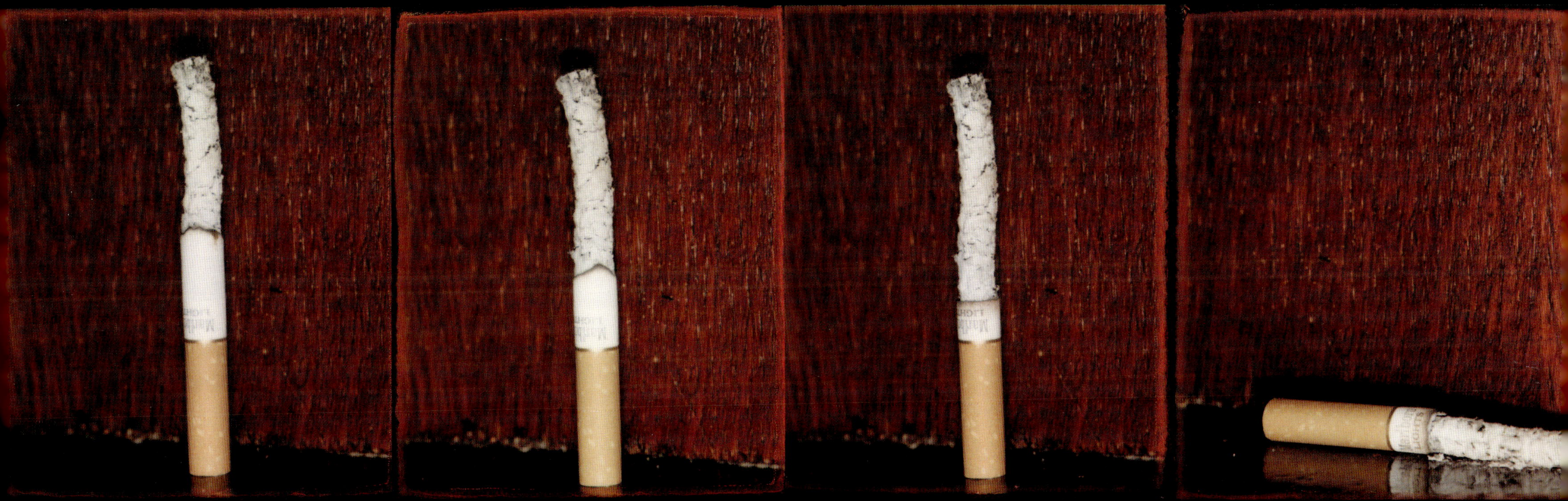

Richard Billingham, artist

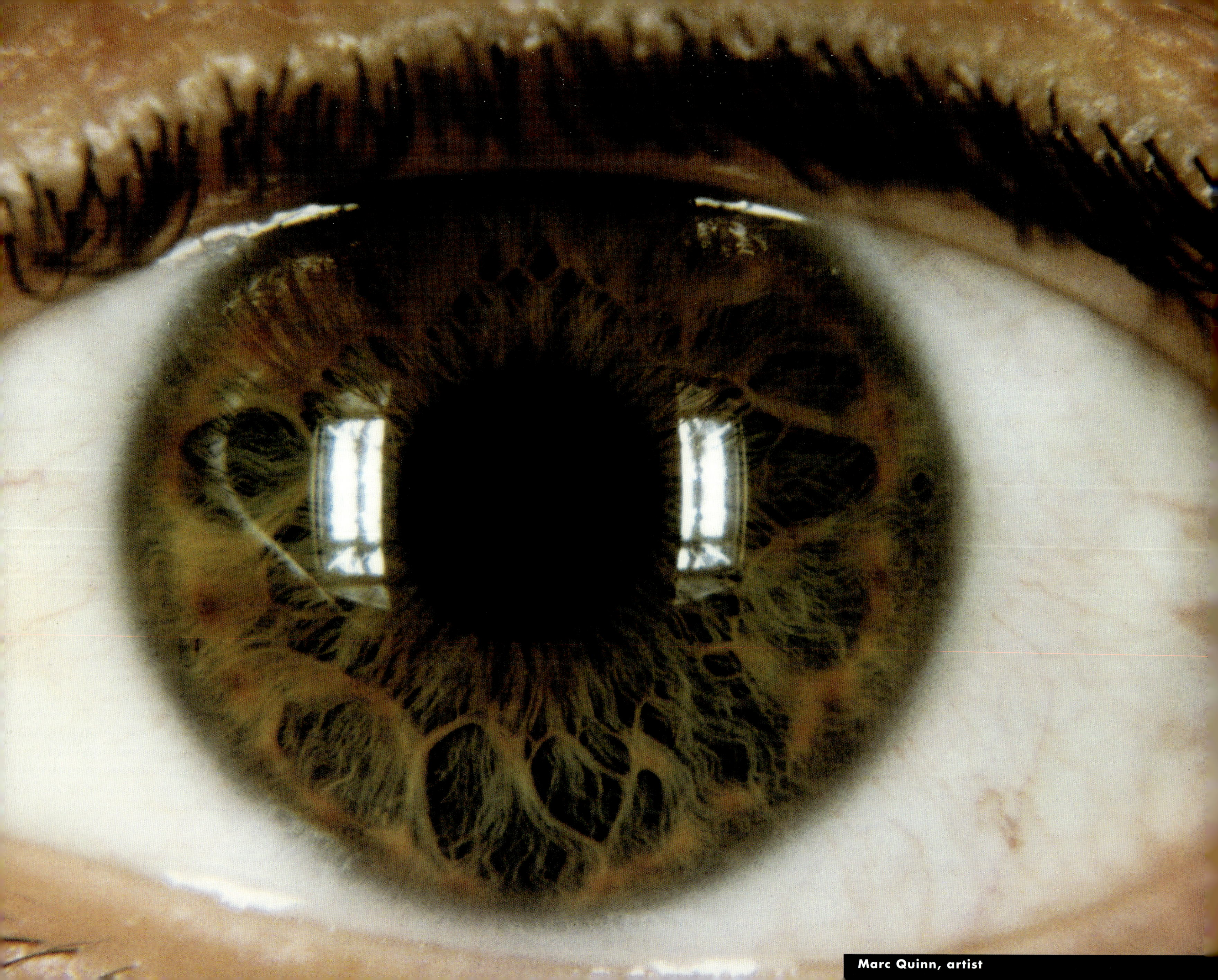

Marc Quinn, artist

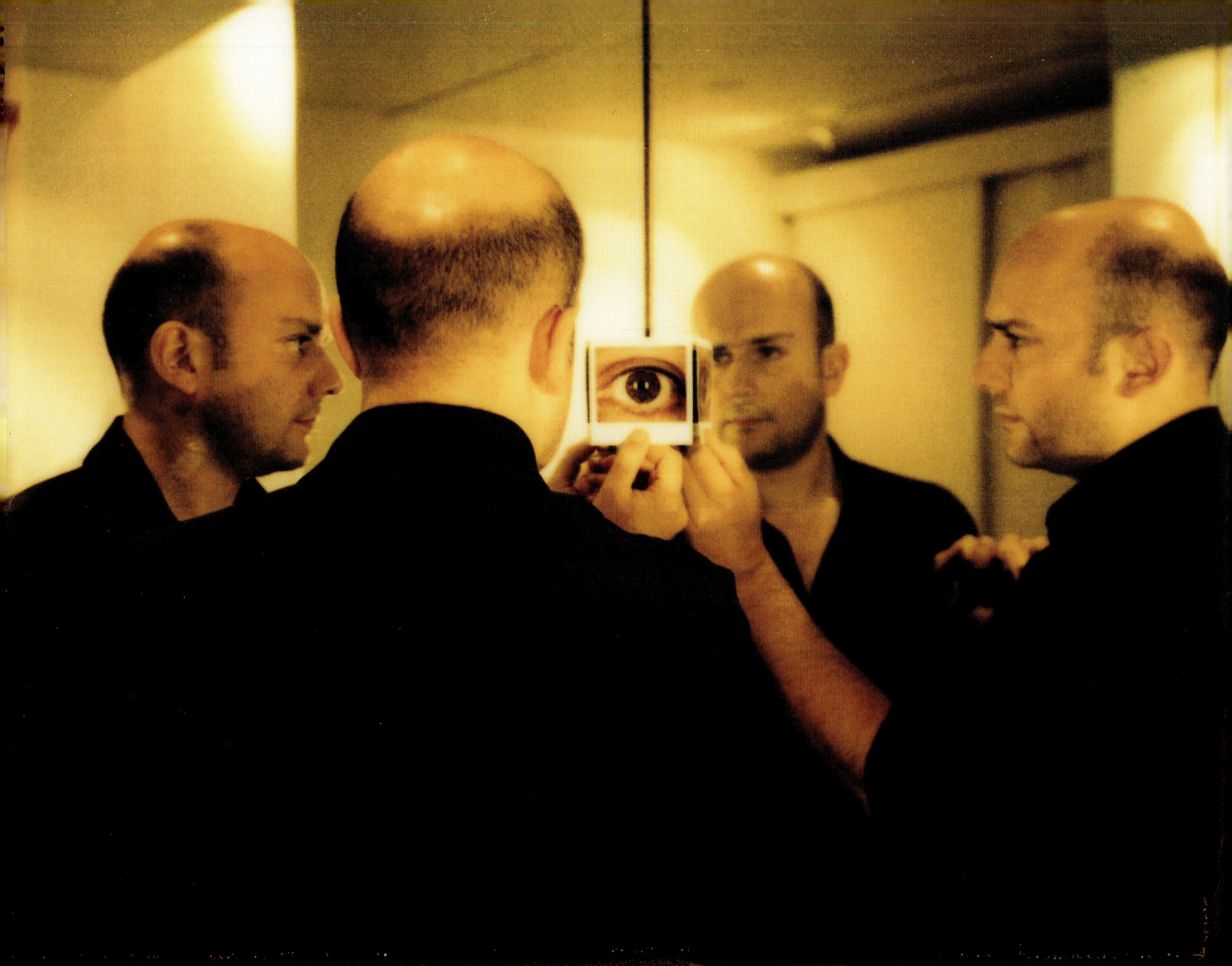

Marc Quinn, artist

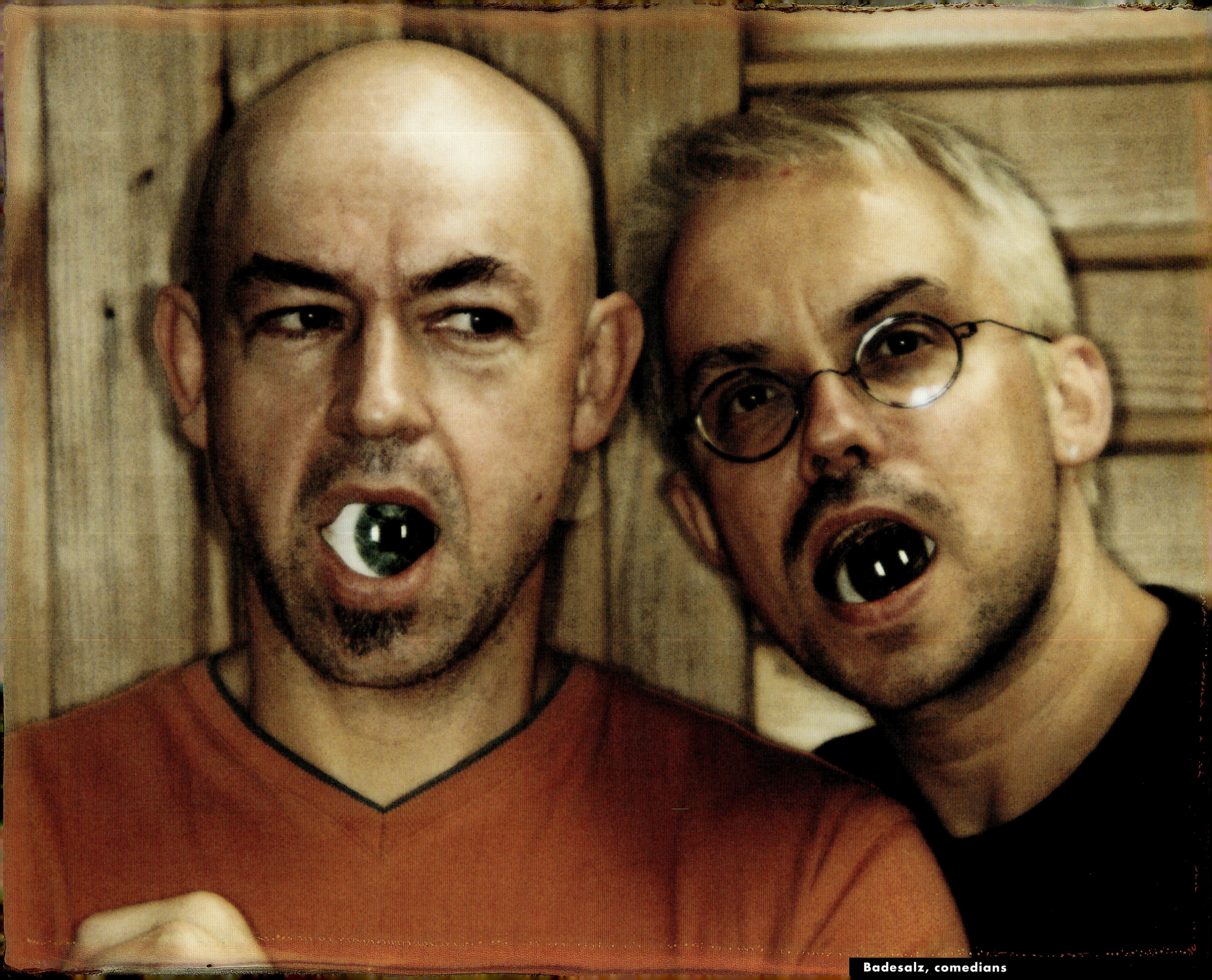

Badesalz, comedians

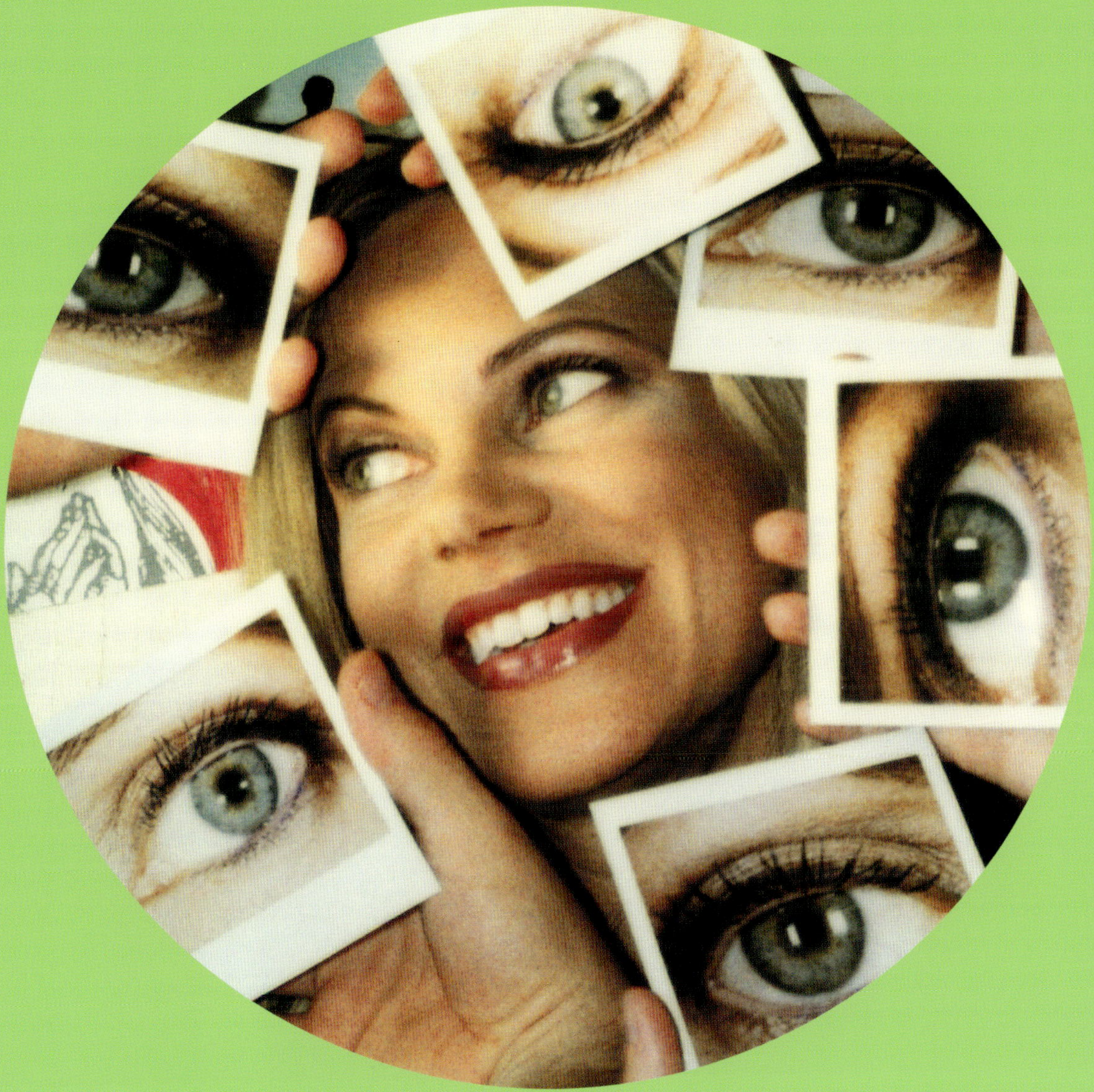

Nina Ruge, moderator

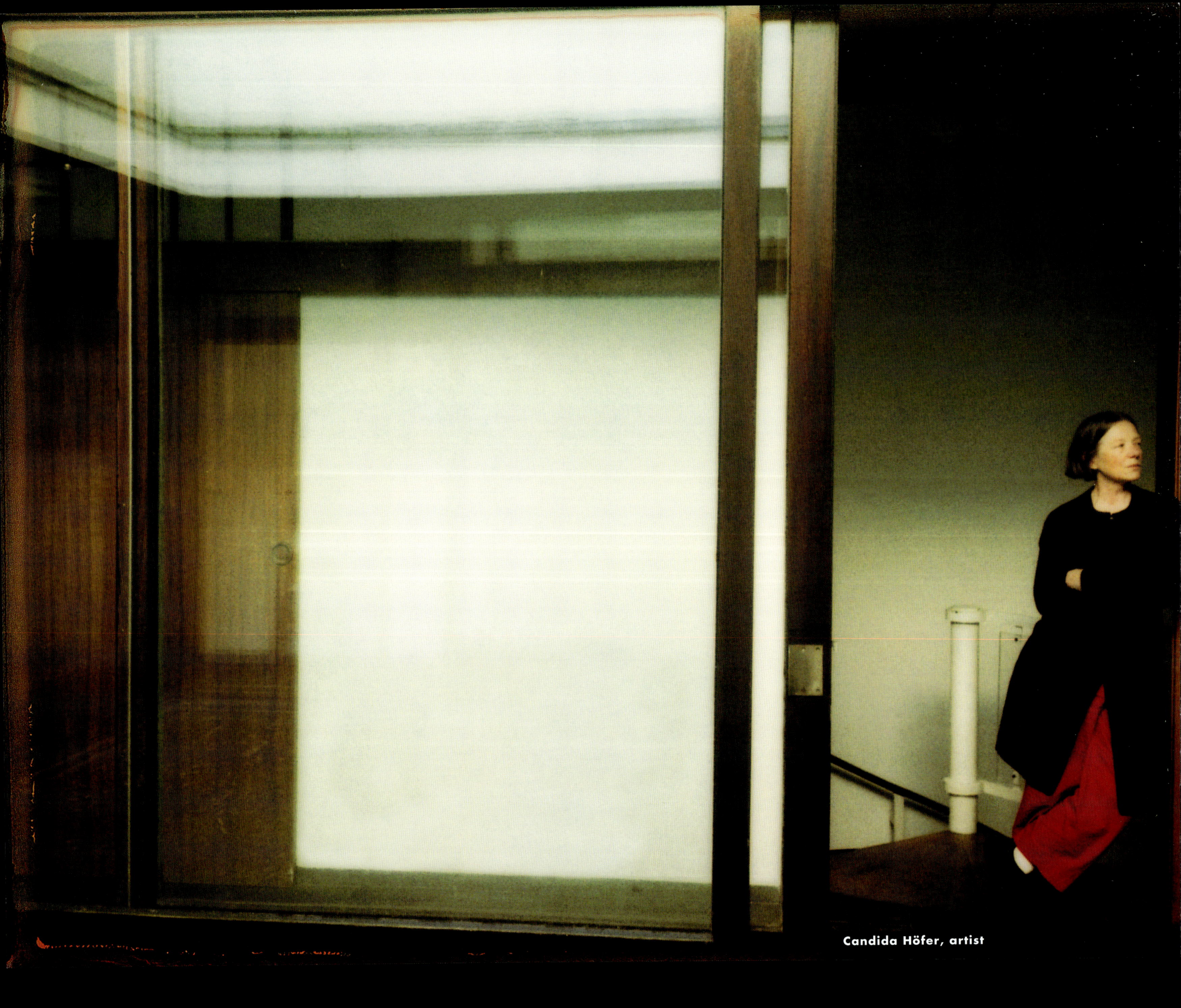

Candida Höfer, artist

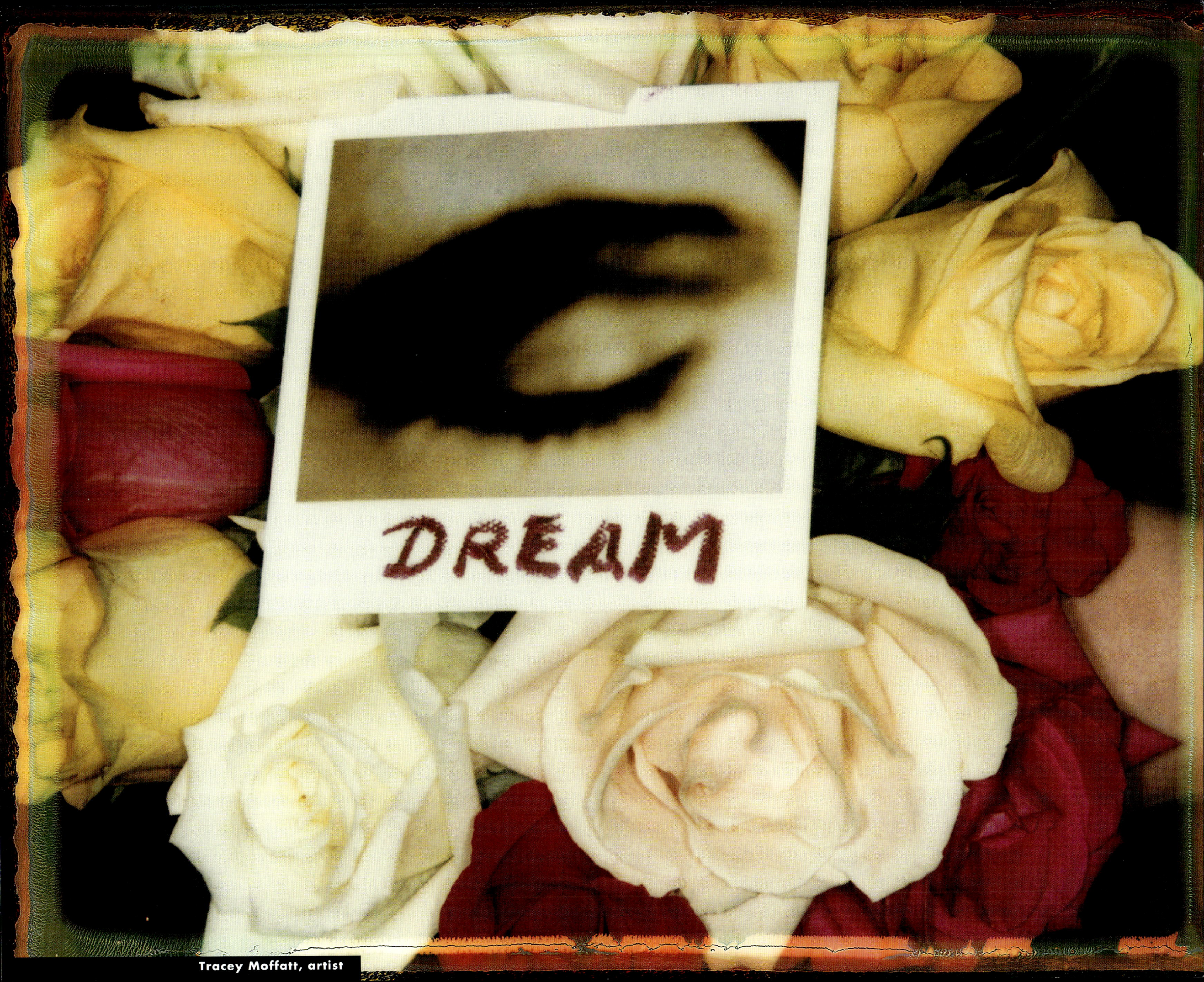

Tracey Moffatt, artist

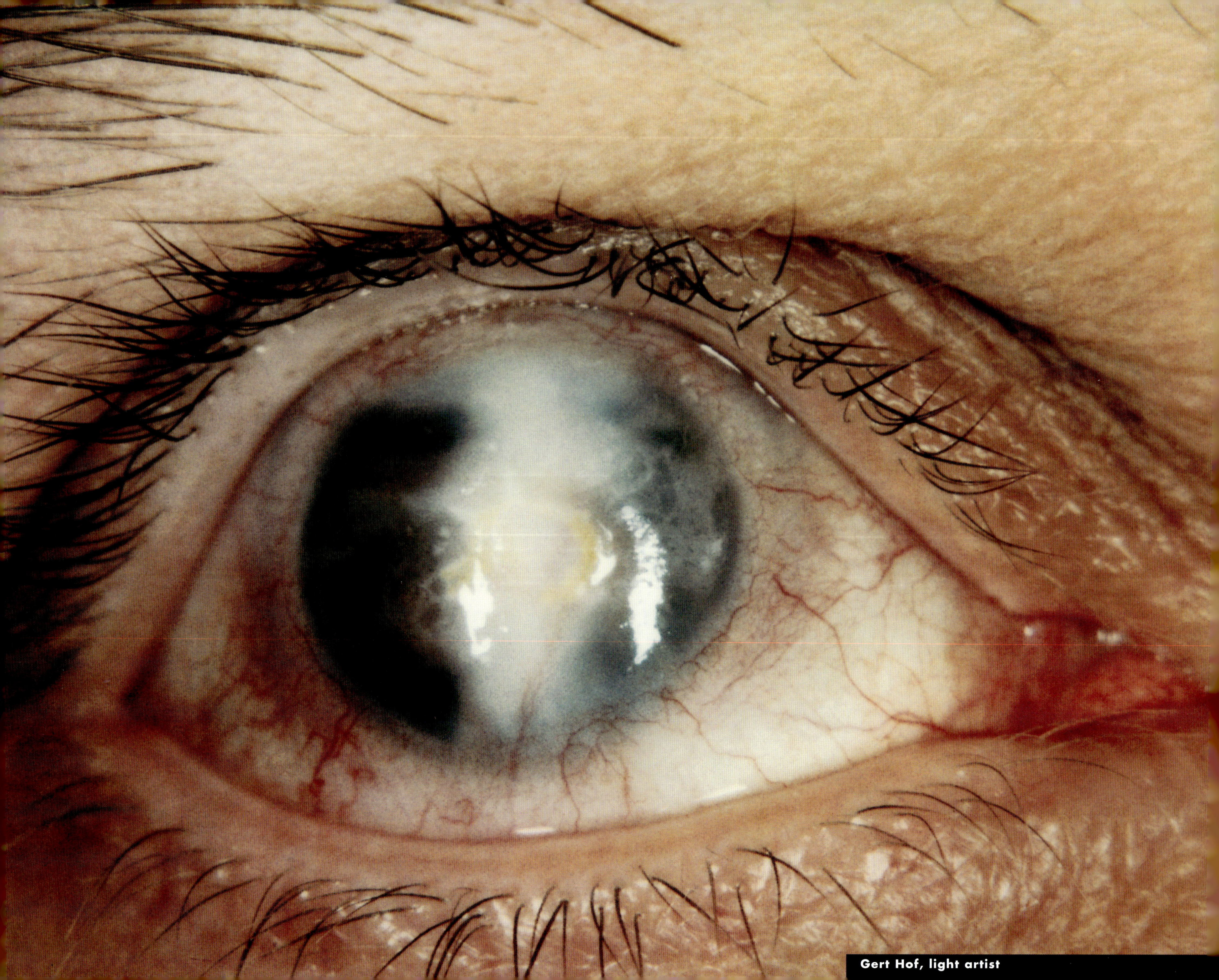

Gert Hof, light artist

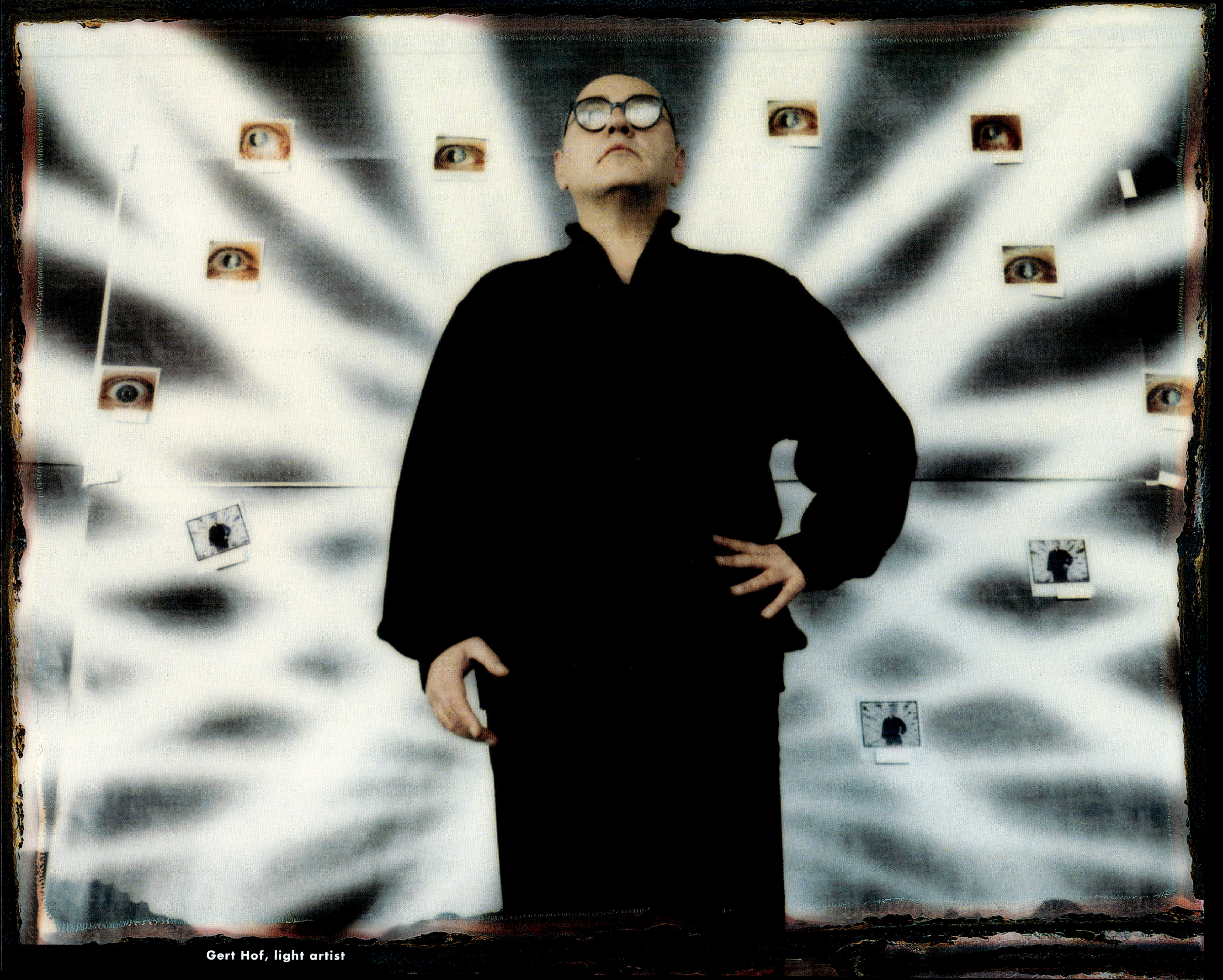

Gert Hof, light artist

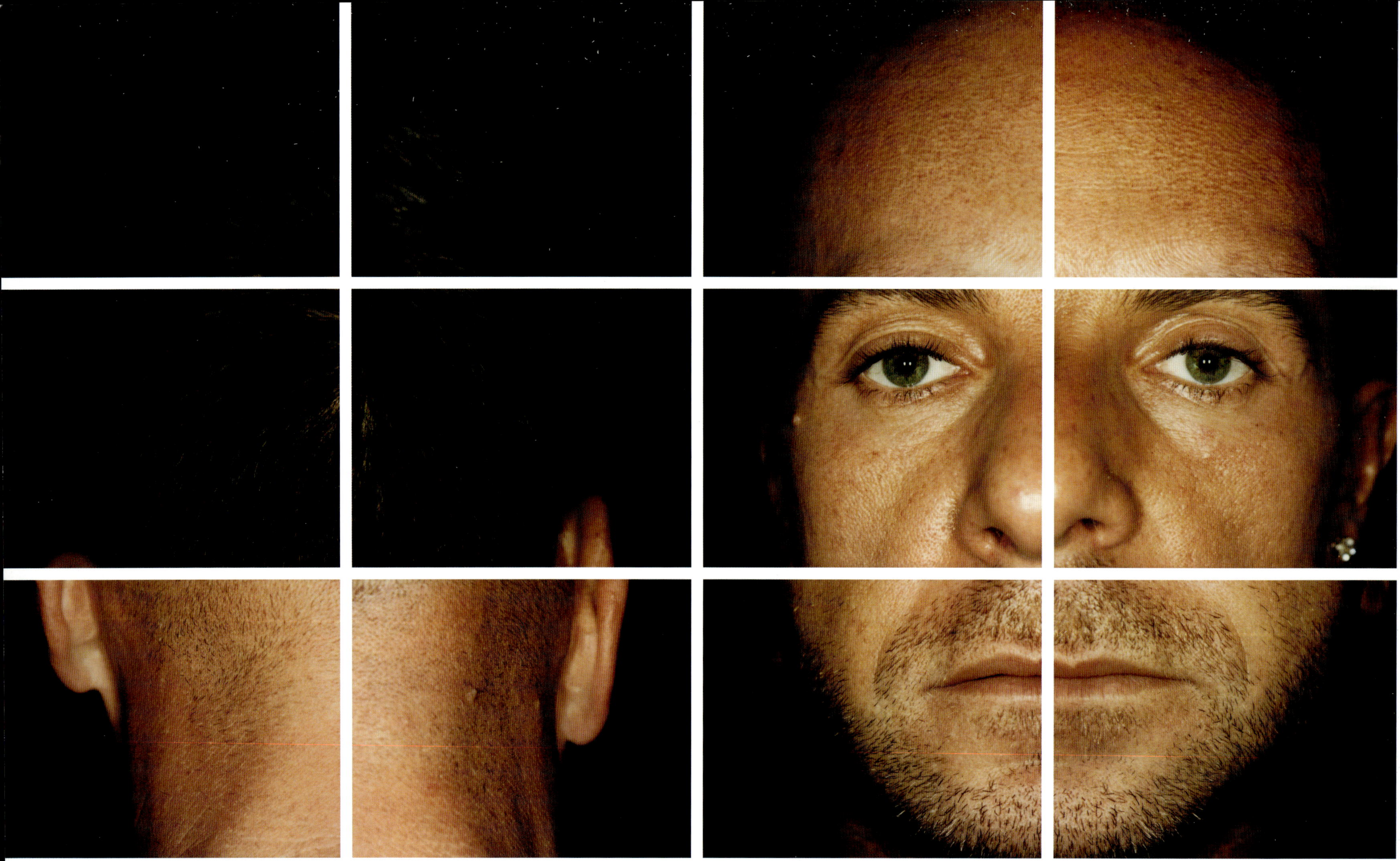

Dolce & Gabbana, designers

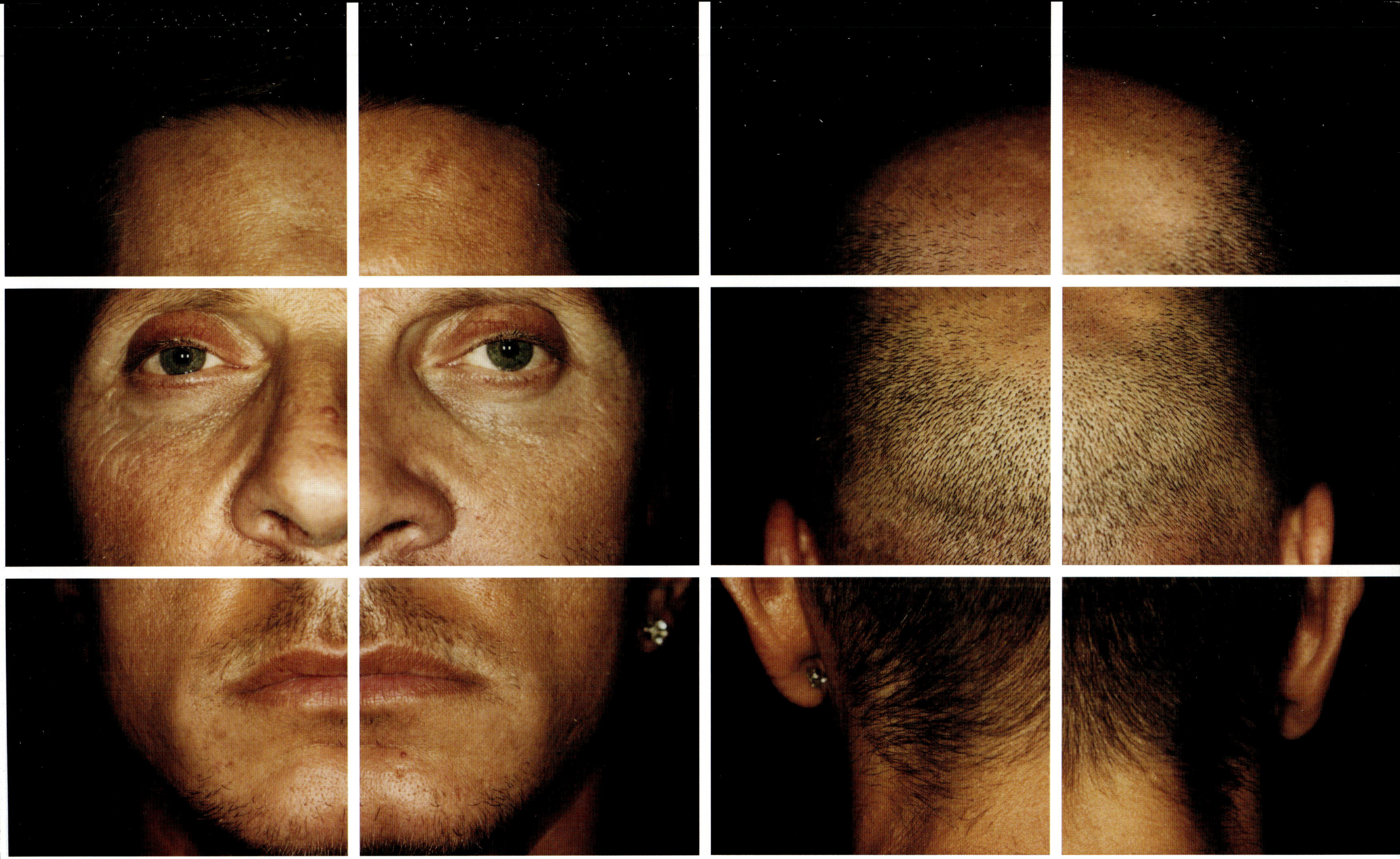

Dolce & Gabbana, designers

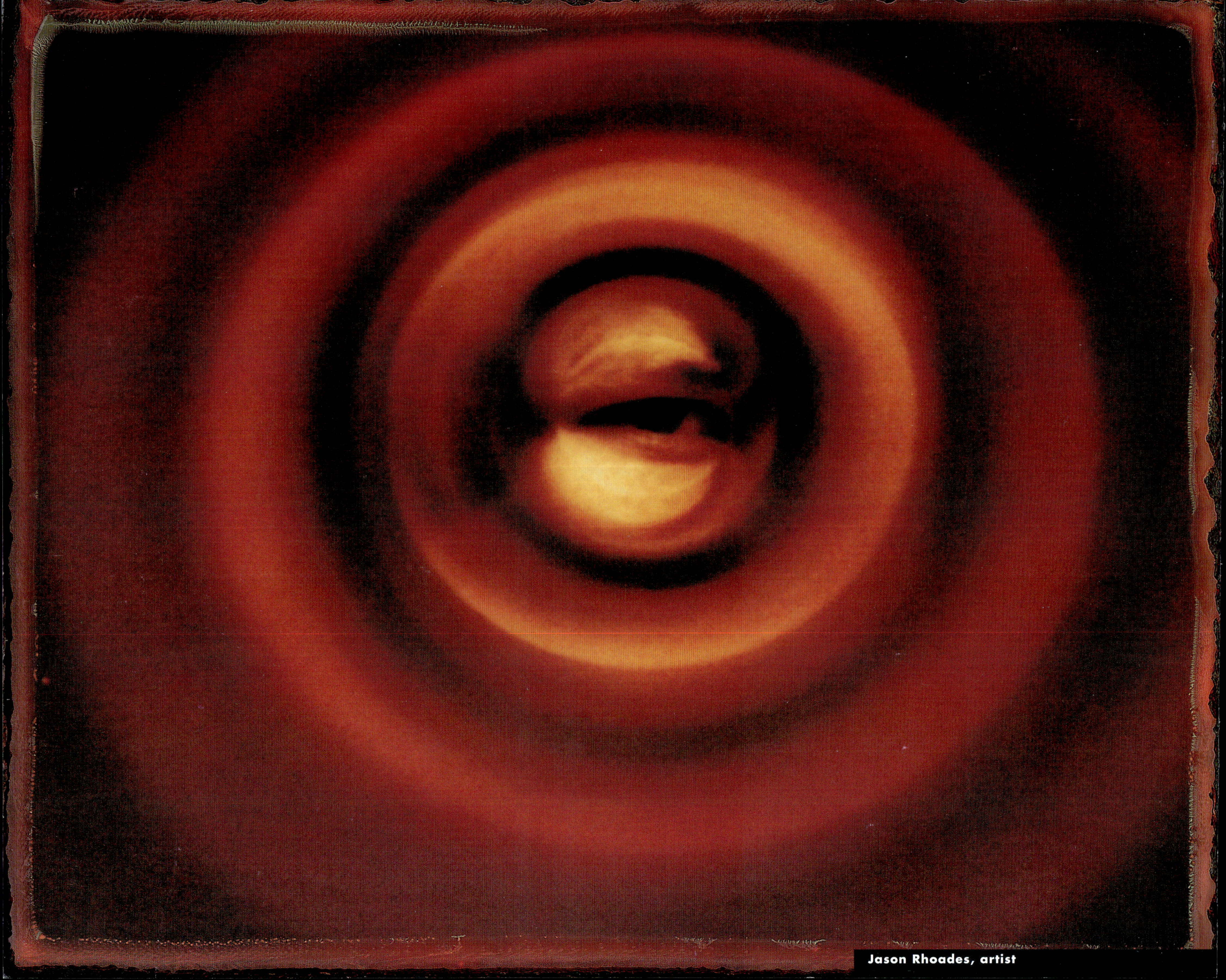

Jason Rhoades, artist

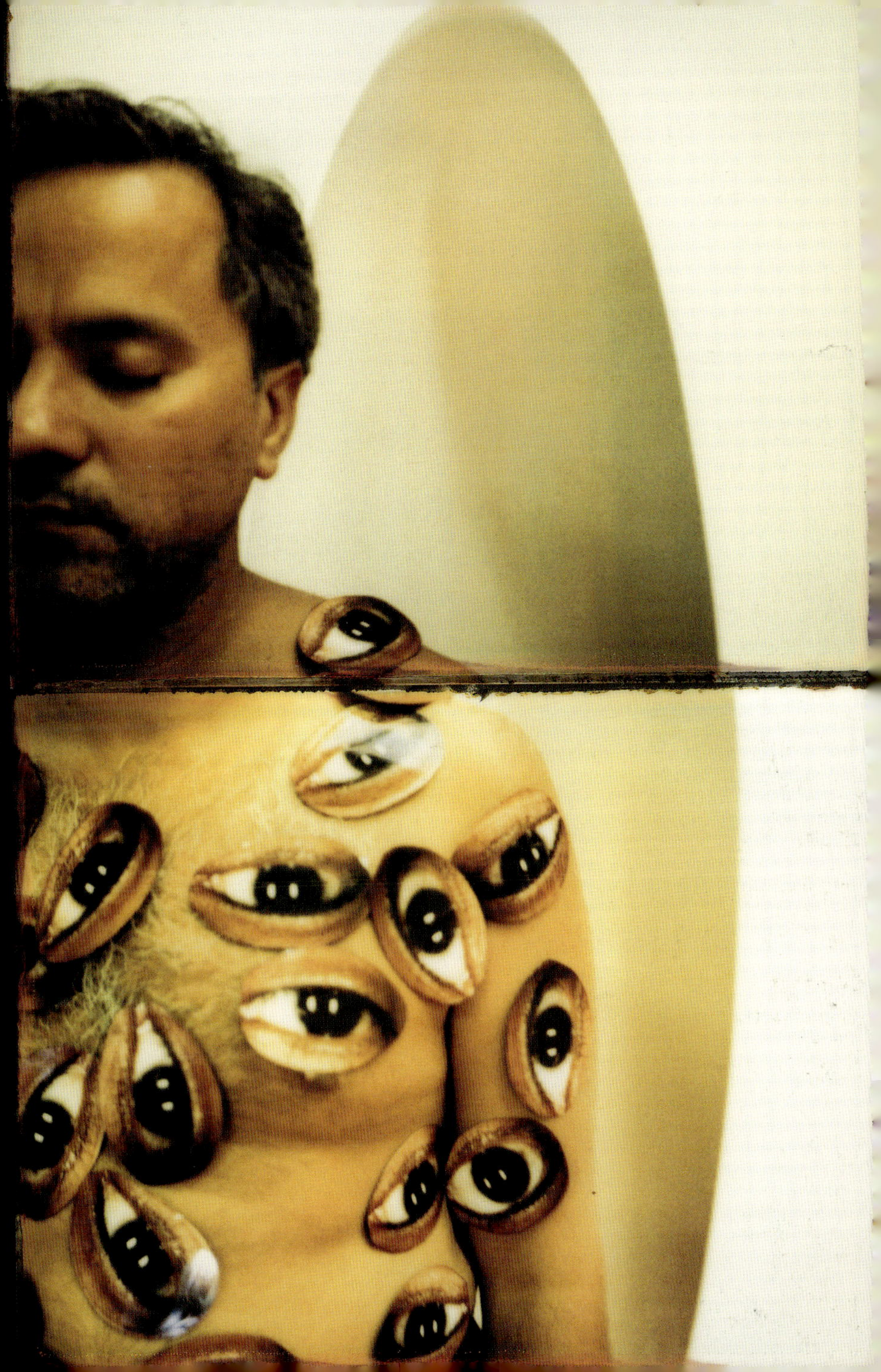

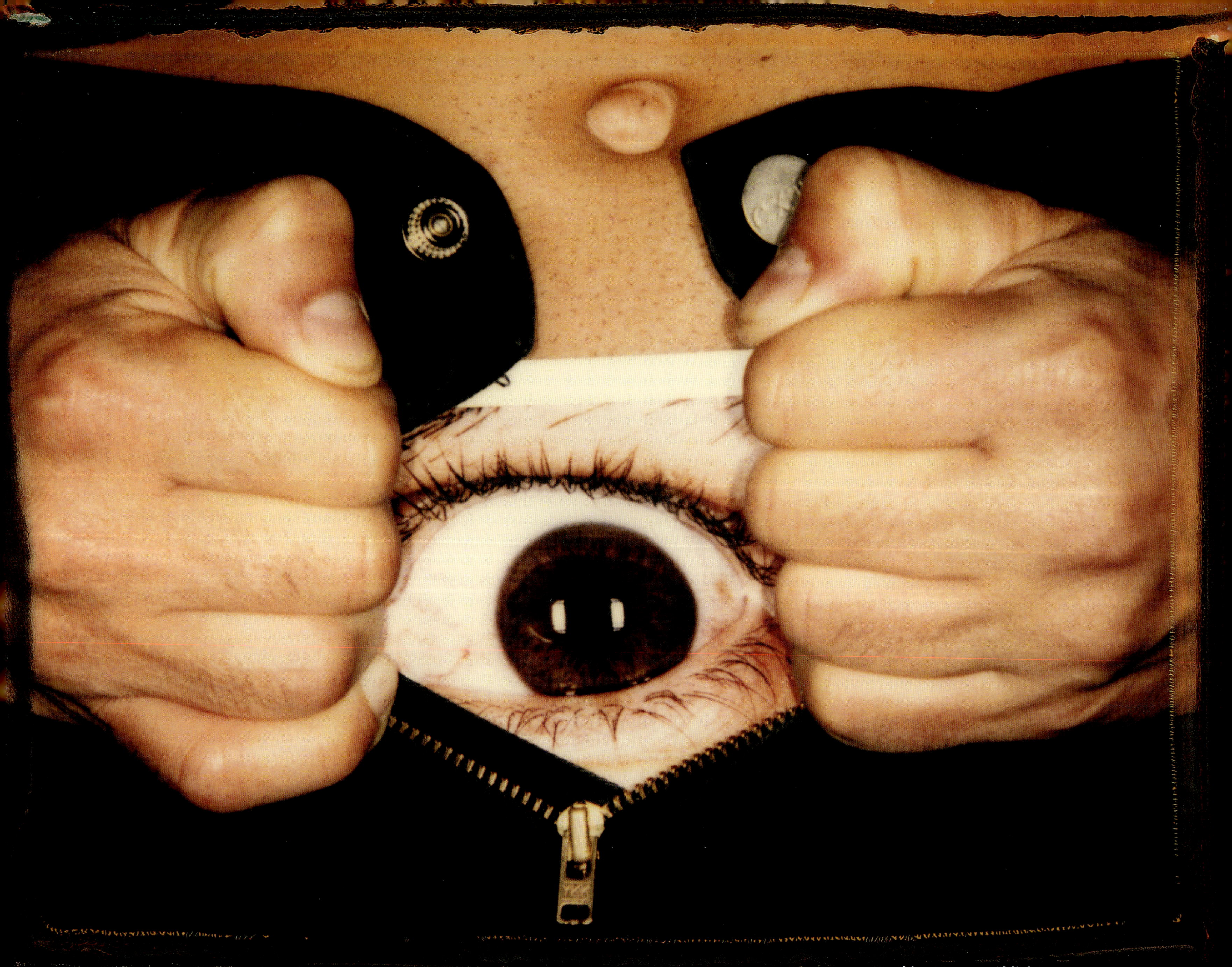

Matthias Reuss, publisher

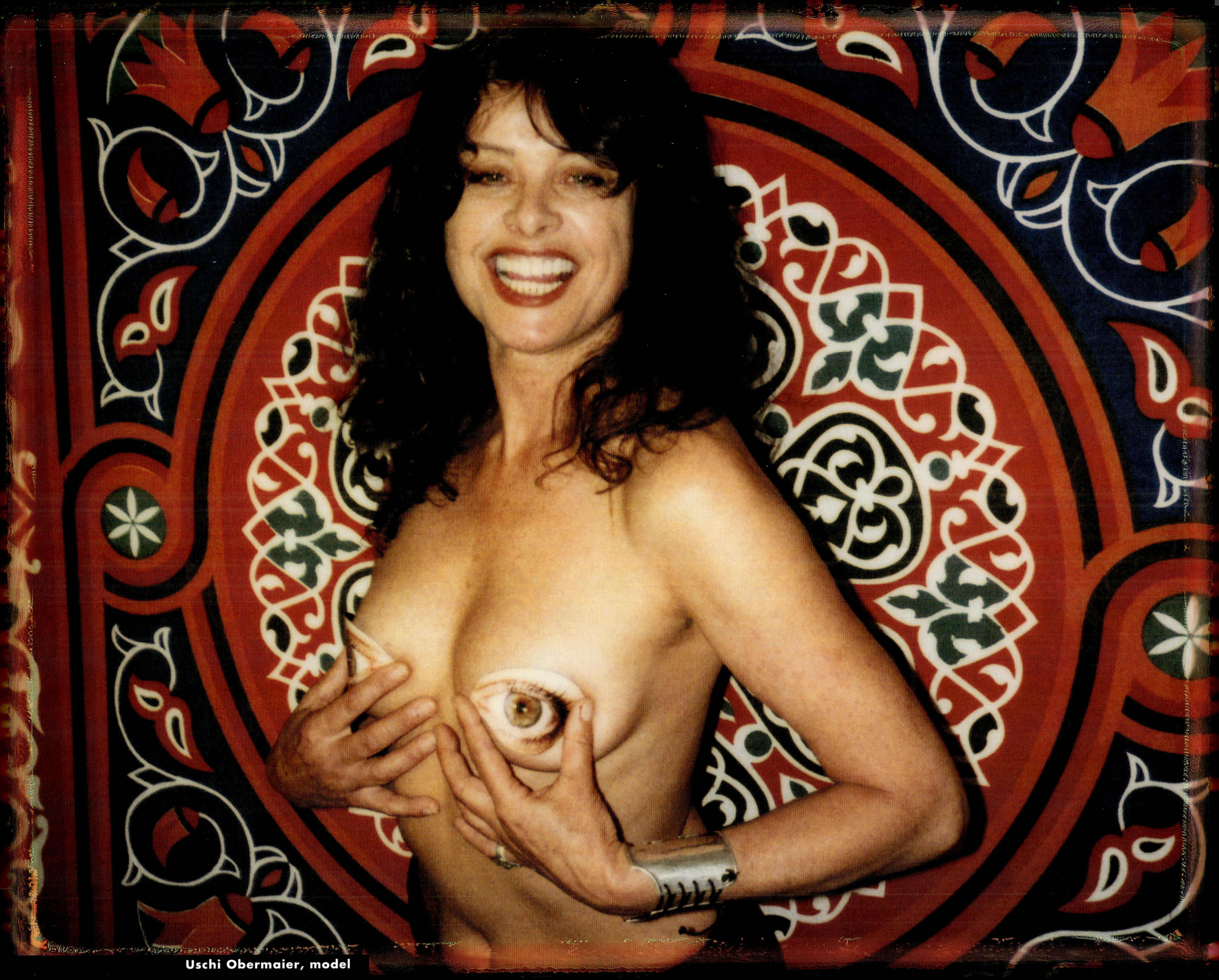

Uschi Obermaier, model

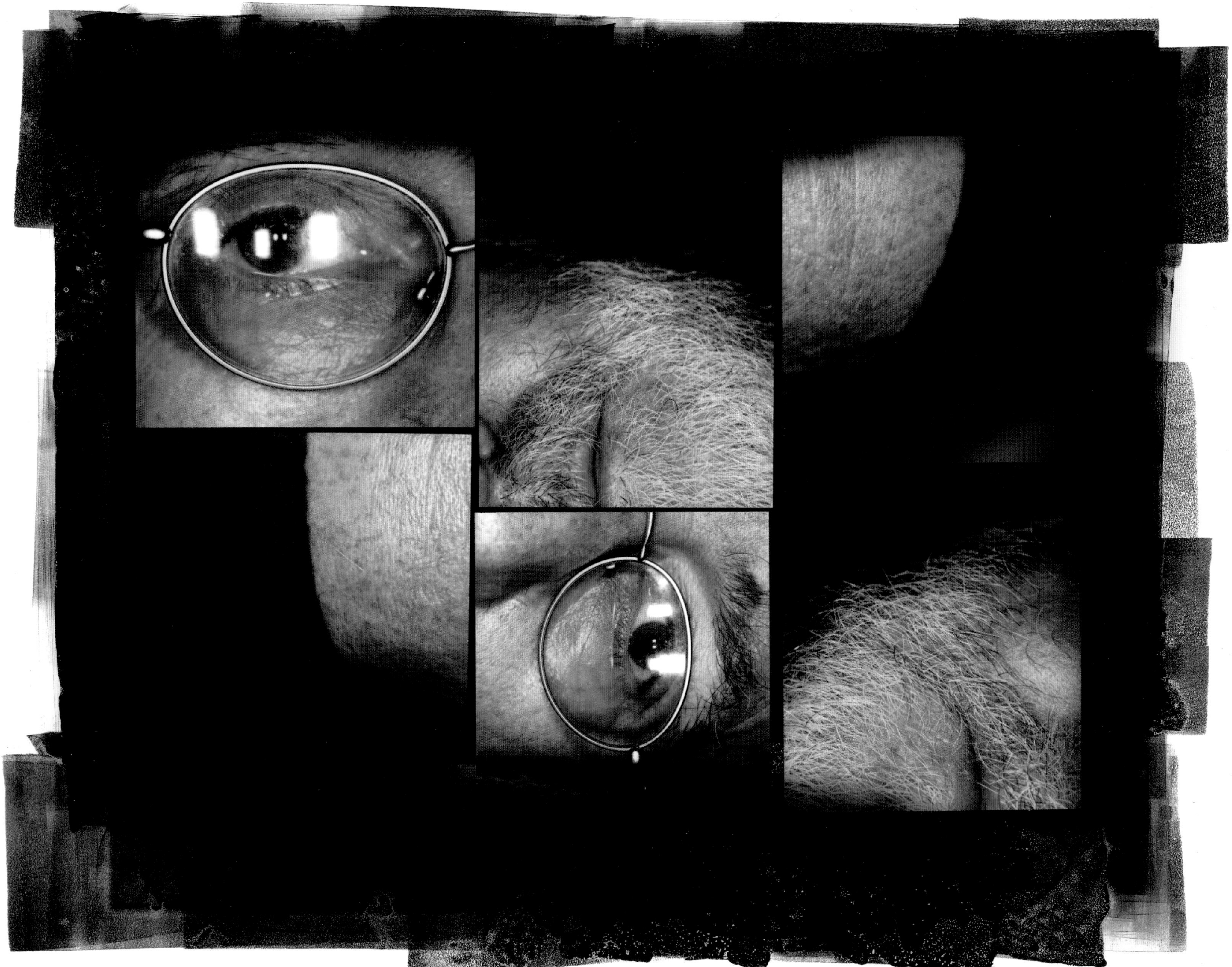

Albert Watson, photographer

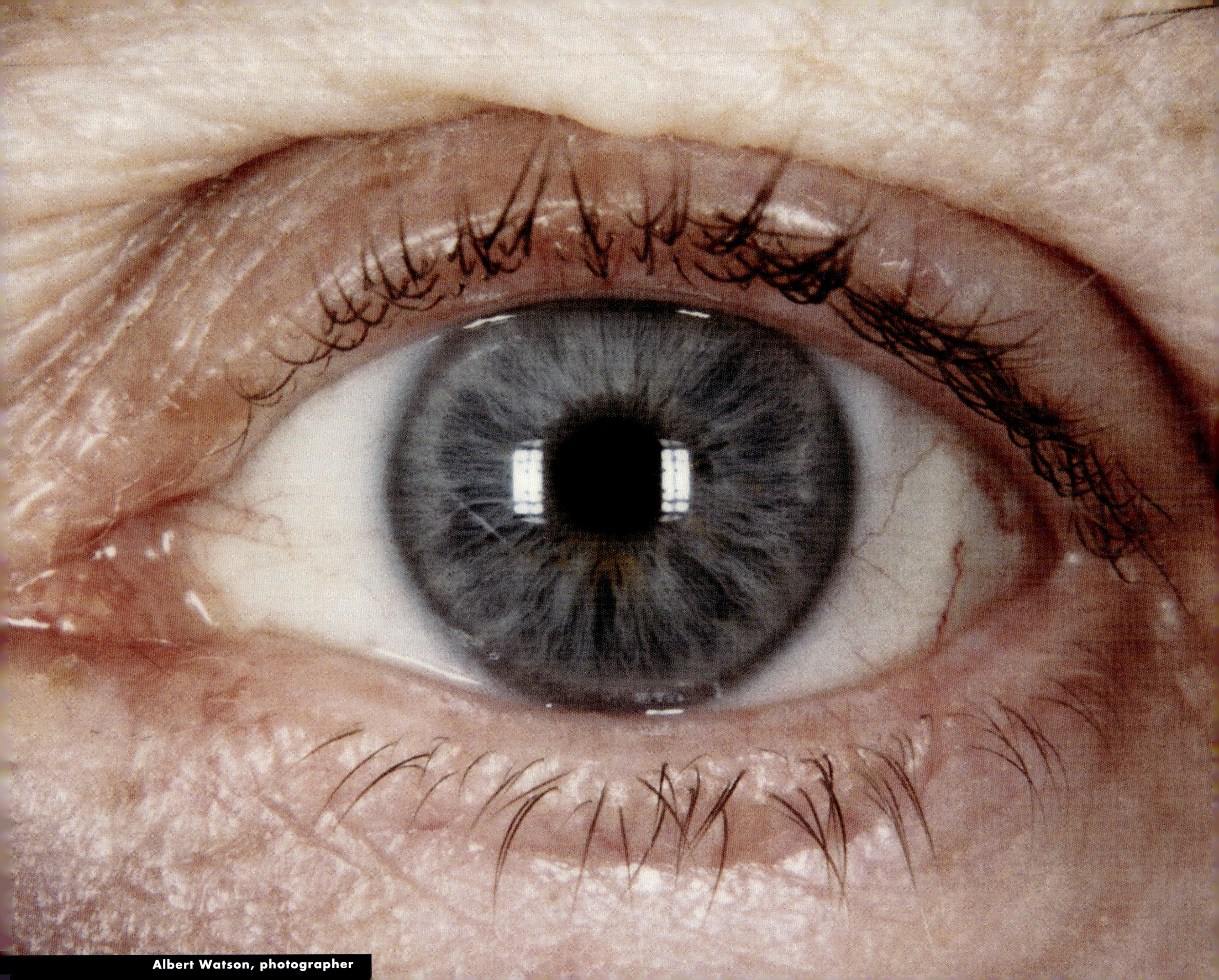

Albert Watson, photographer

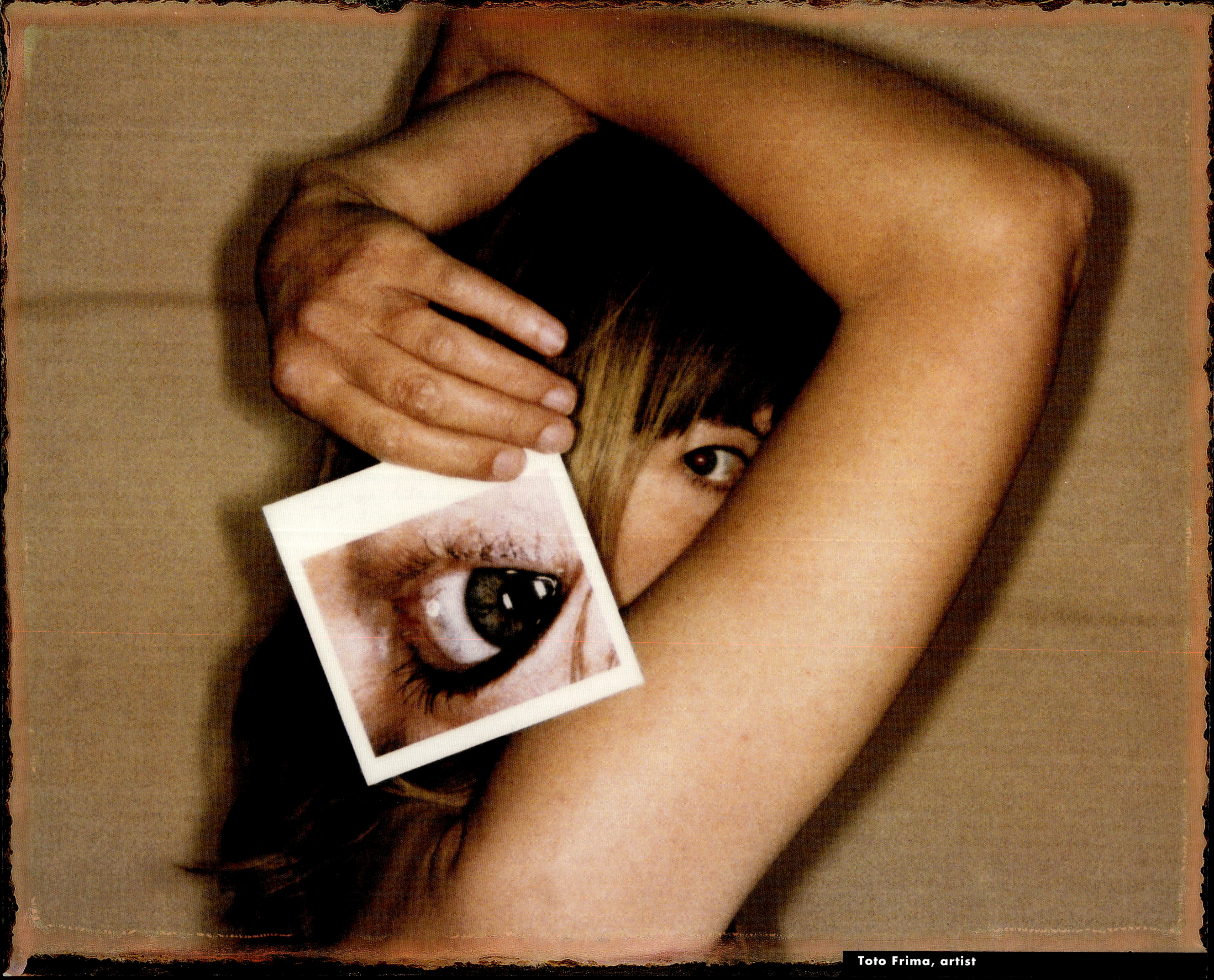

Toto Frima, artist

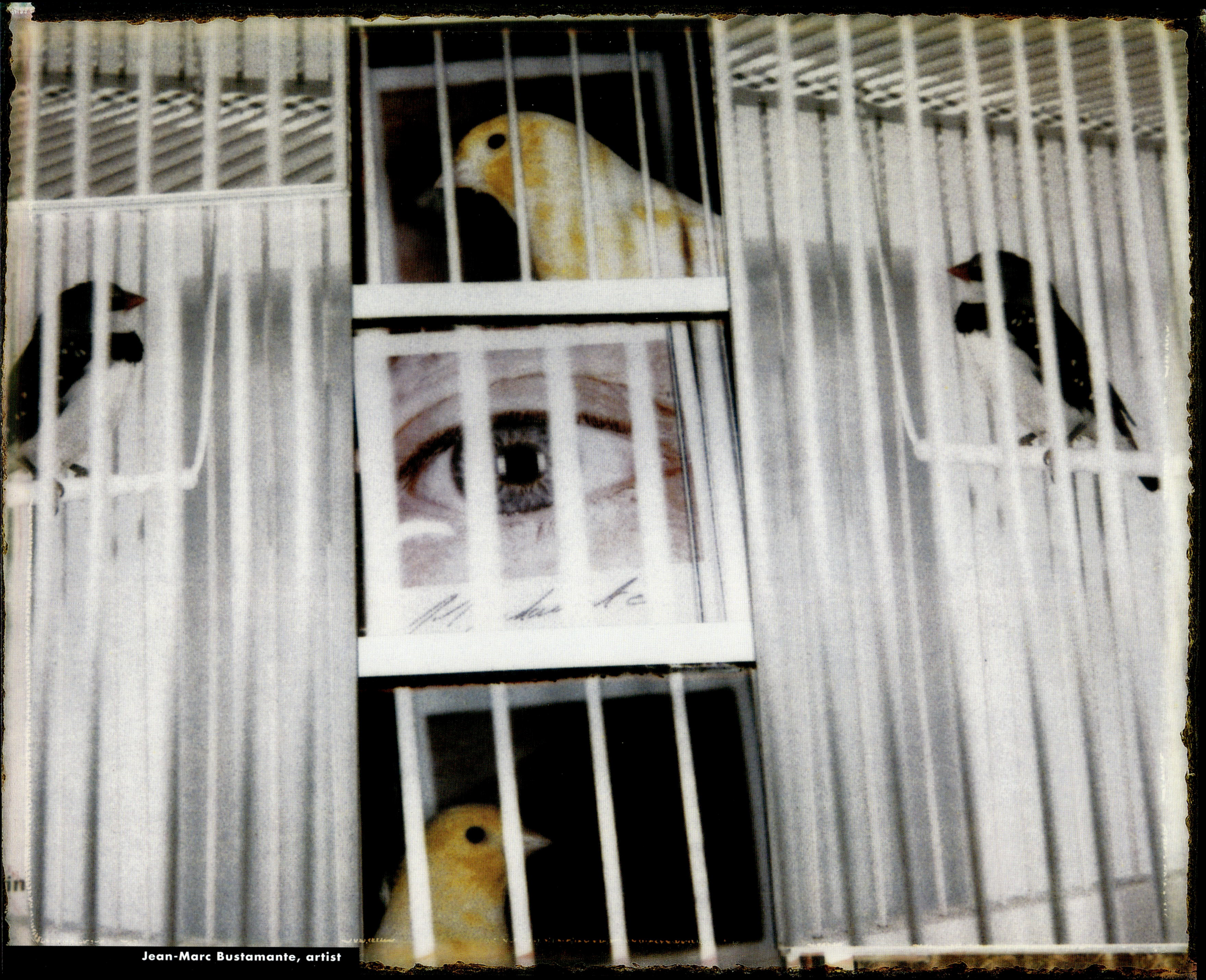

Jean-Marc Bustamante, artist

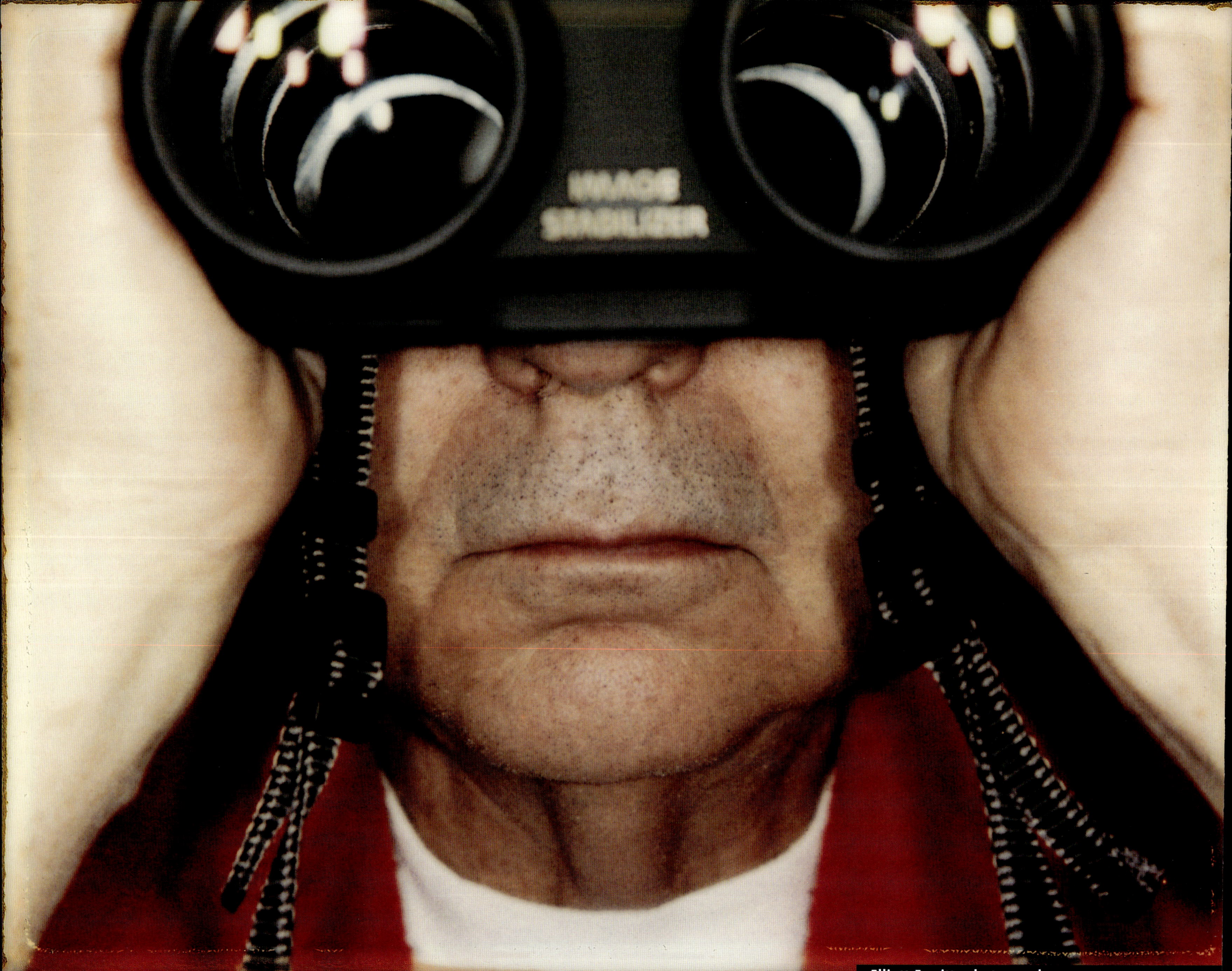

Elliott Erwitt, photographer

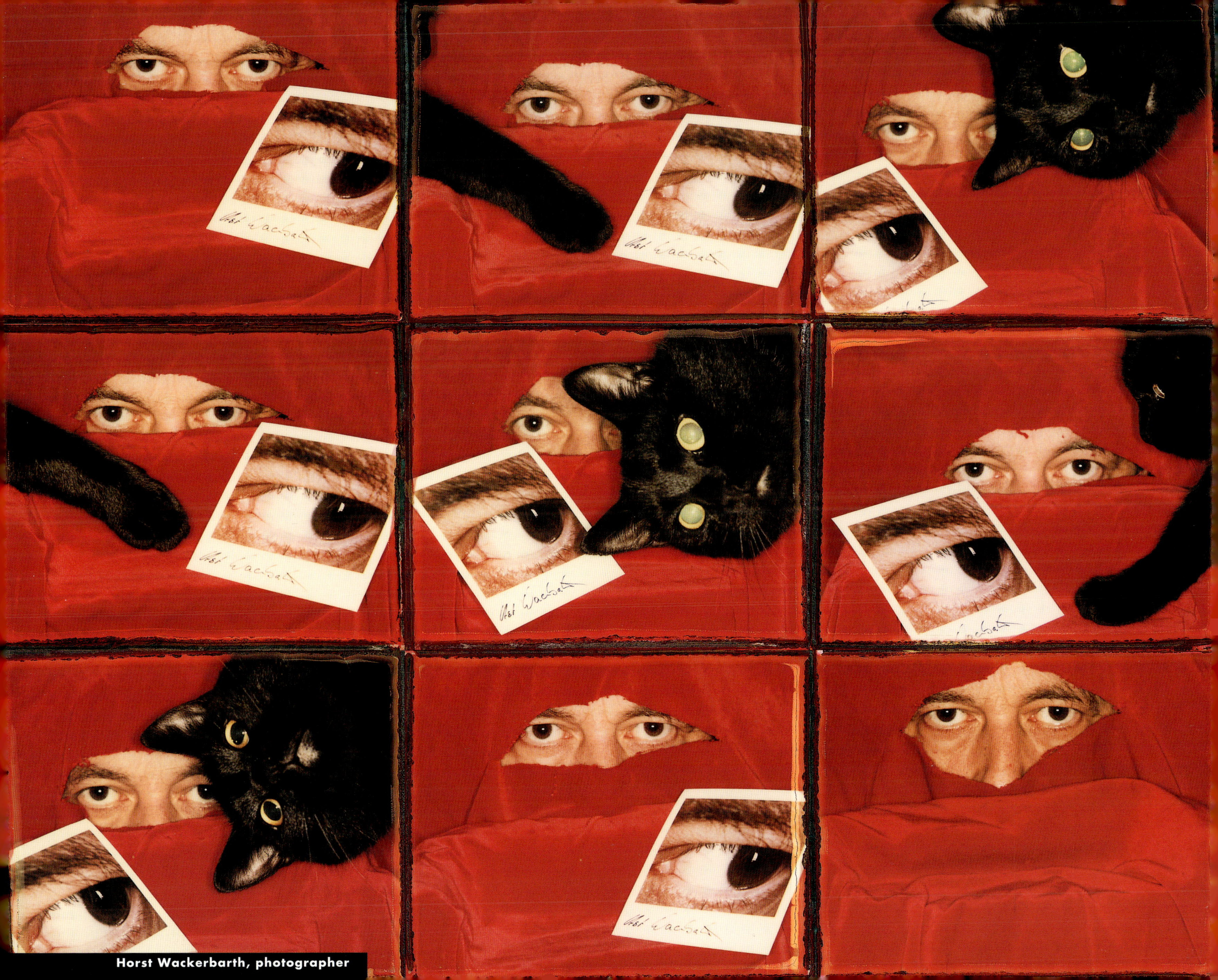

Horst Wackerbarth, photographer

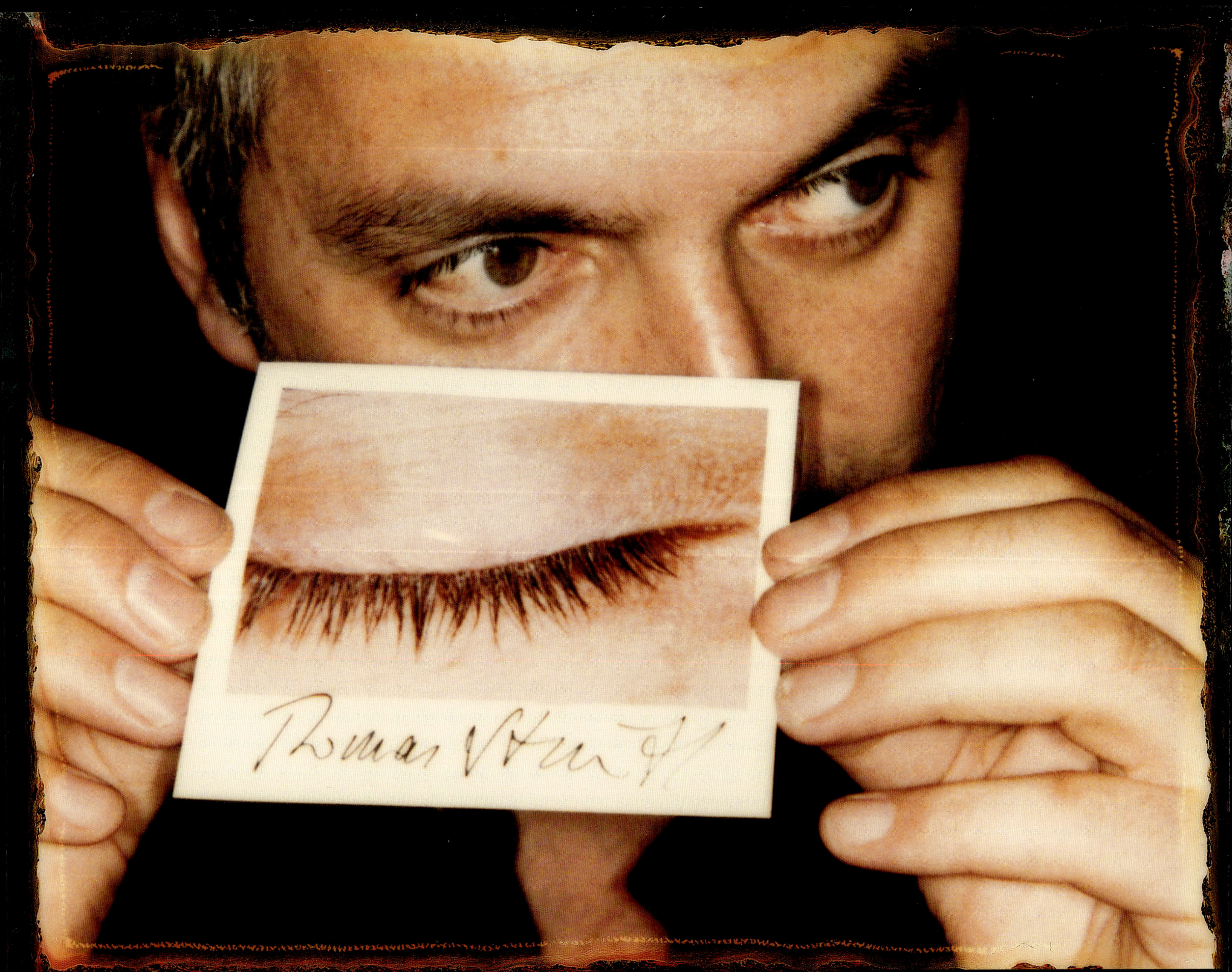

Thomas Struth, artist

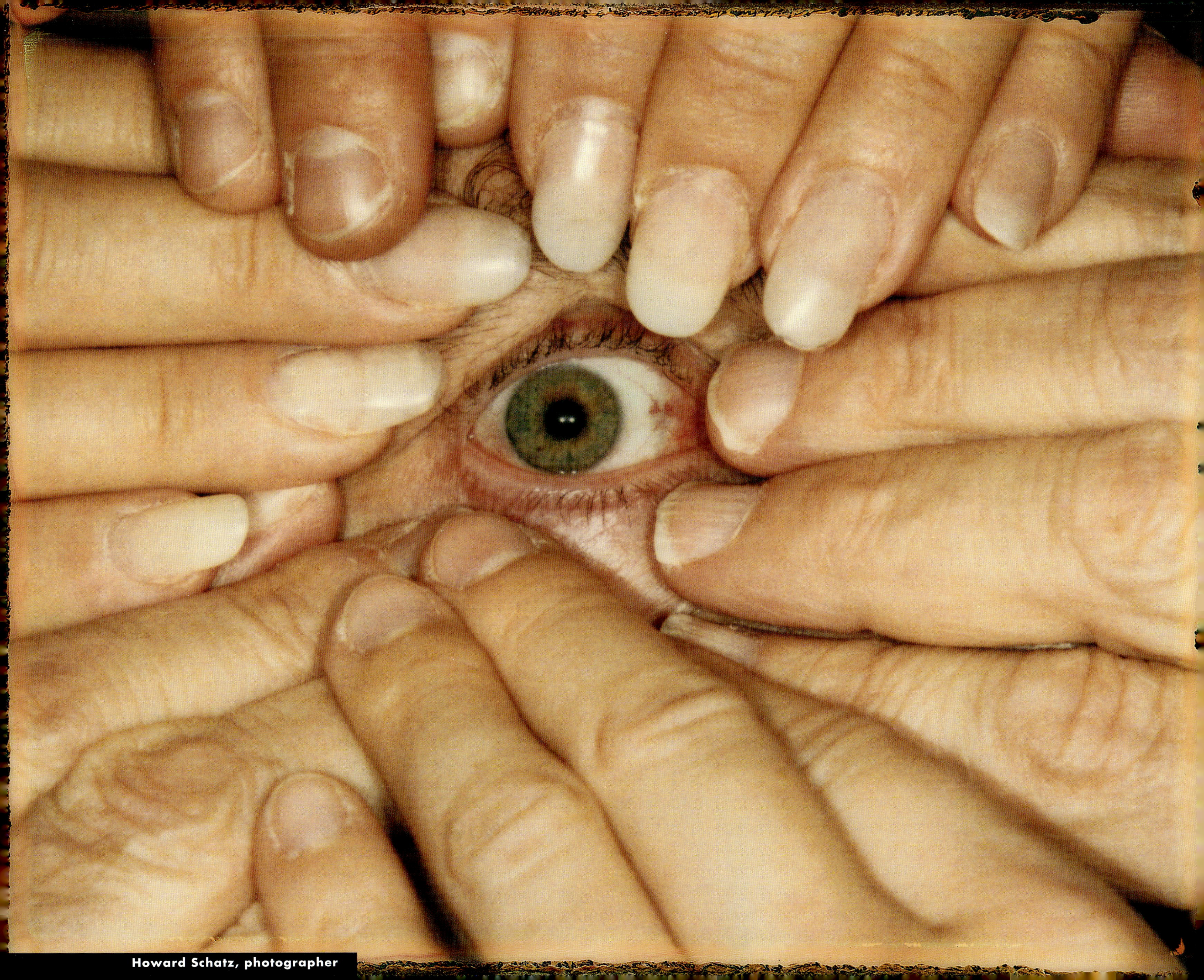

Howard Schatz, photographer

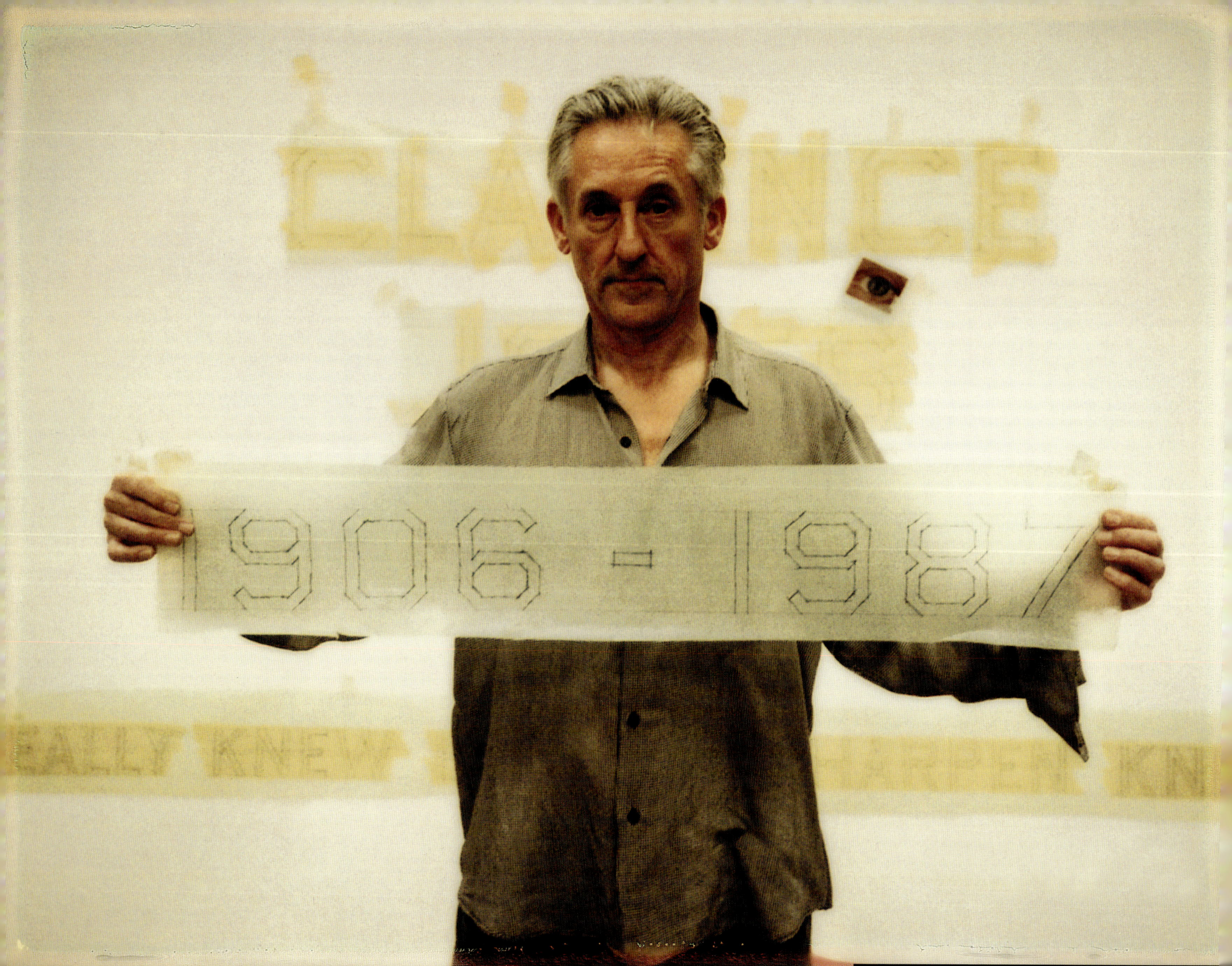

Ed Ruscha, artist

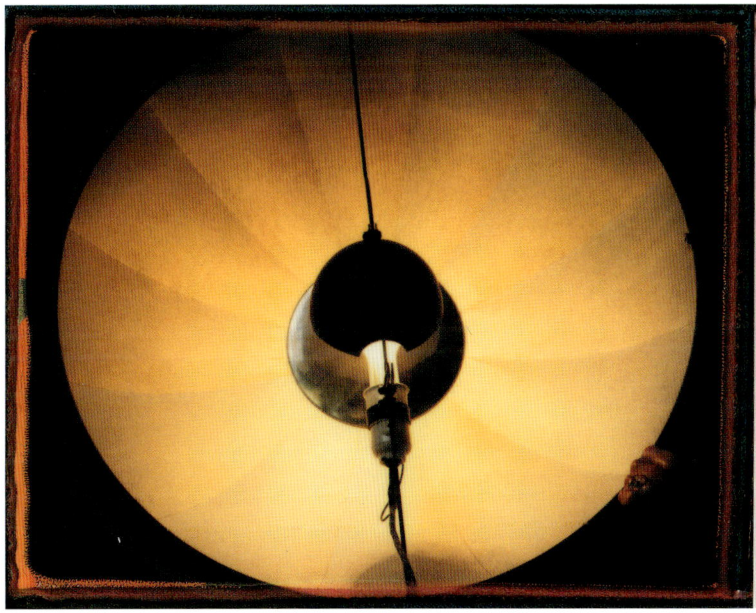
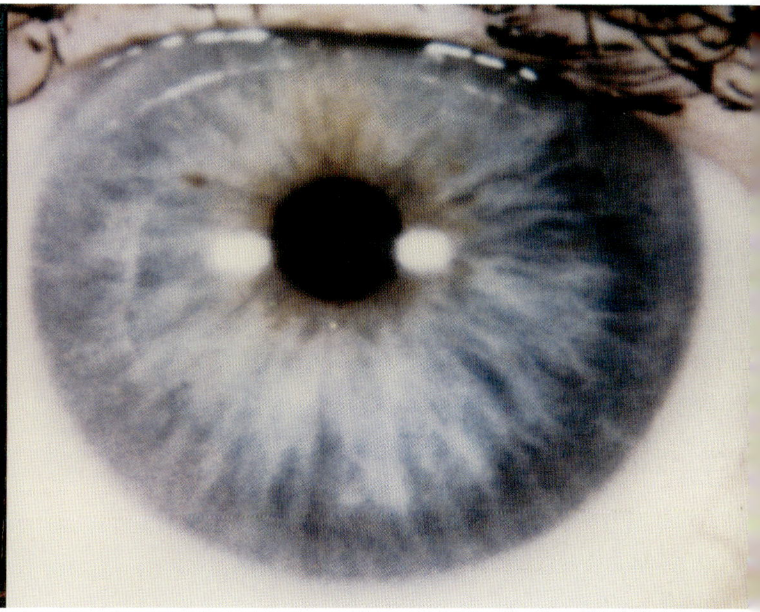

Andrée Putman, designer

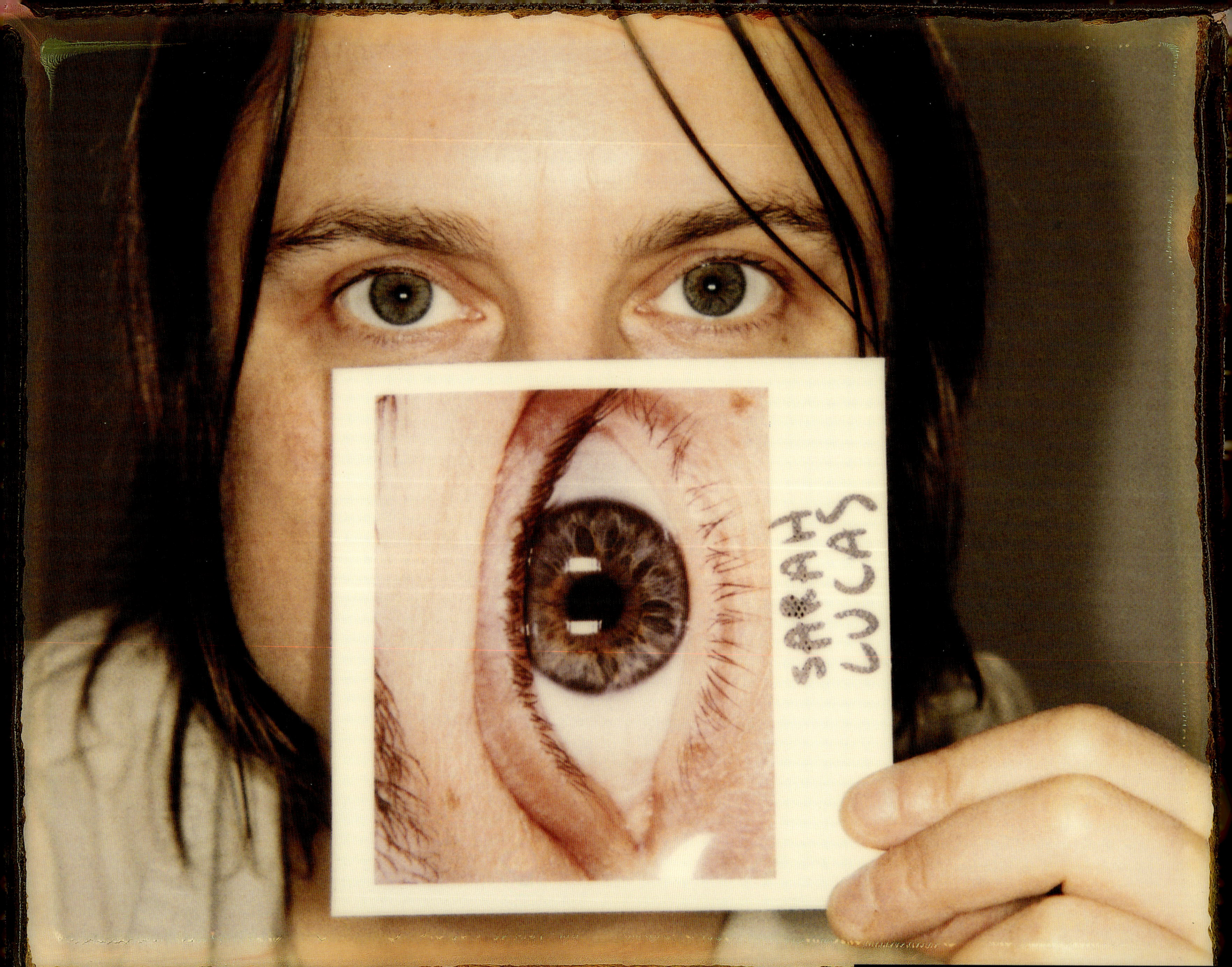

Sarah Lucas, artist

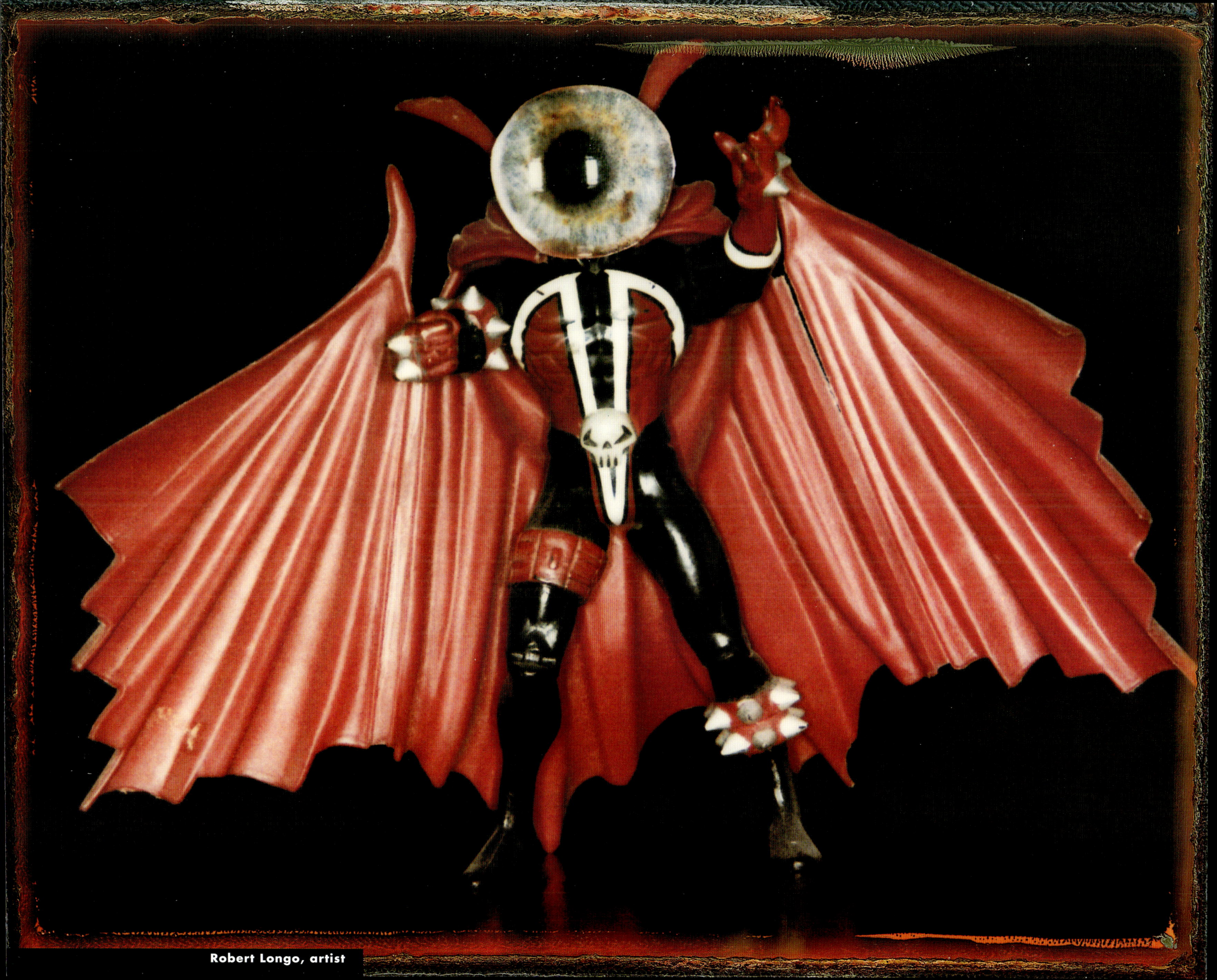

Robert Longo, artist

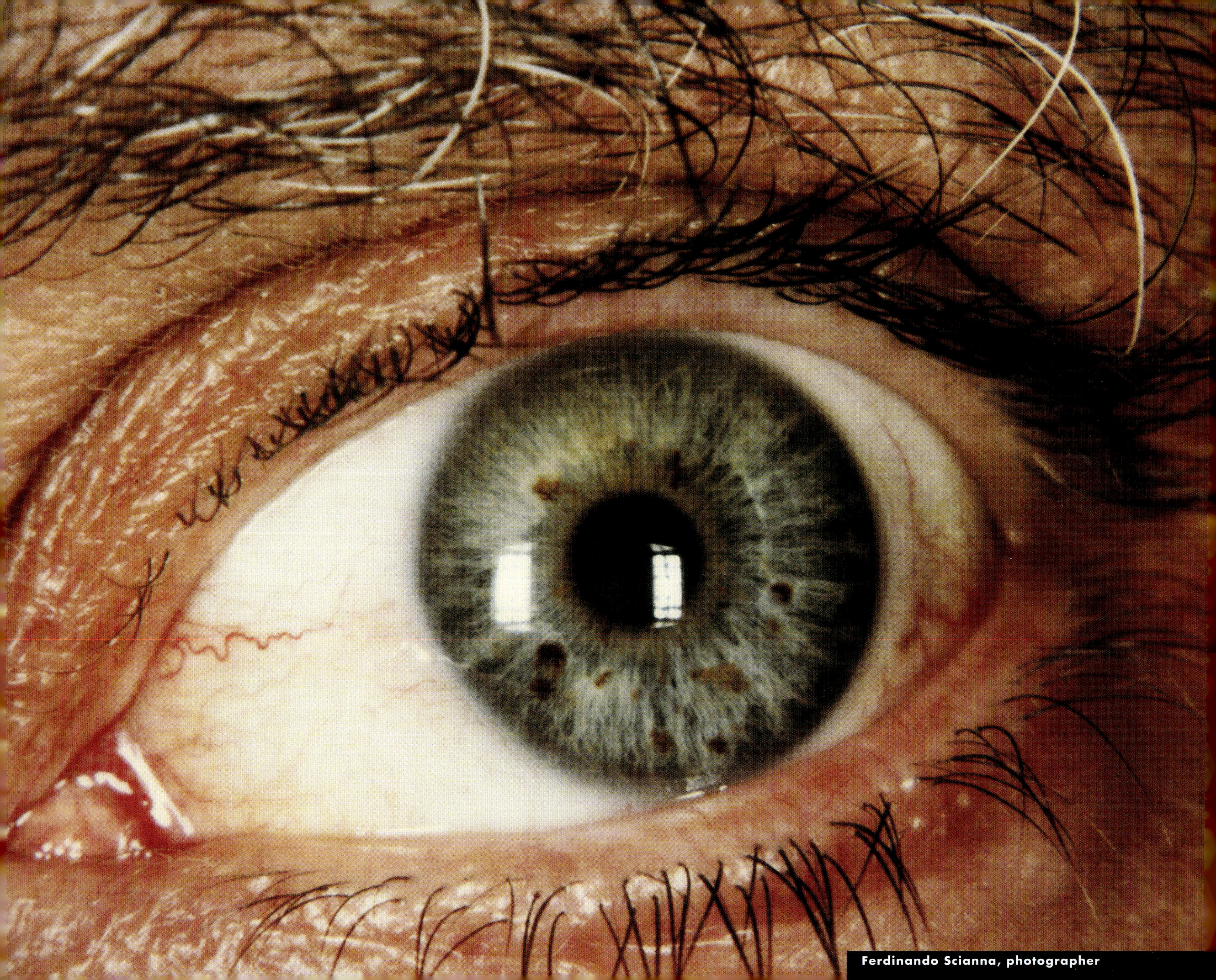

Ferdinando Scianna, photographer

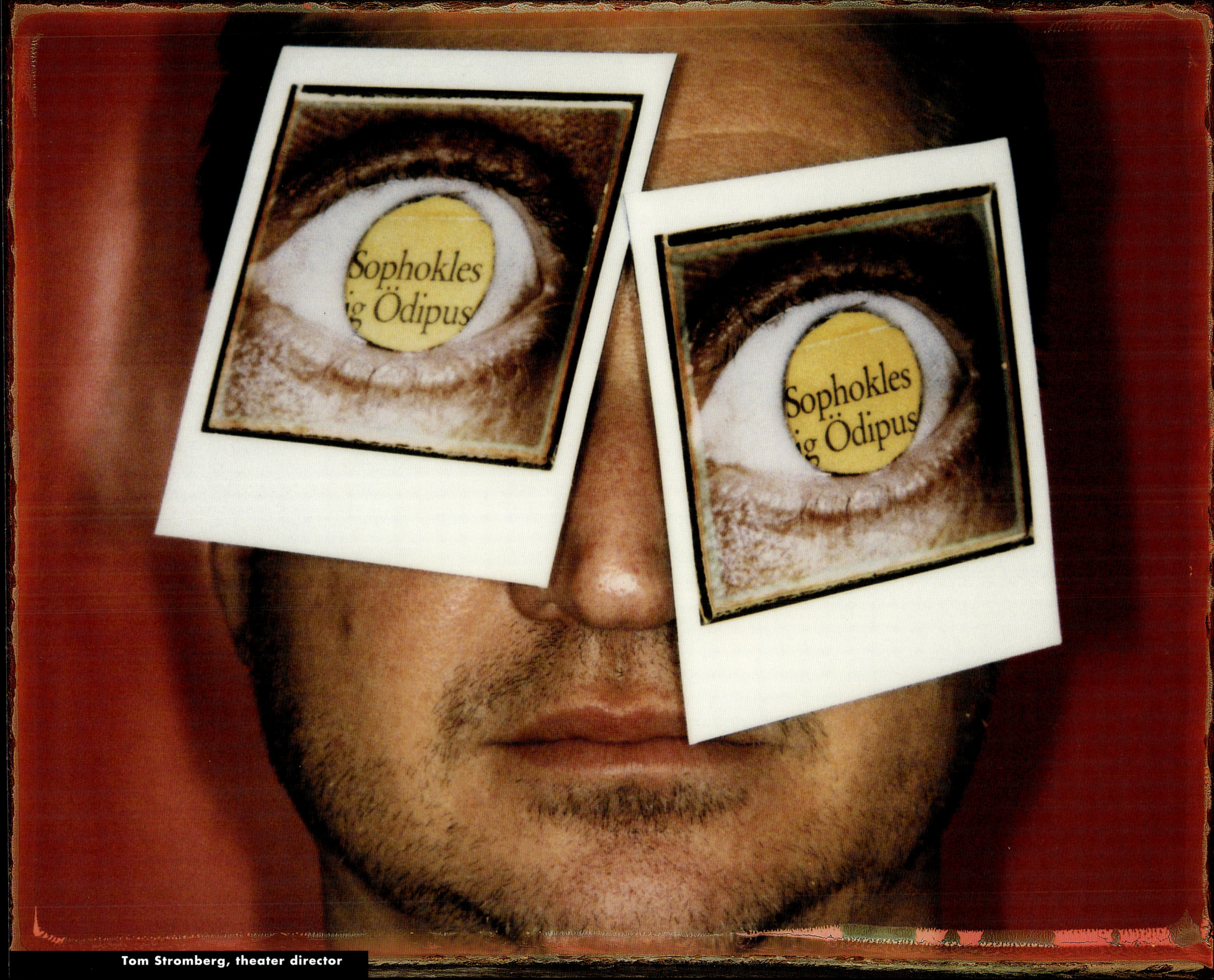

Tom Stromberg, theater director

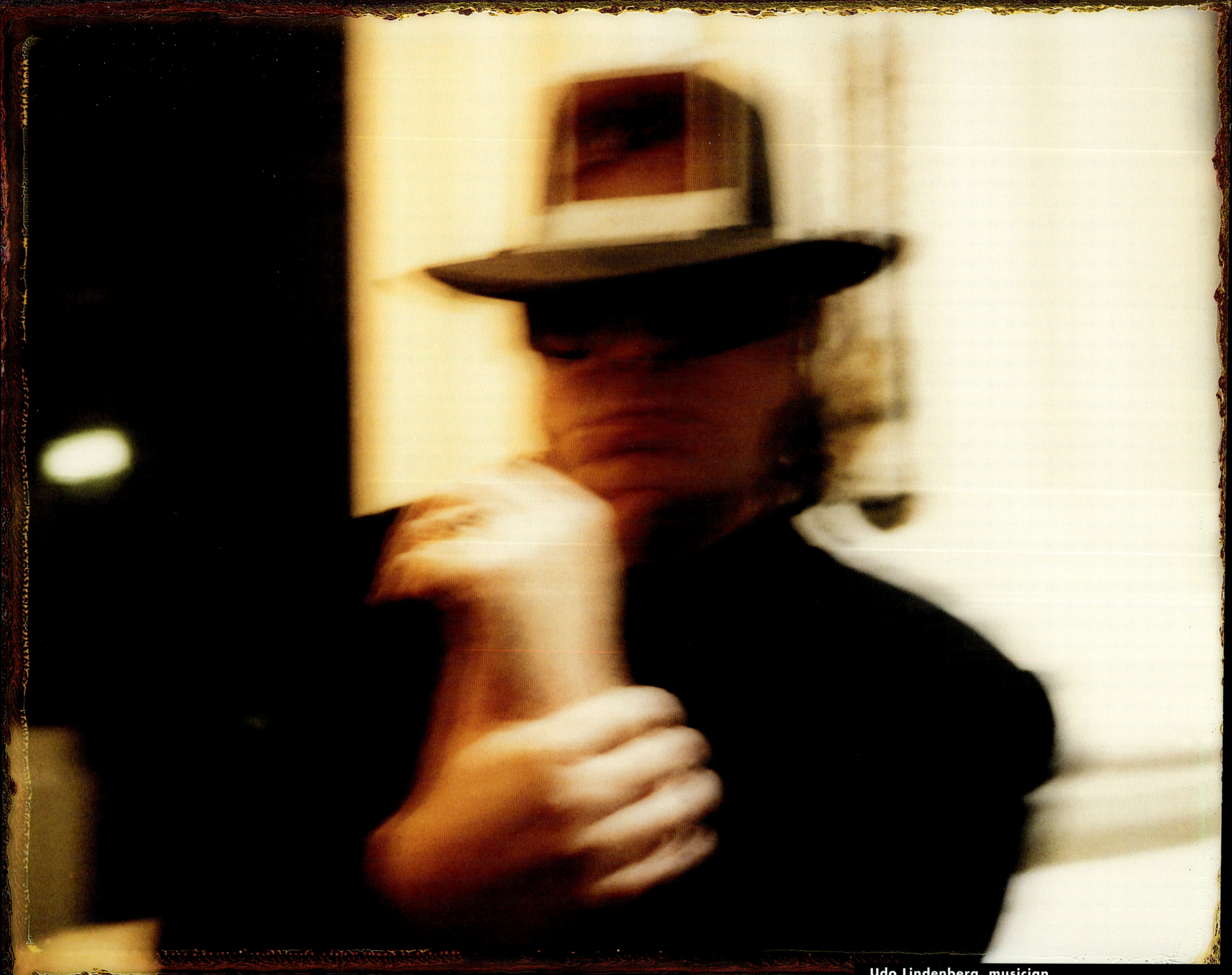

Udo Lindenberg, musician

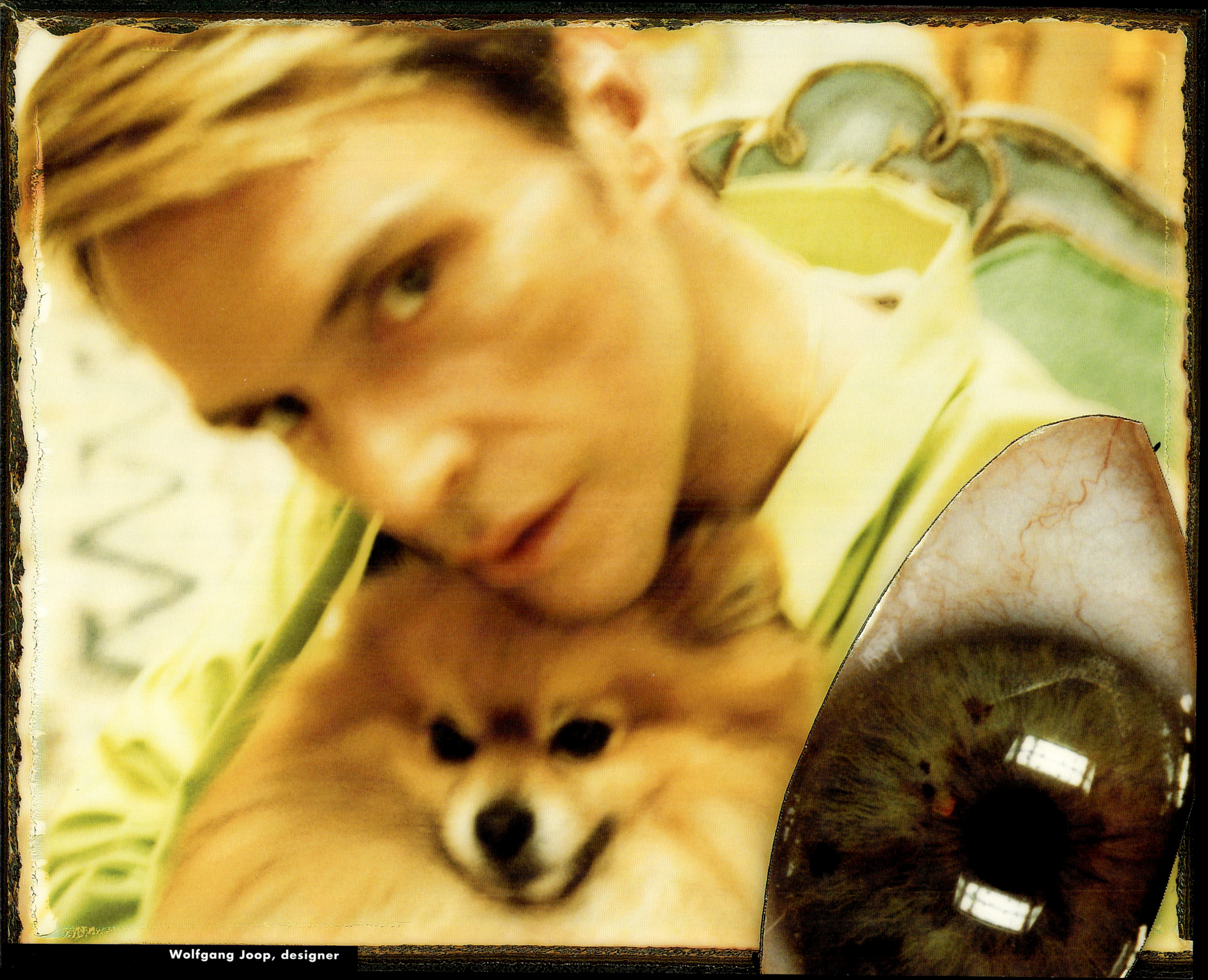

Wolfgang Joop, designer

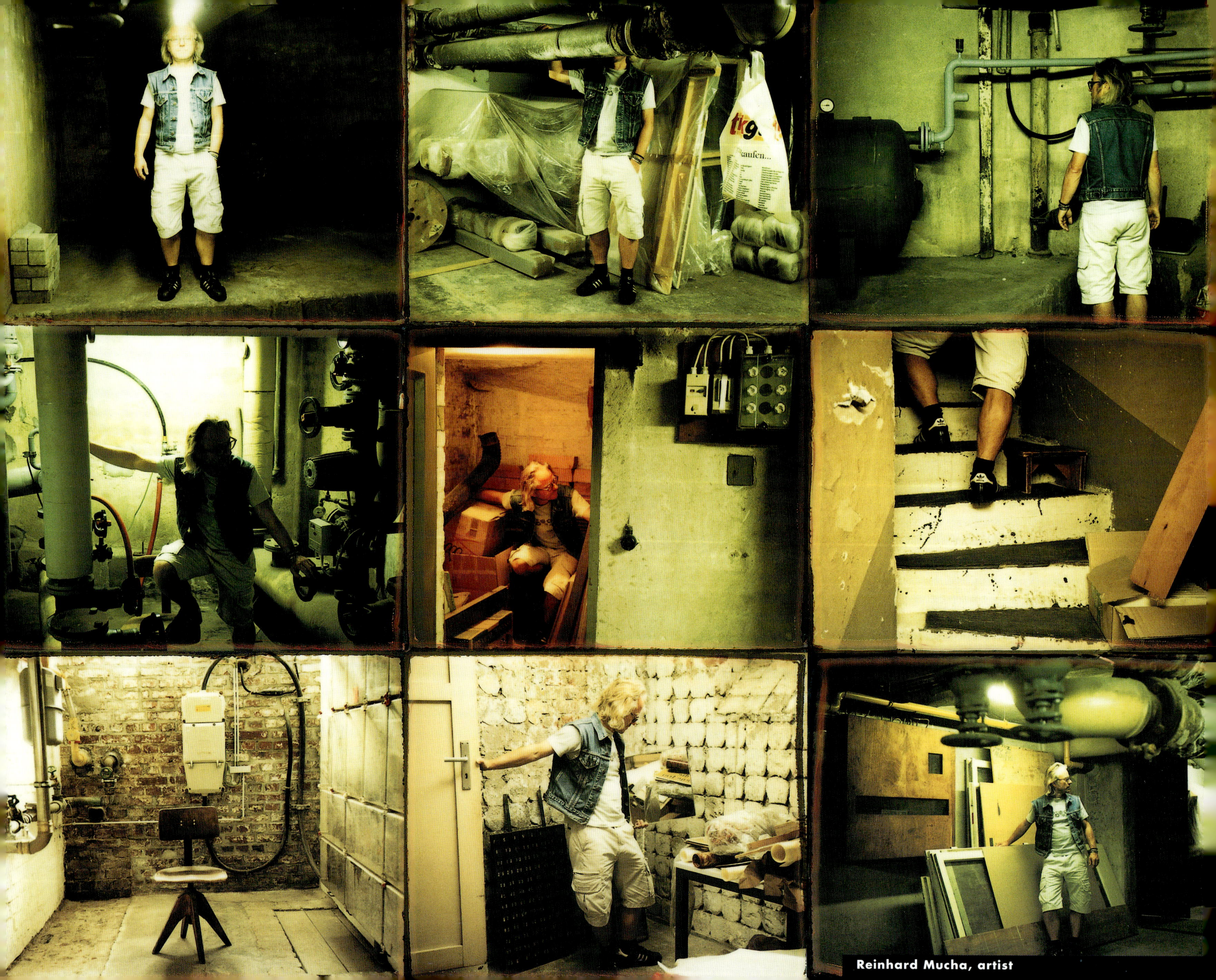

Reinhard Mucha, artist

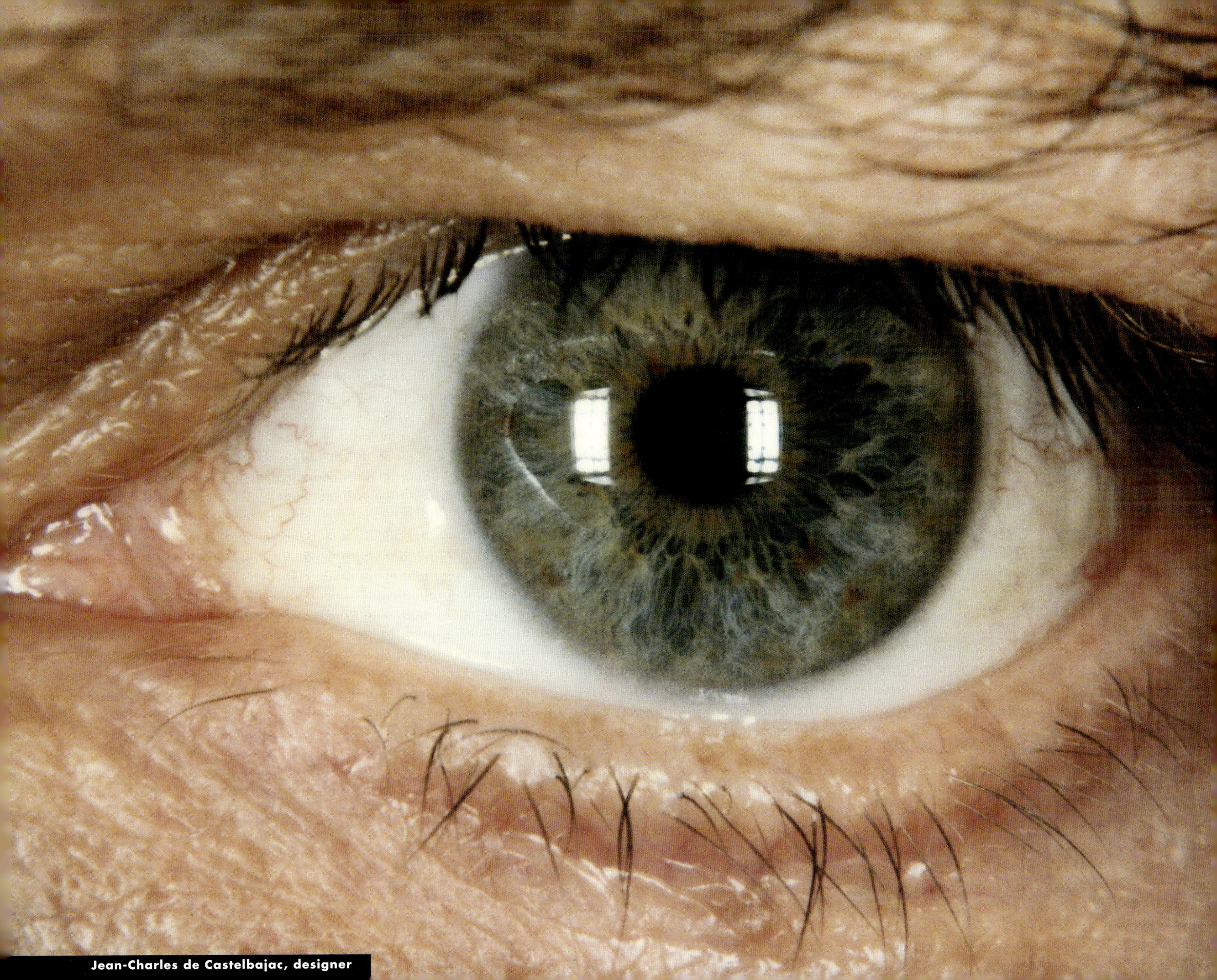

Jean-Charles de Castelbajac, designer

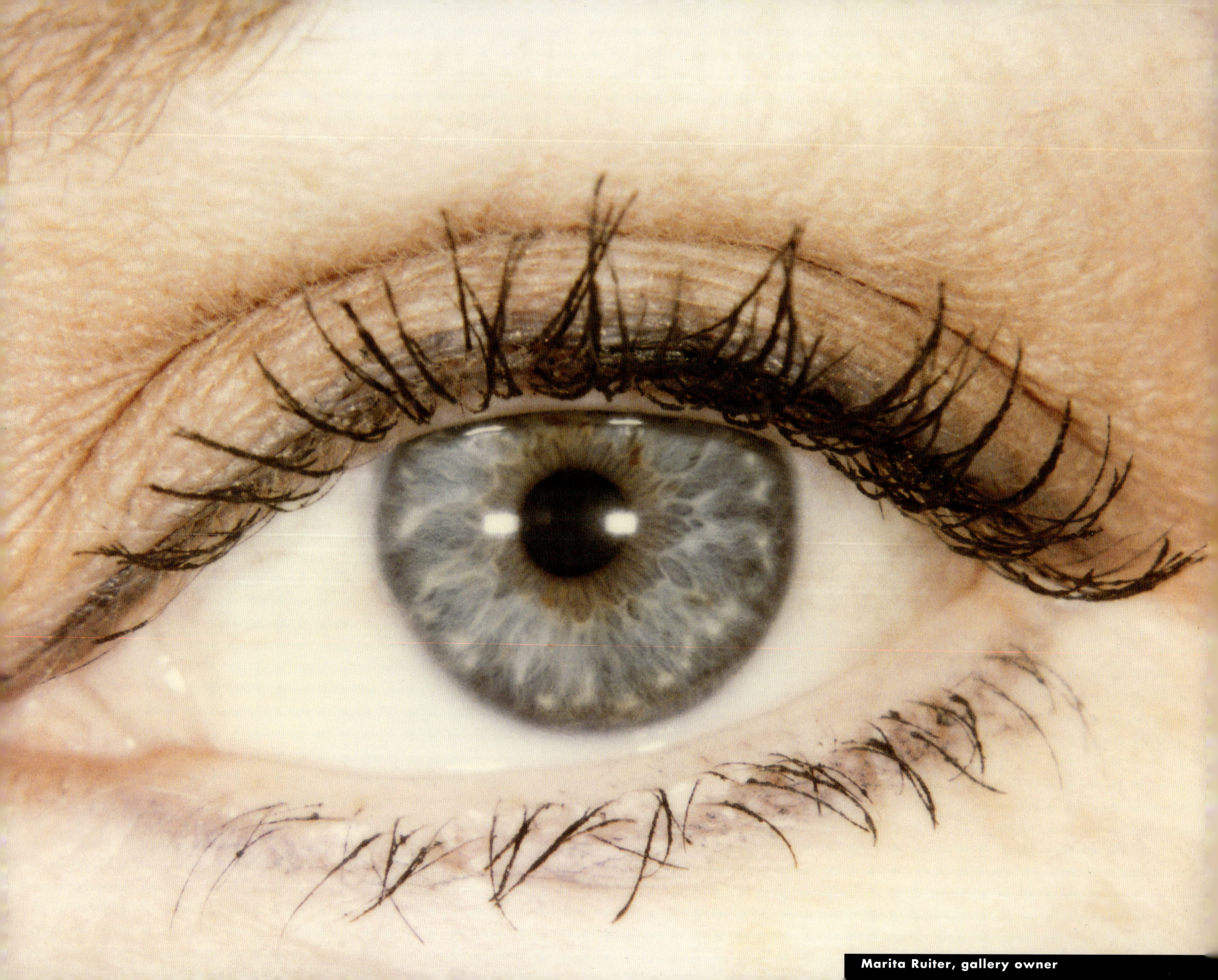

Marita Ruiter, gallery owner

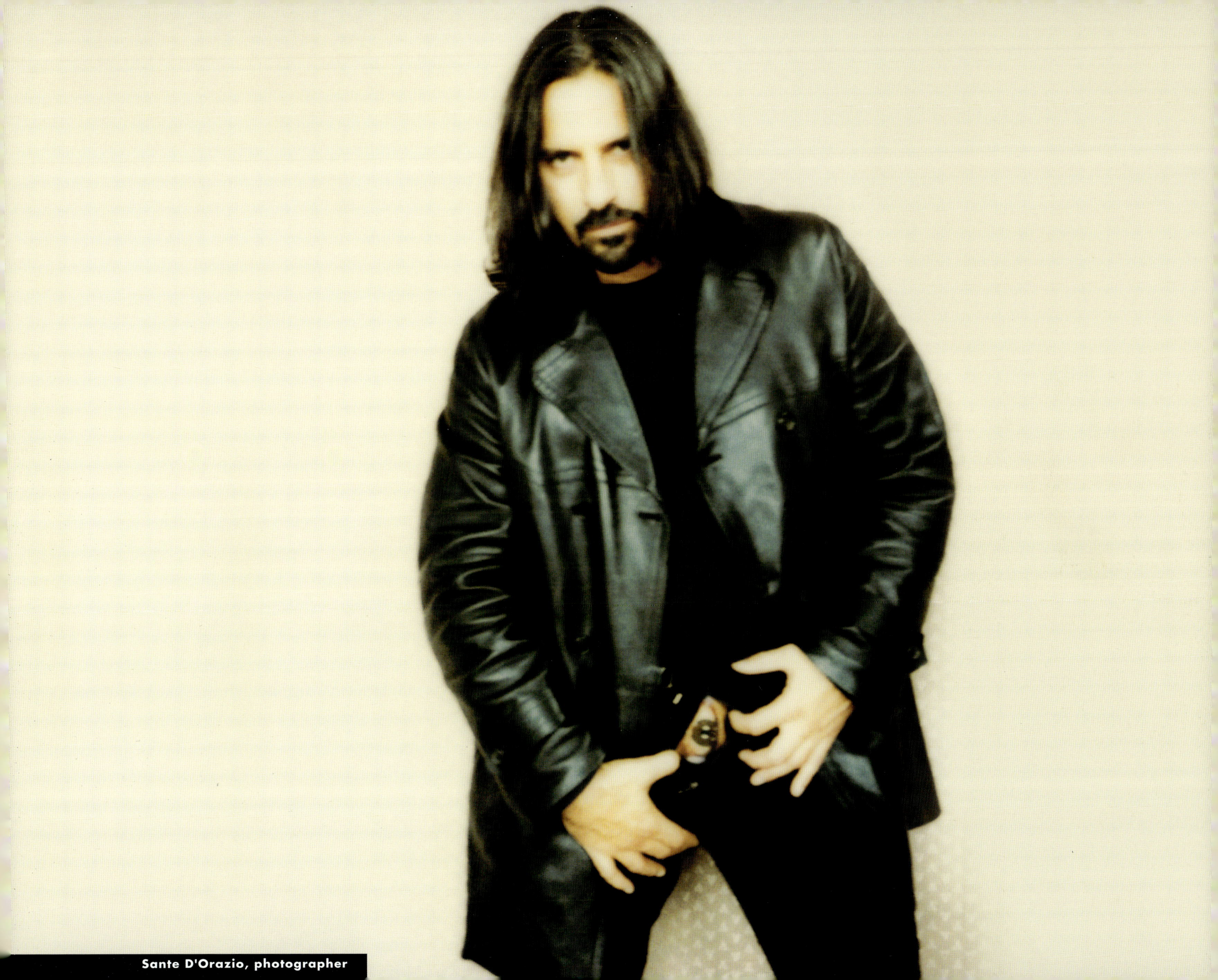

Sante D'Orazio, photographer

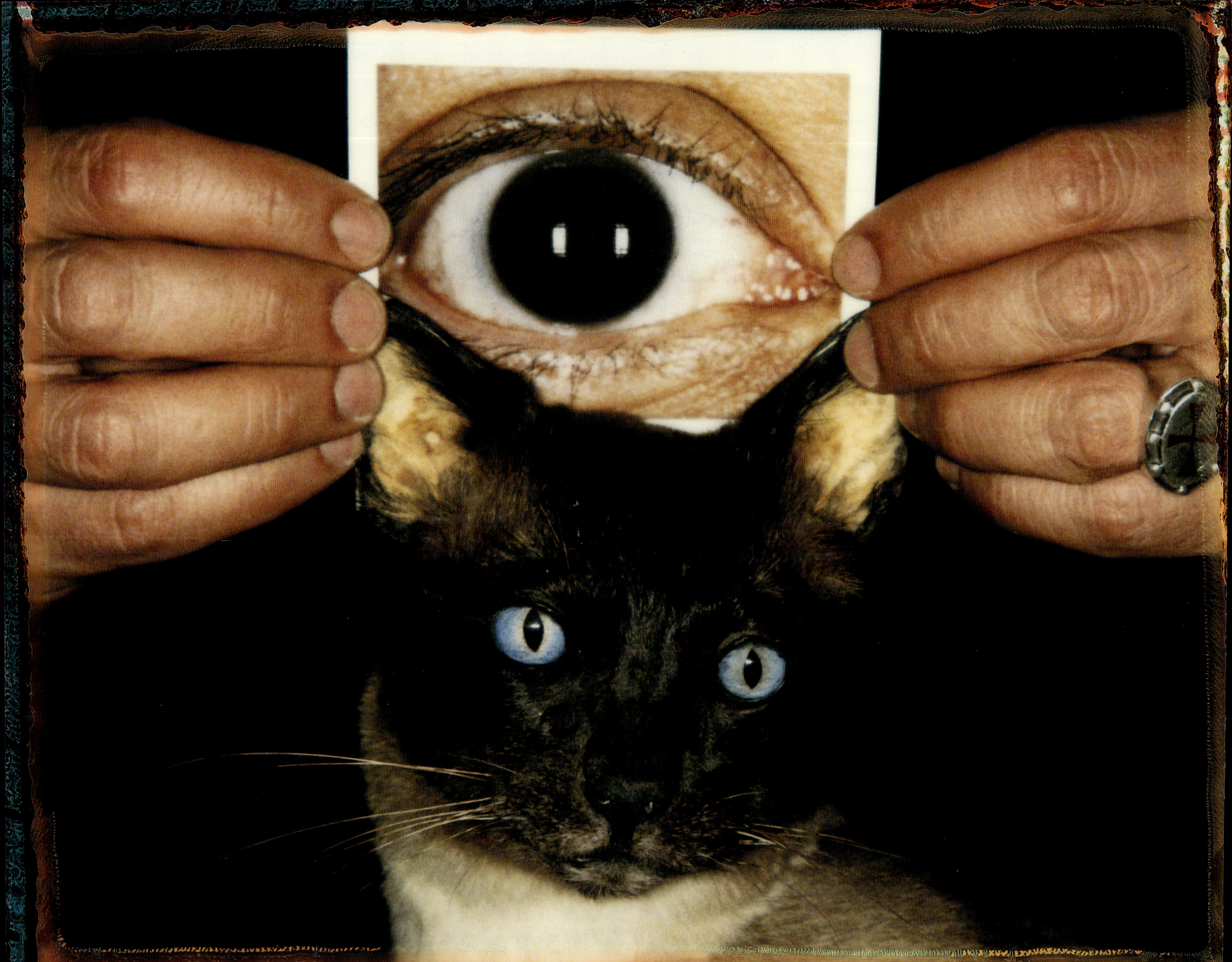

Andres Serrano, artist

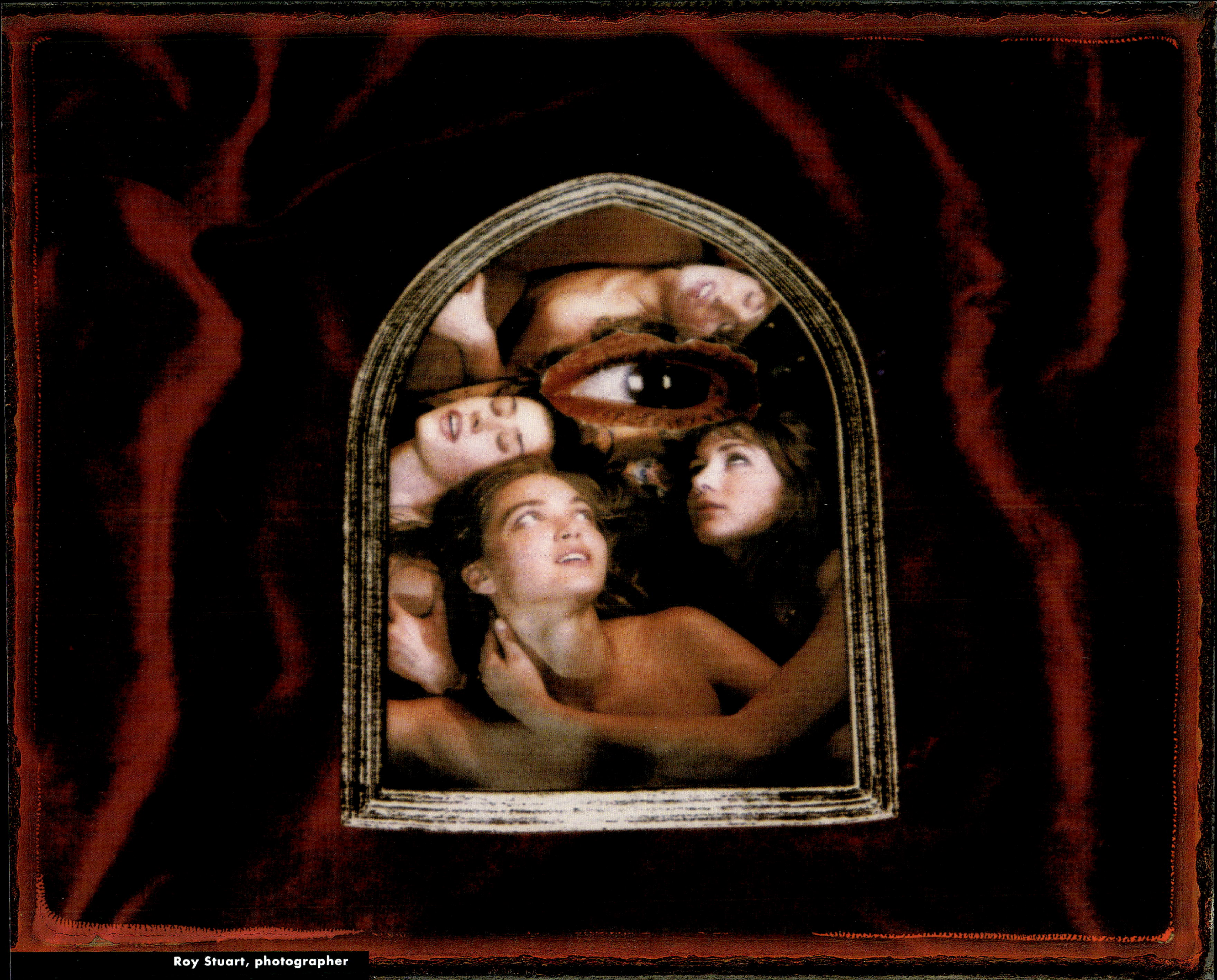

Roy Stuart, photographer

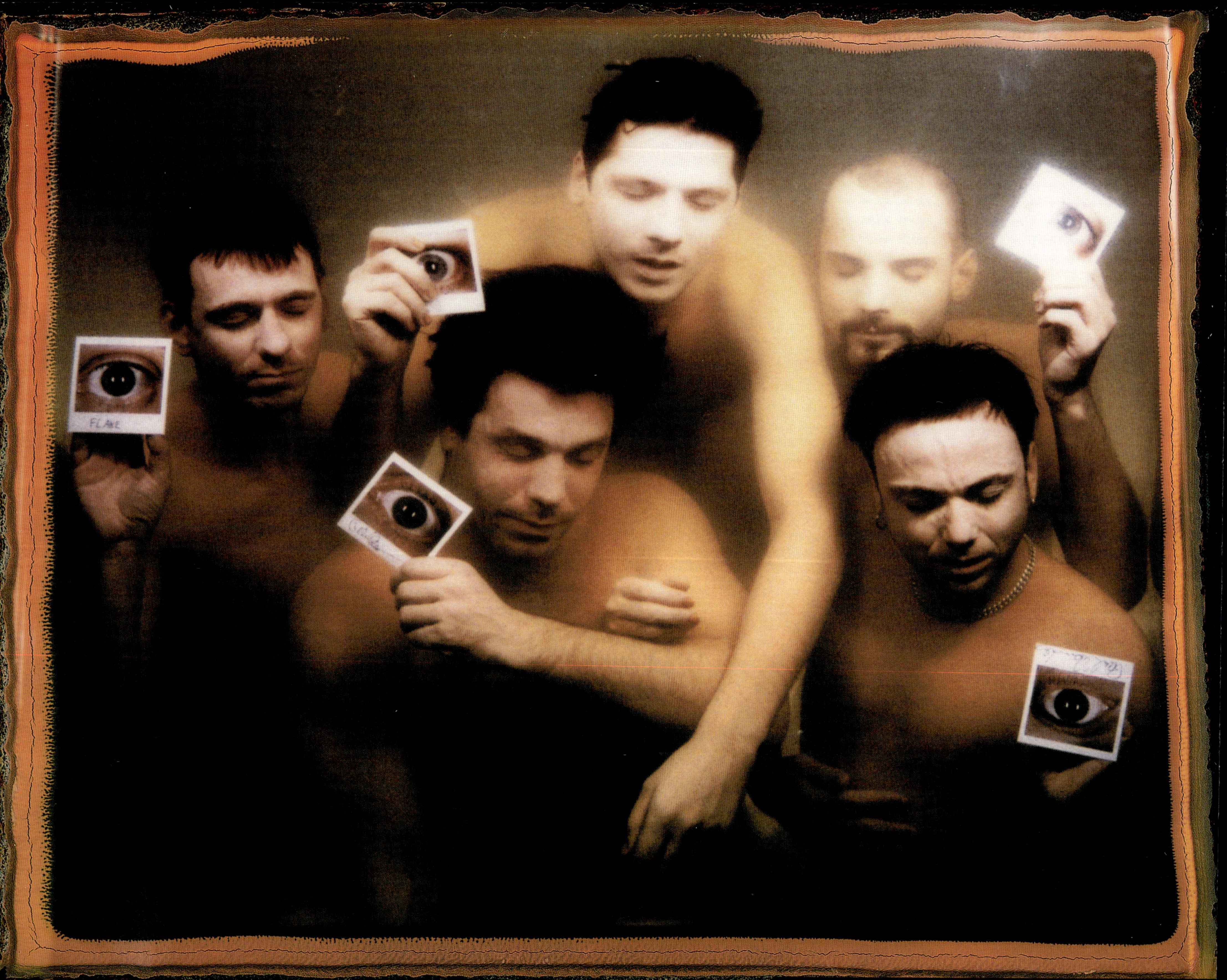

Flake, Till, Christoph, Olli, Paul of Rammstein, musicians

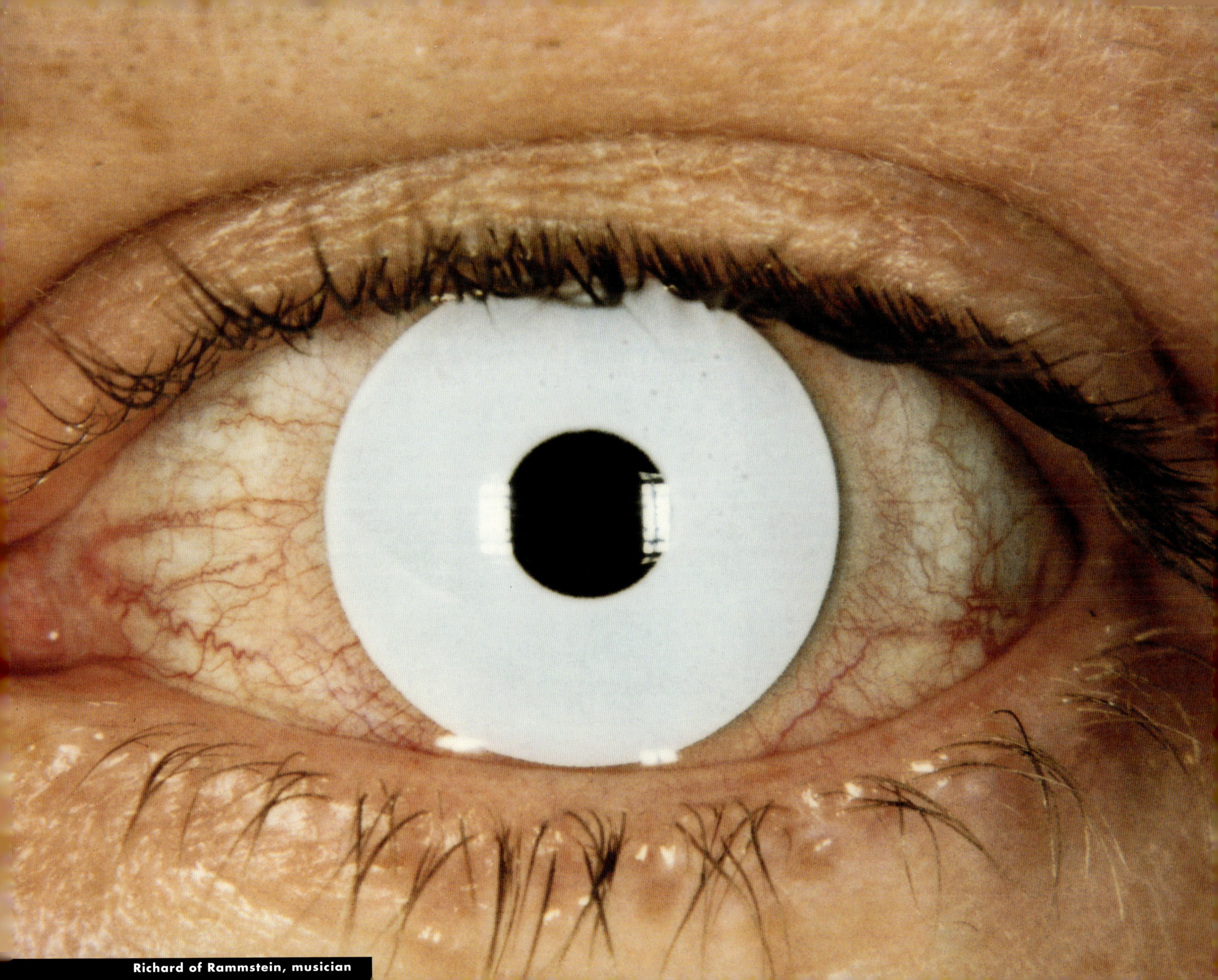

Richard of Rammstein, musician

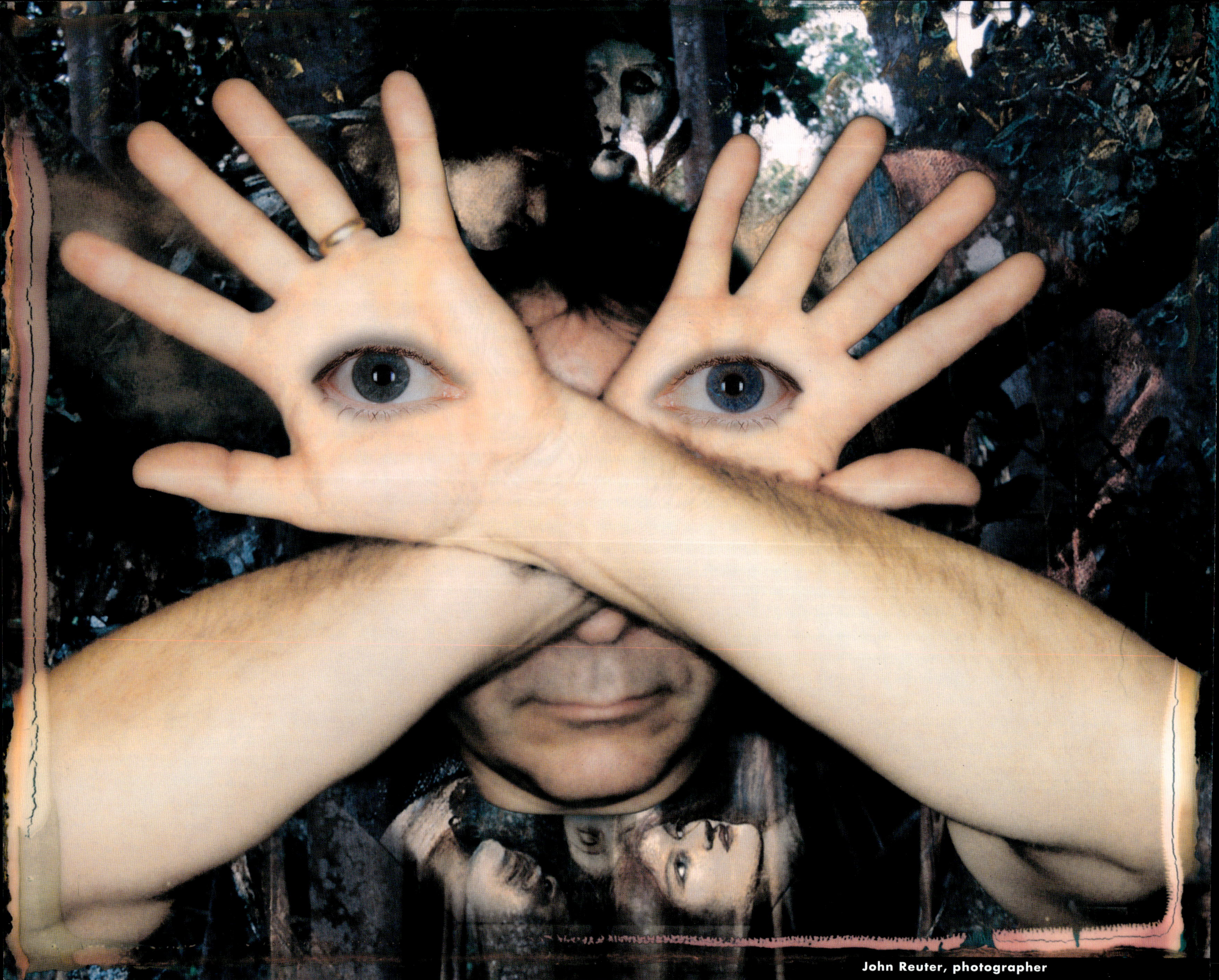

John Reuter, photographer

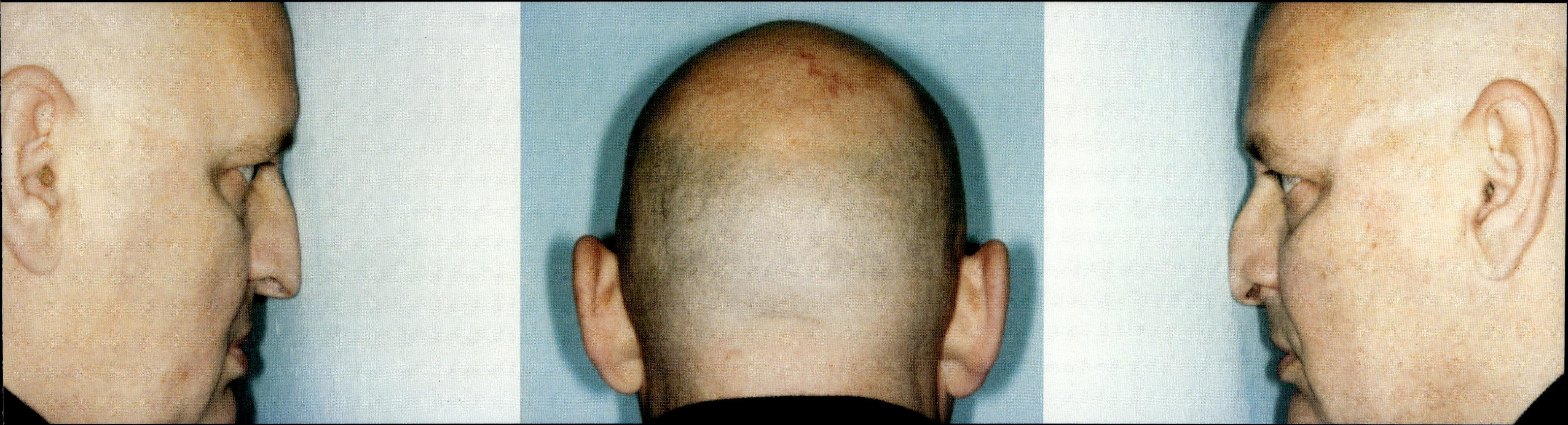

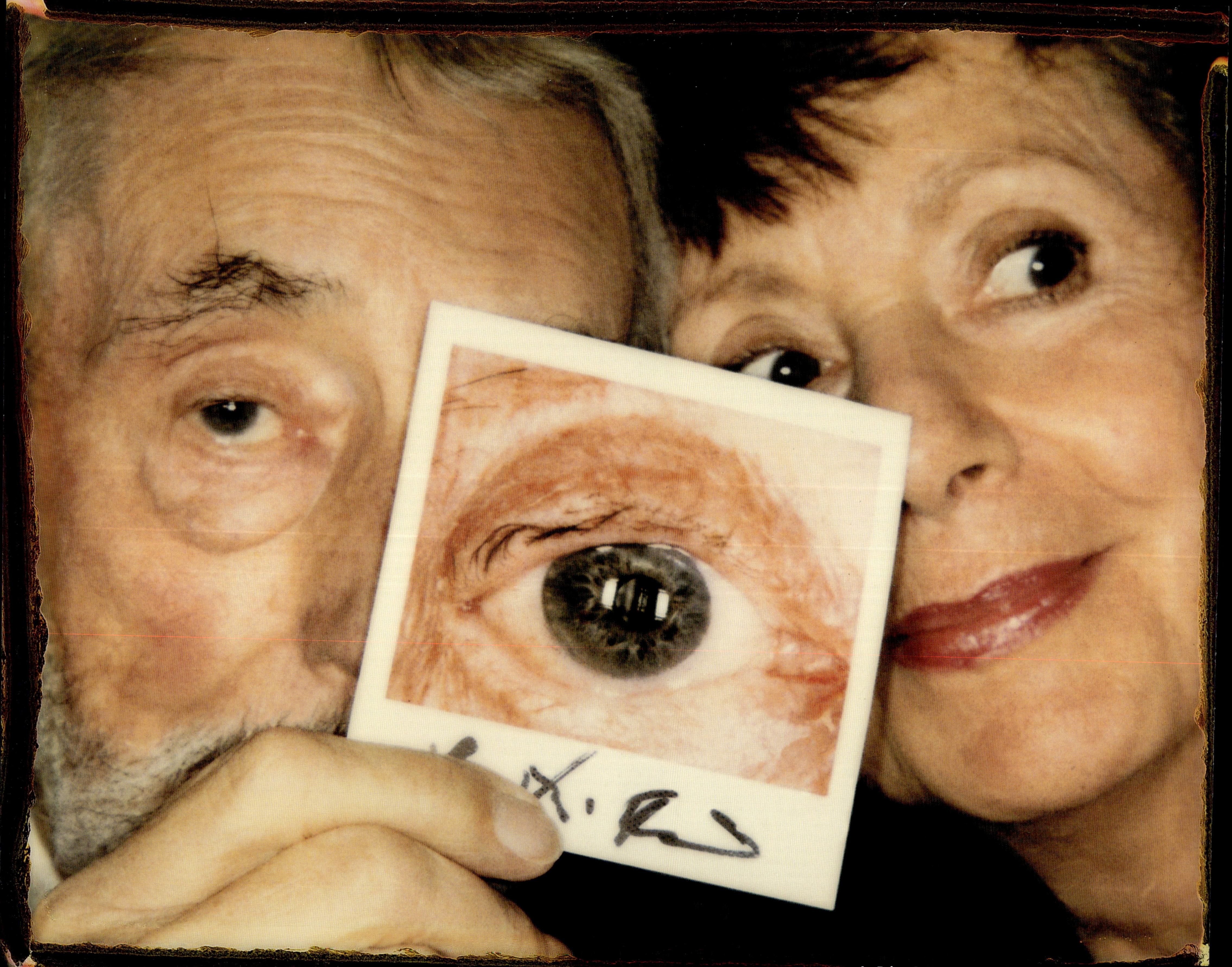

L. Fritz & Renate Gruber, collectors

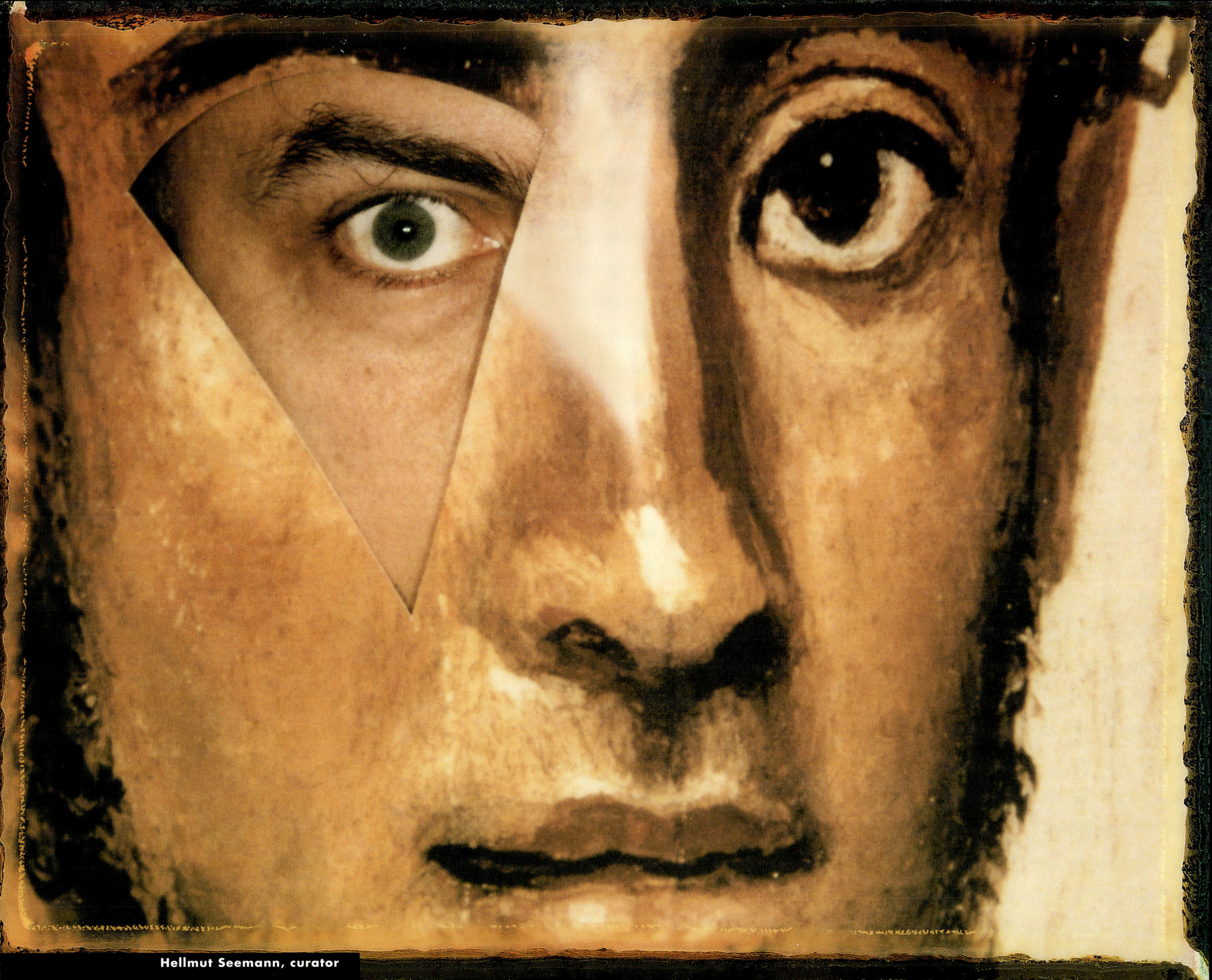

Hellmut Seemann, curator

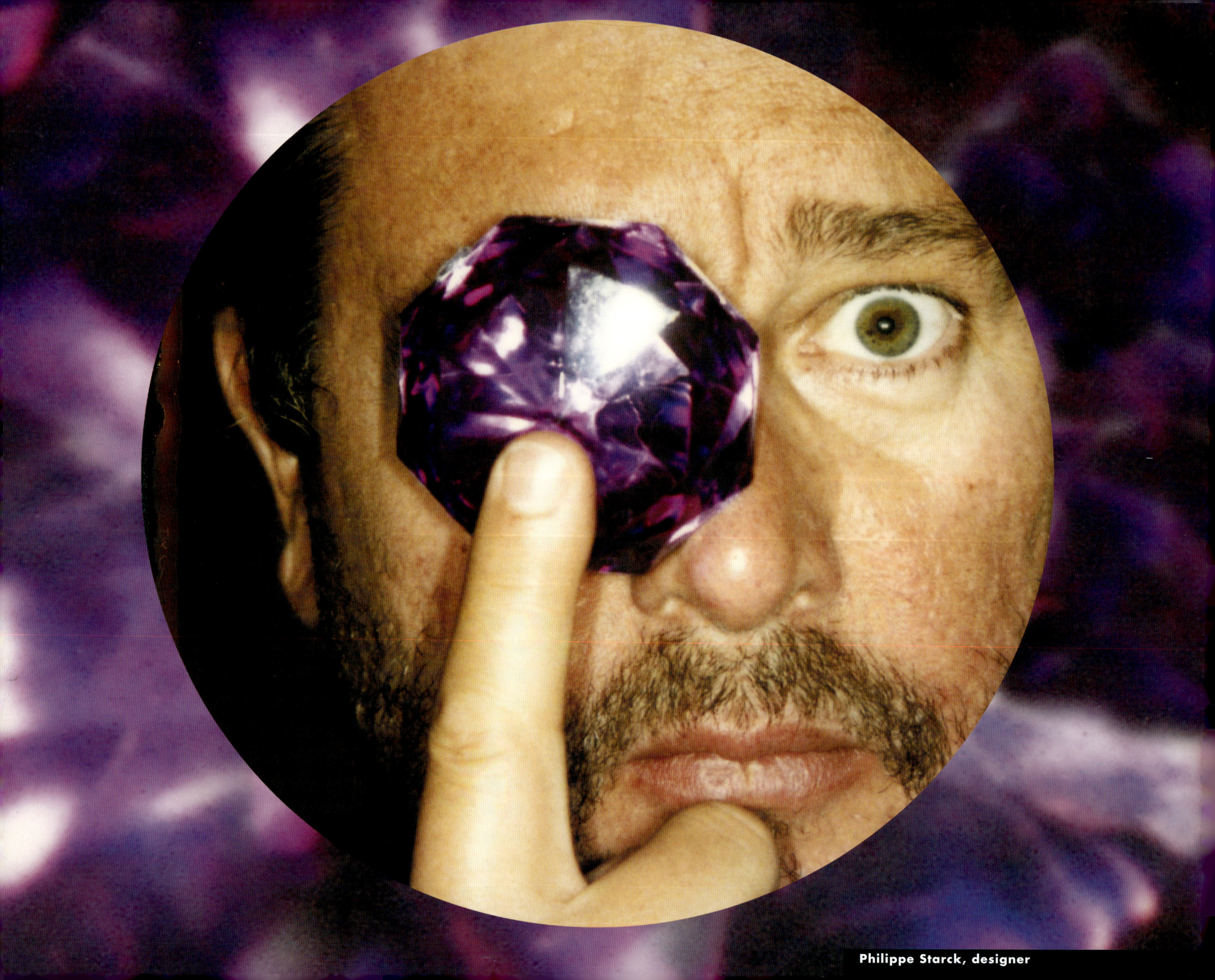

Philippe Starck, designer

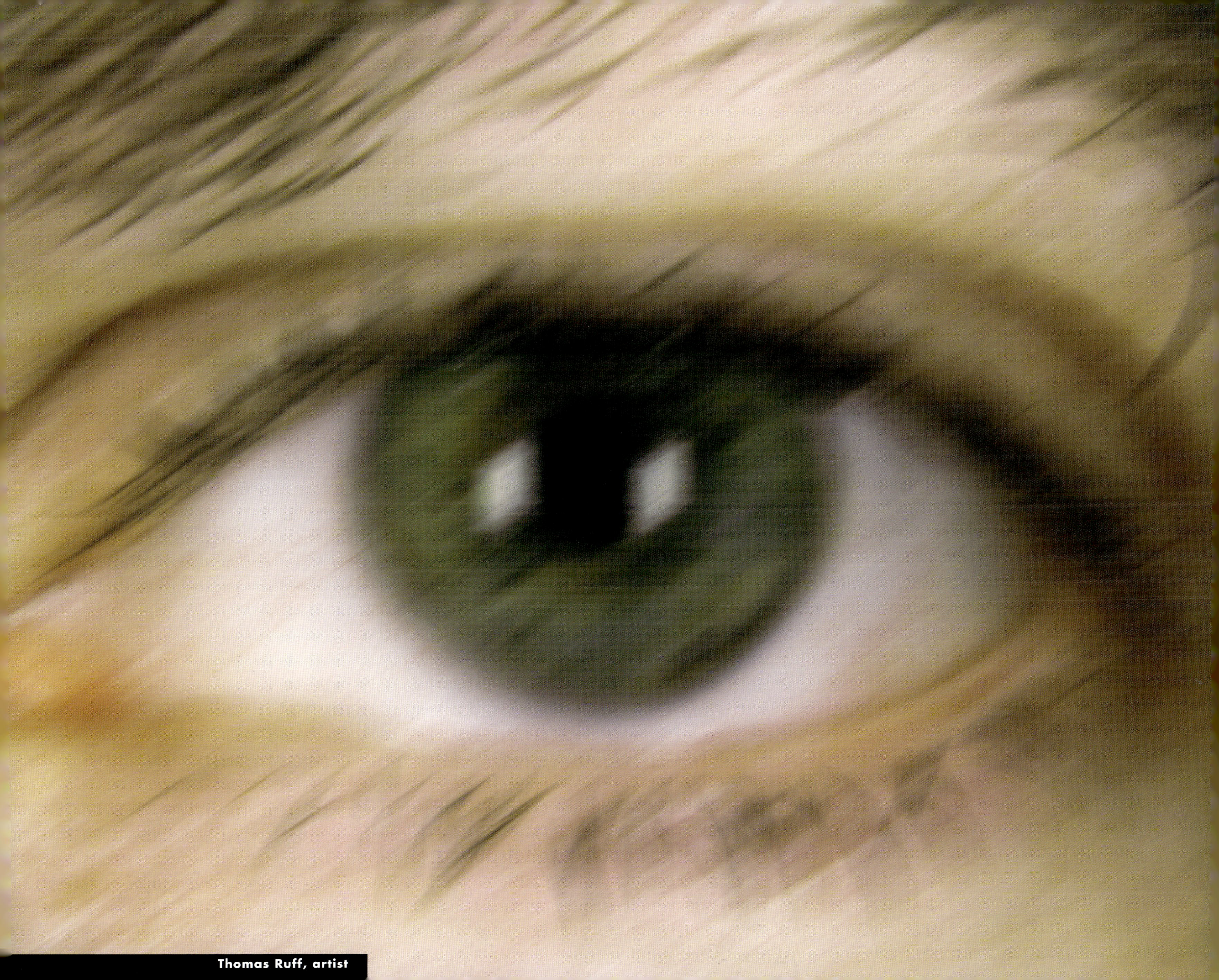

Thomas Ruff, artist

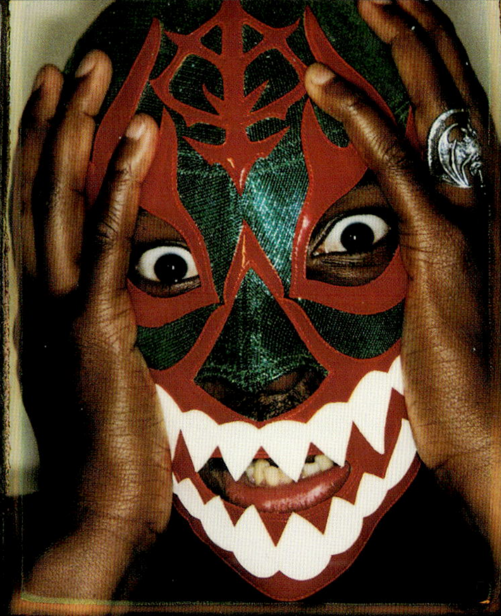
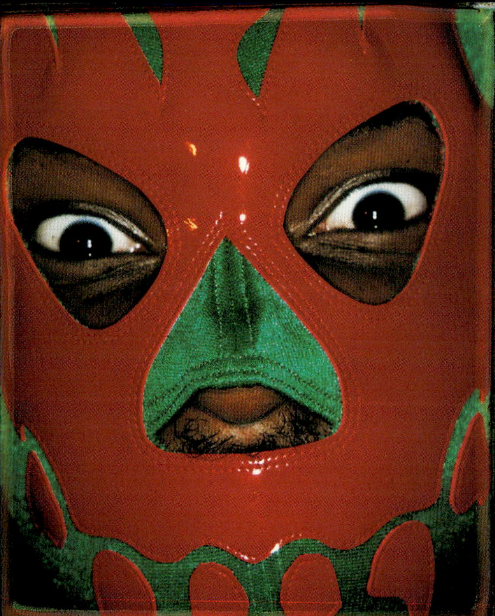
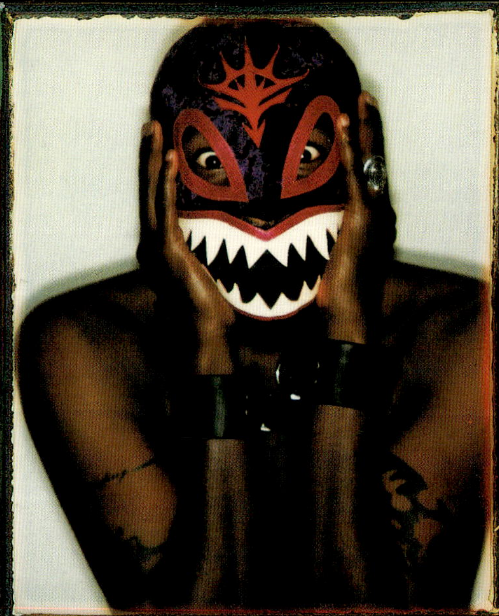
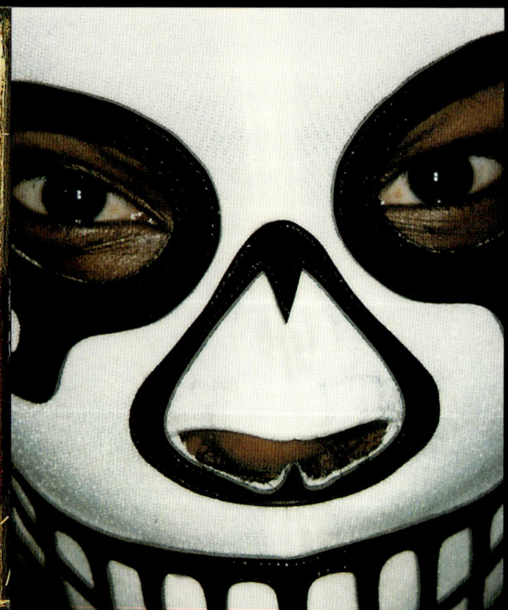
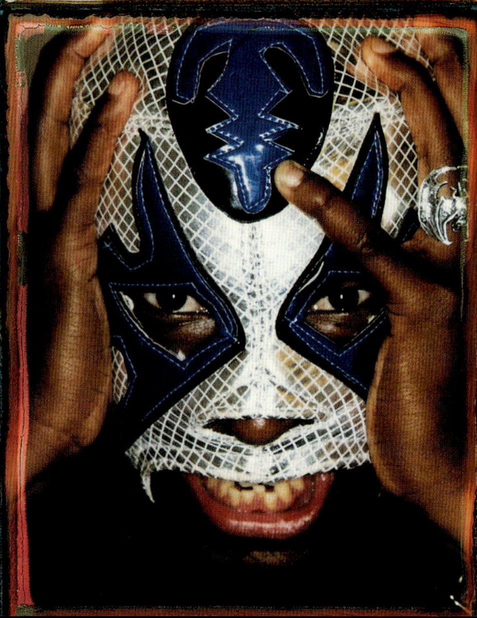

Carlos Batts, photograph

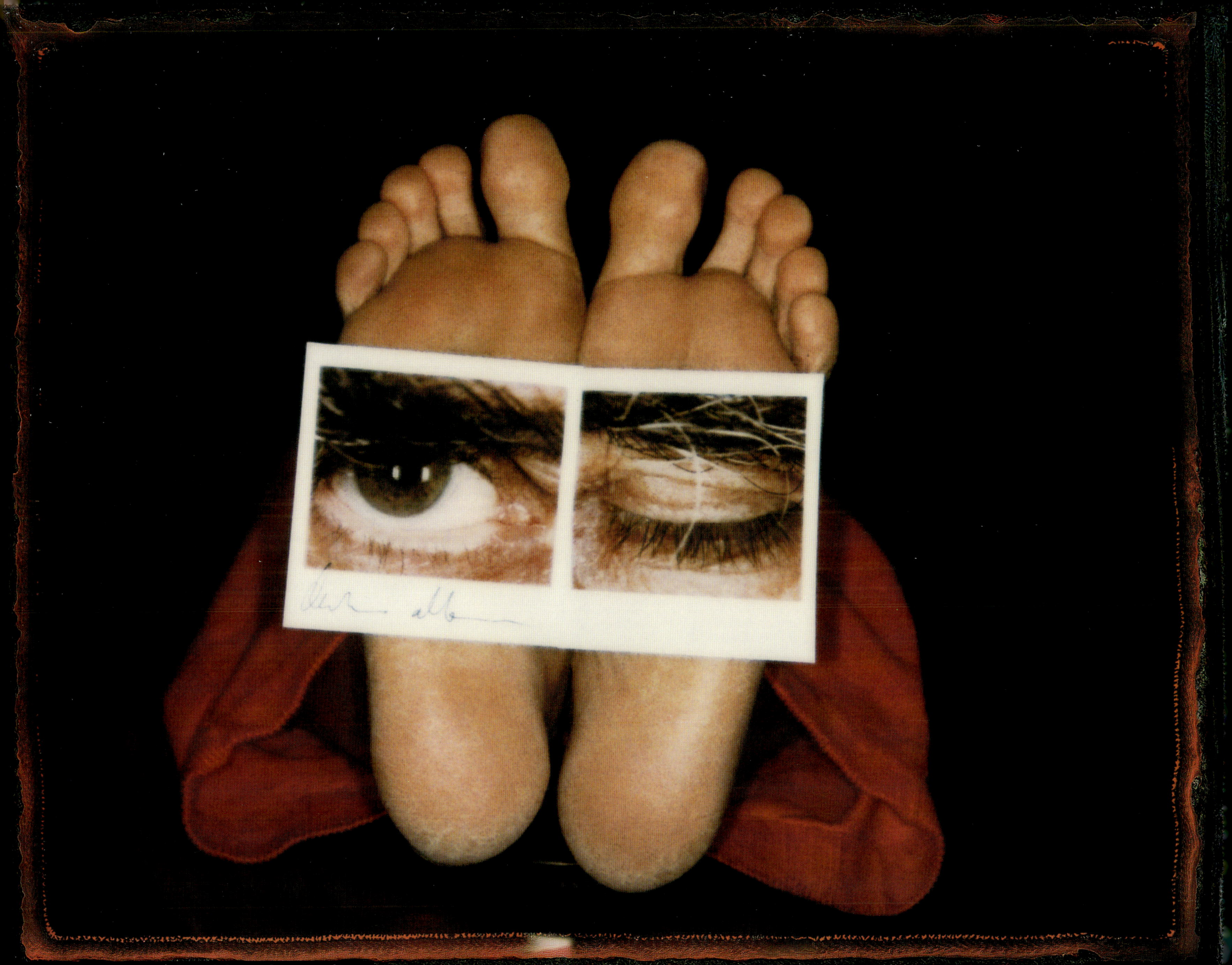

Christian von Alvensleben, photographer

Harald Szeemann, curator

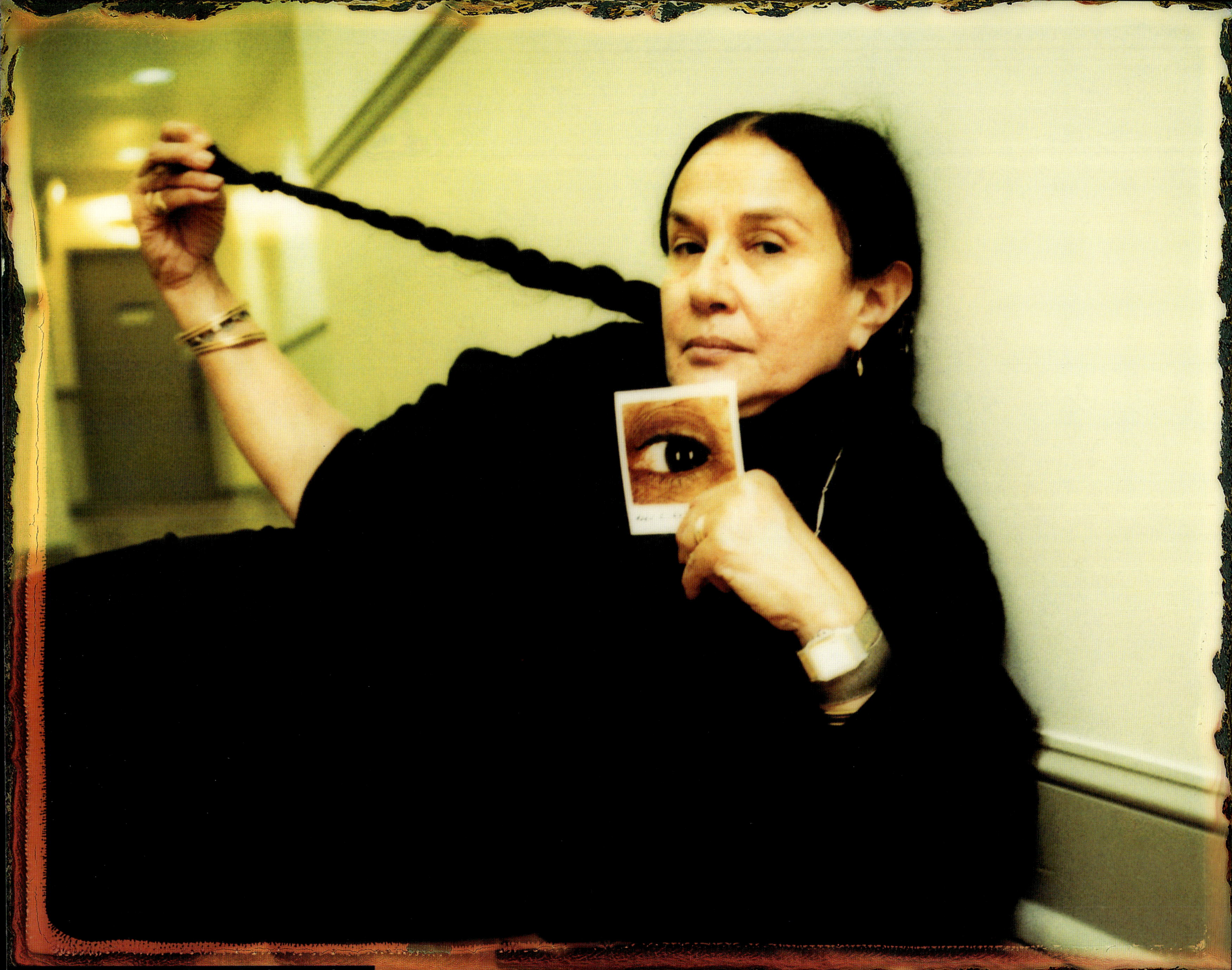

Mary Ellen Mark, photographer

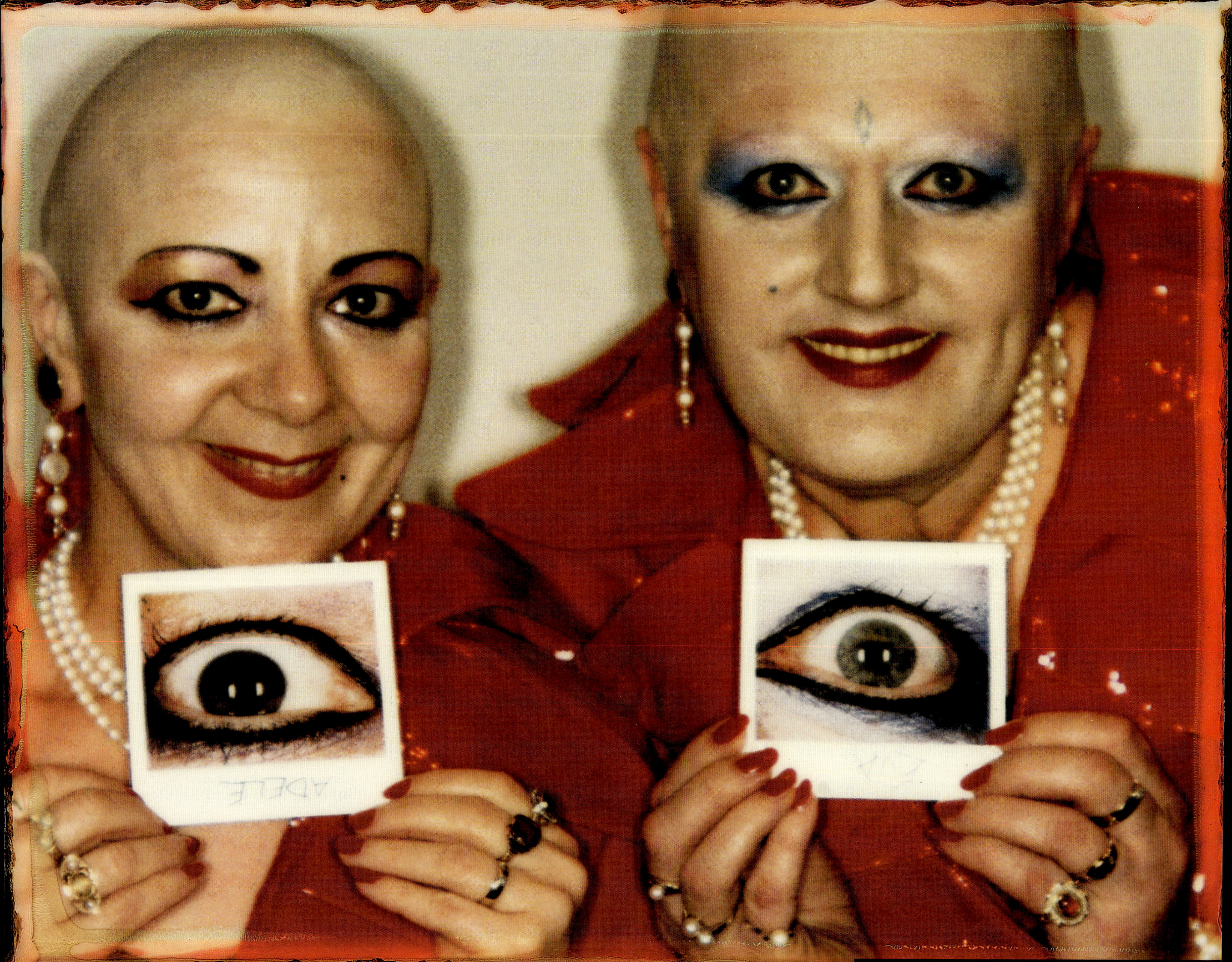

Eva & Adele, artists

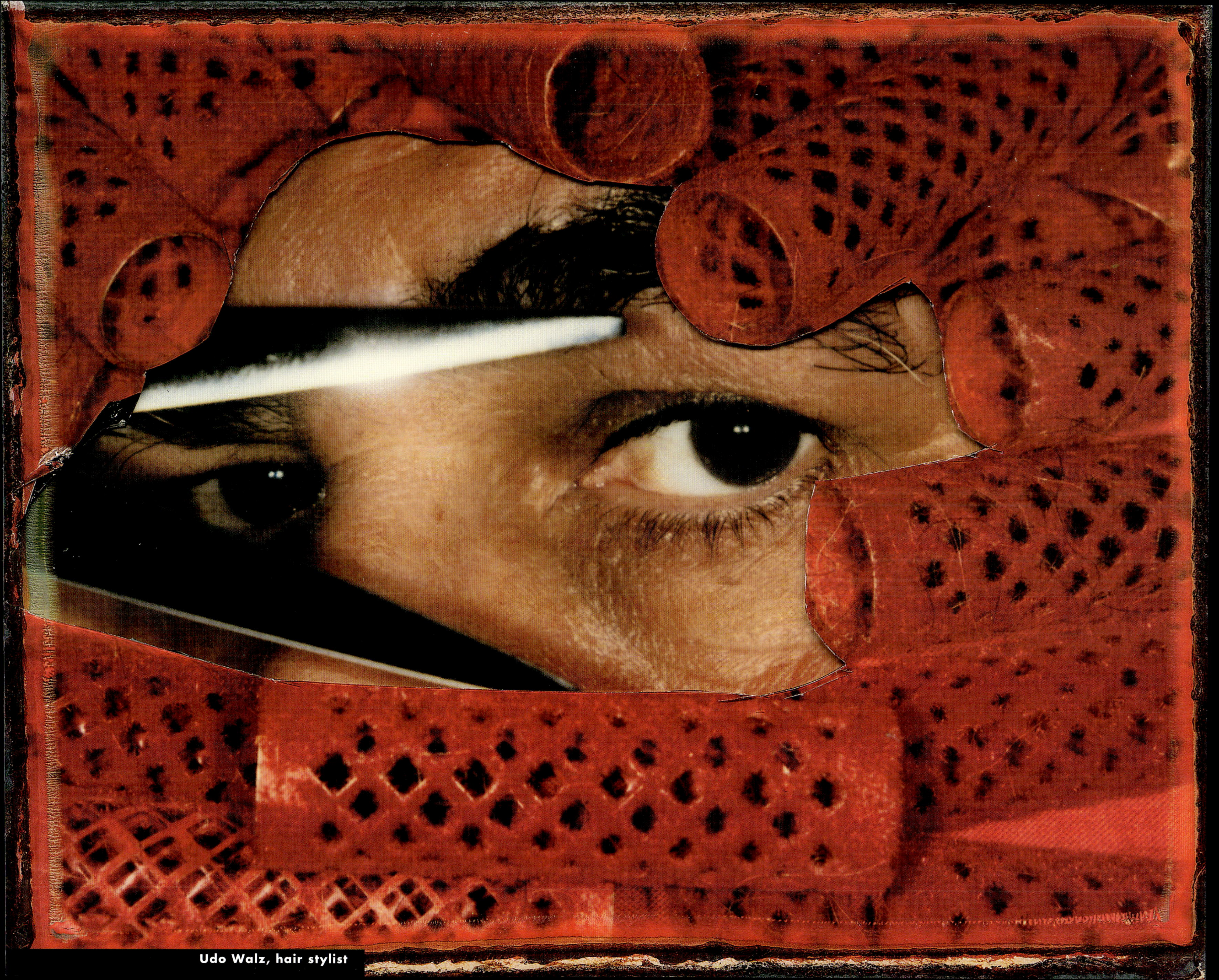

Udo Walz, hair stylist

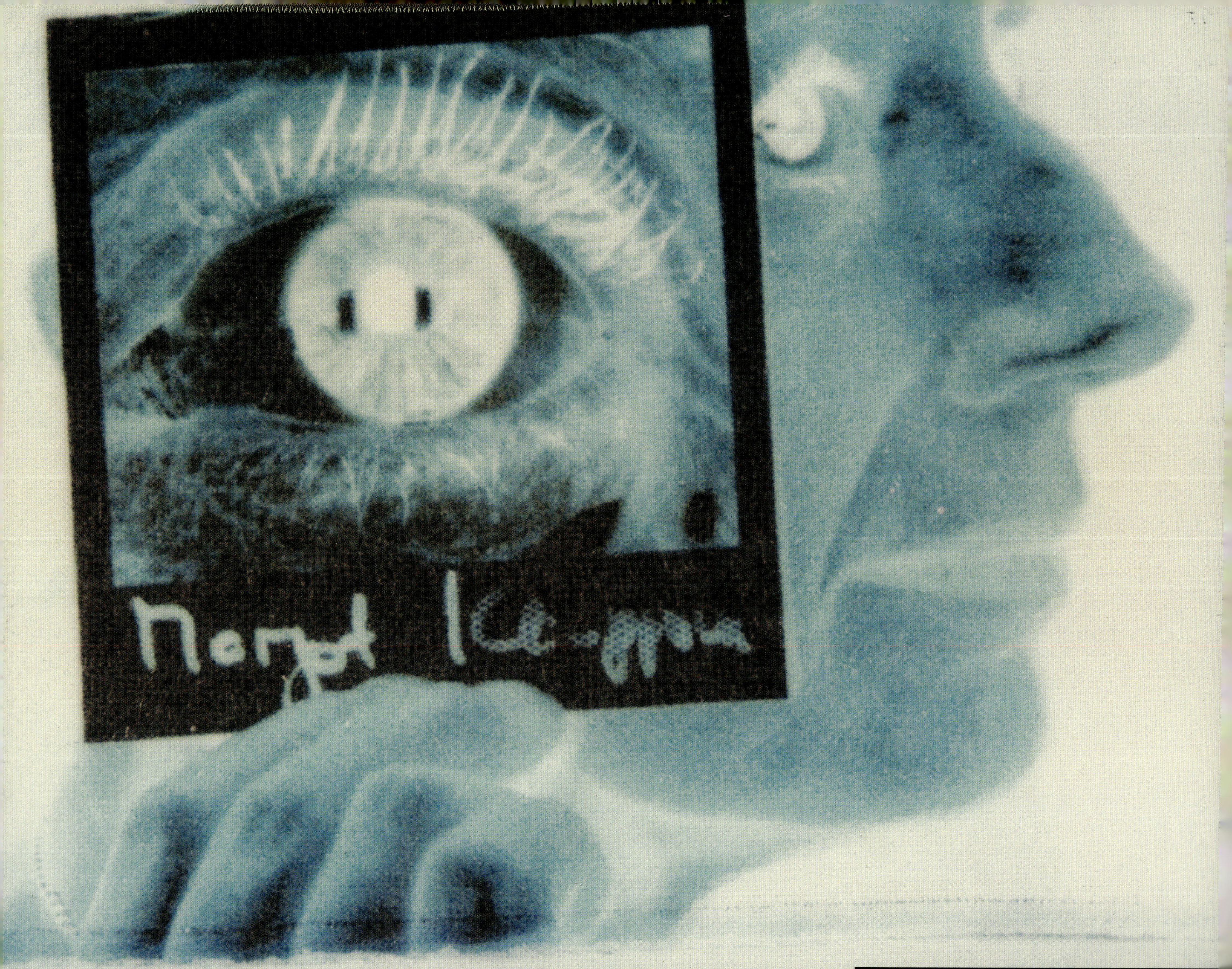

Margot Klingsporn, director of photography

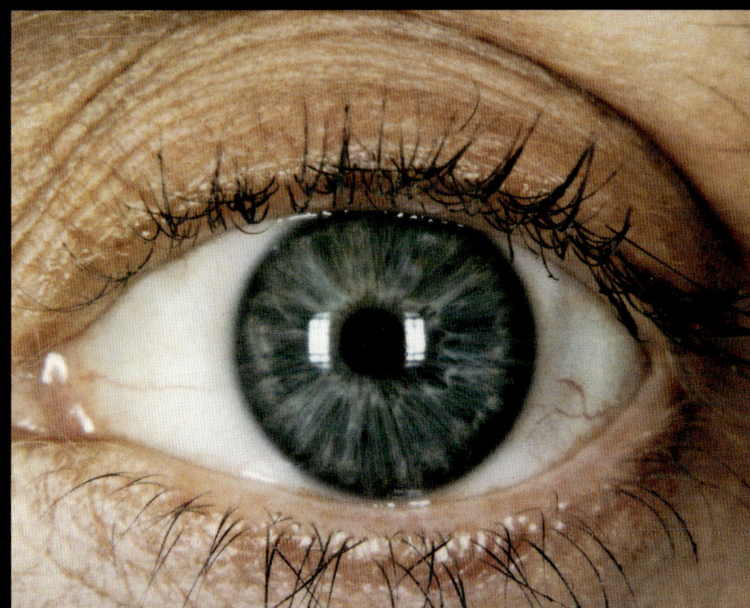

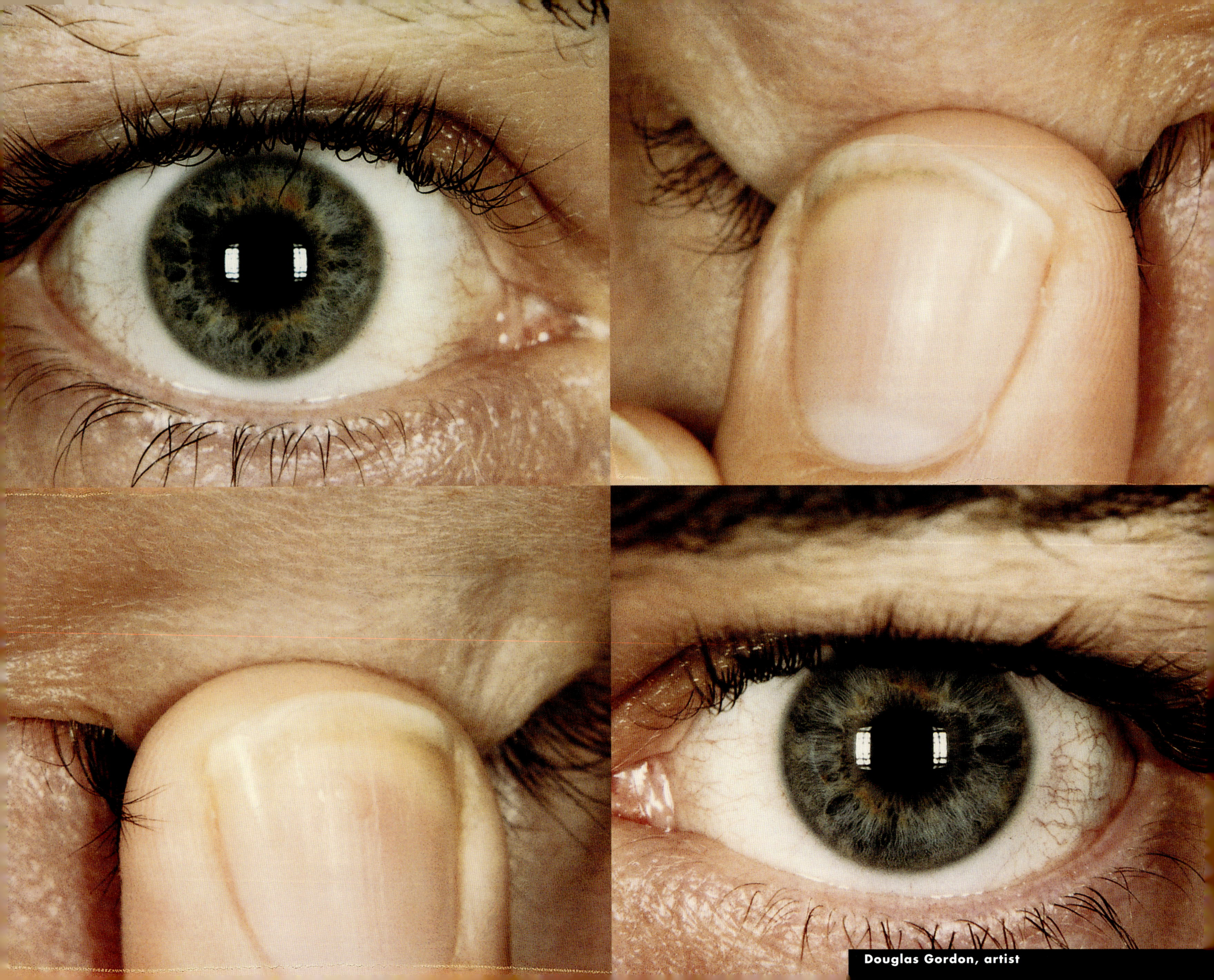

Douglas Gordon, artist

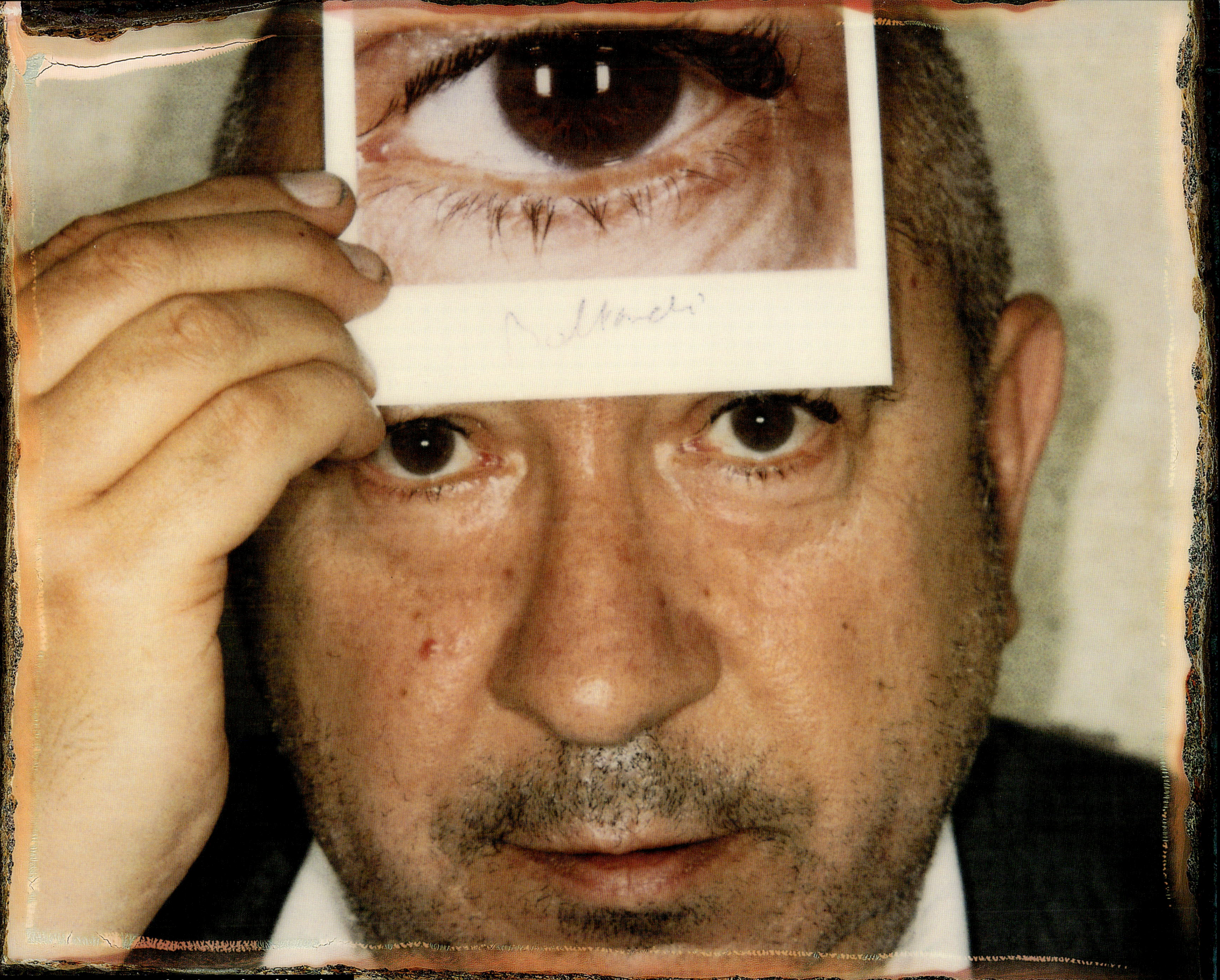

Christian Boltanski, artist

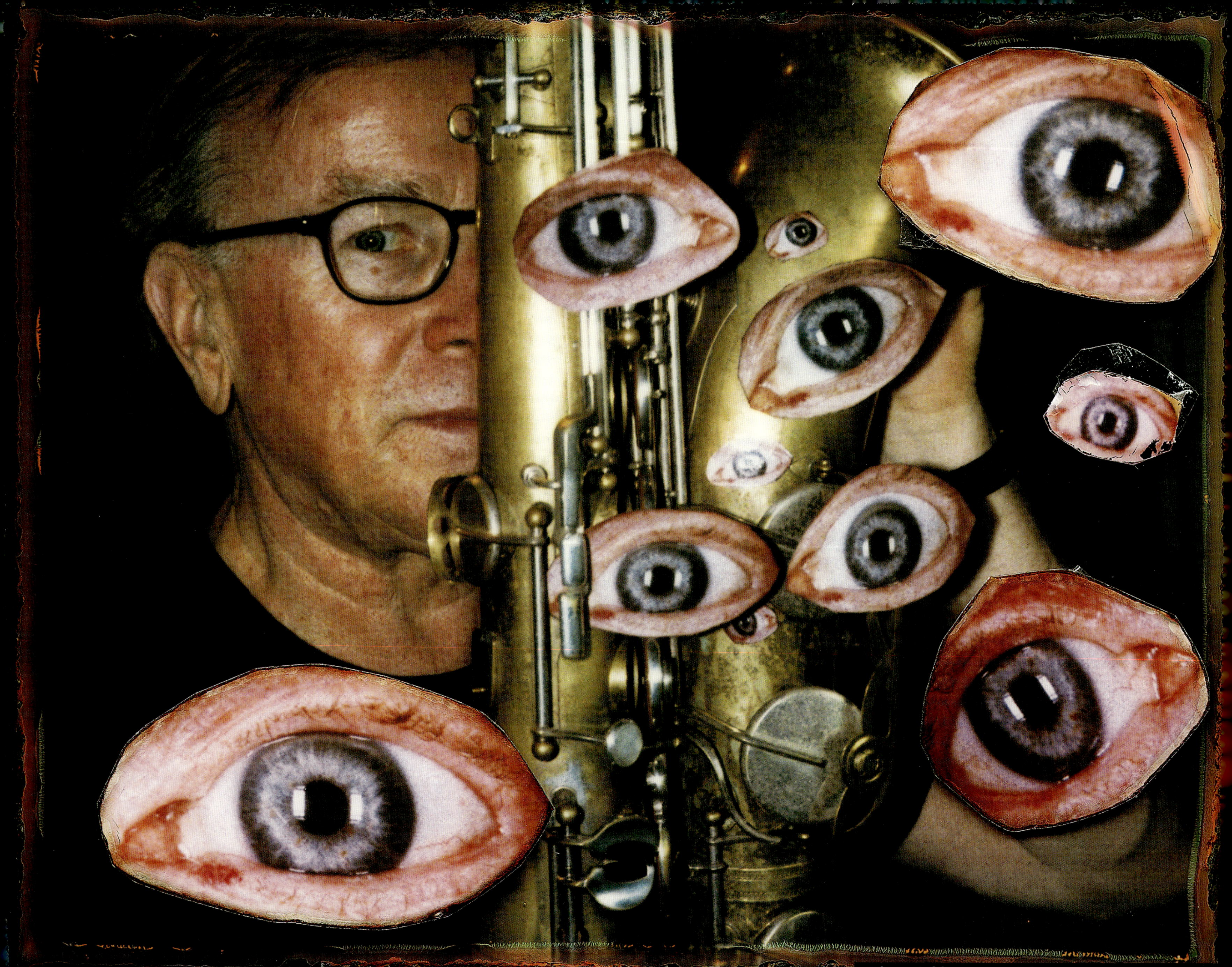

Klaus Doldinger, musician

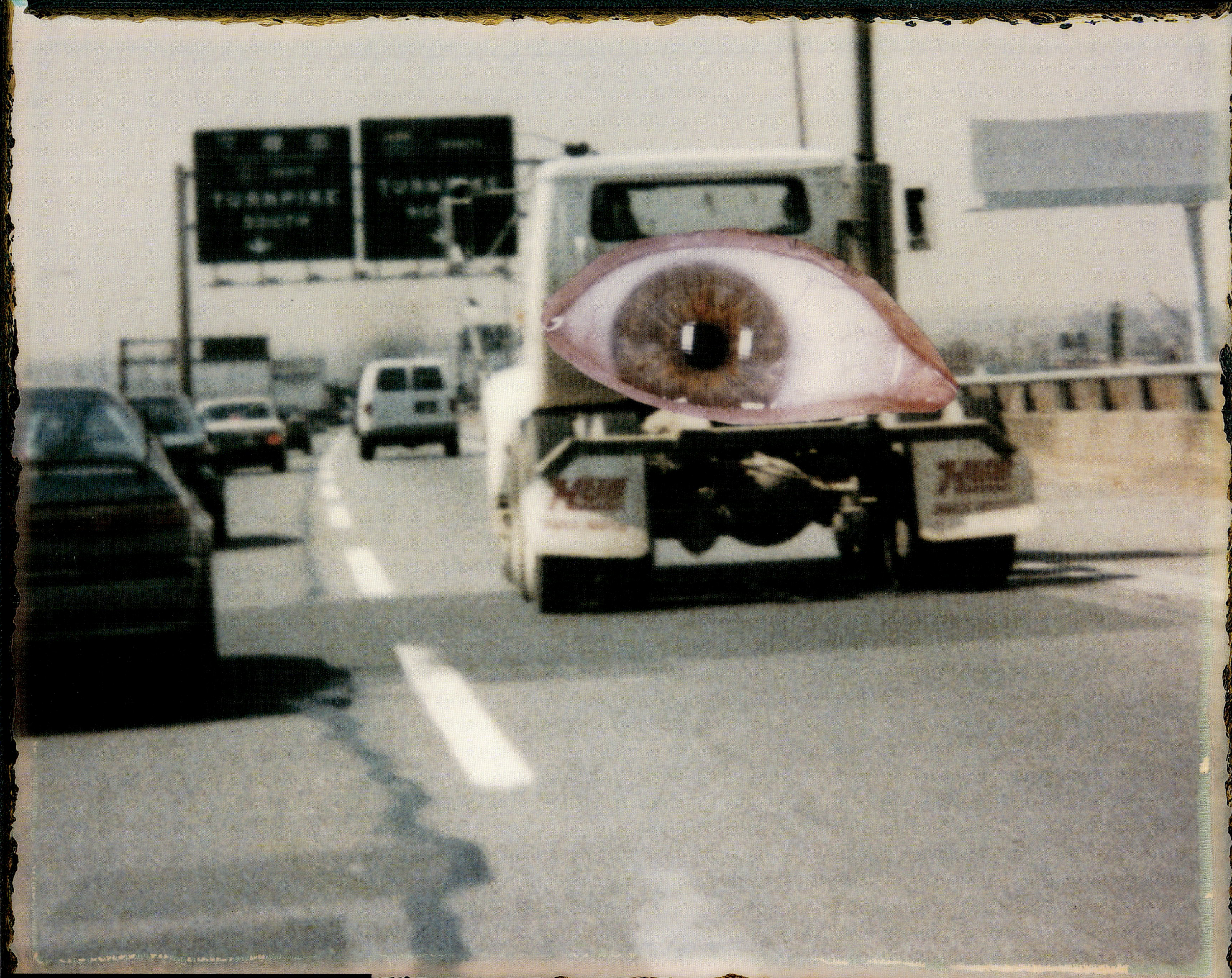

Martha Rosler, artist

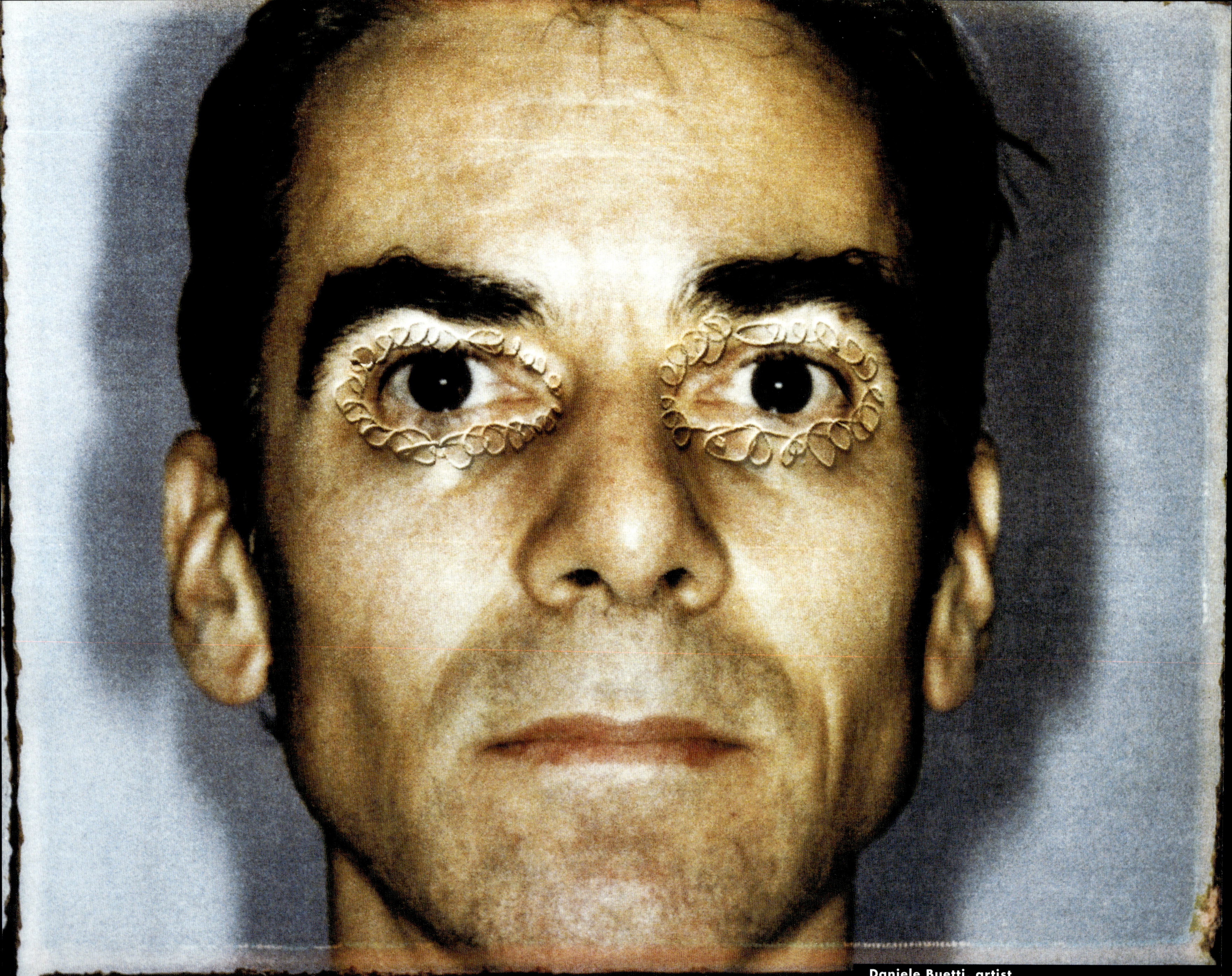

Daniele Buetti, artist

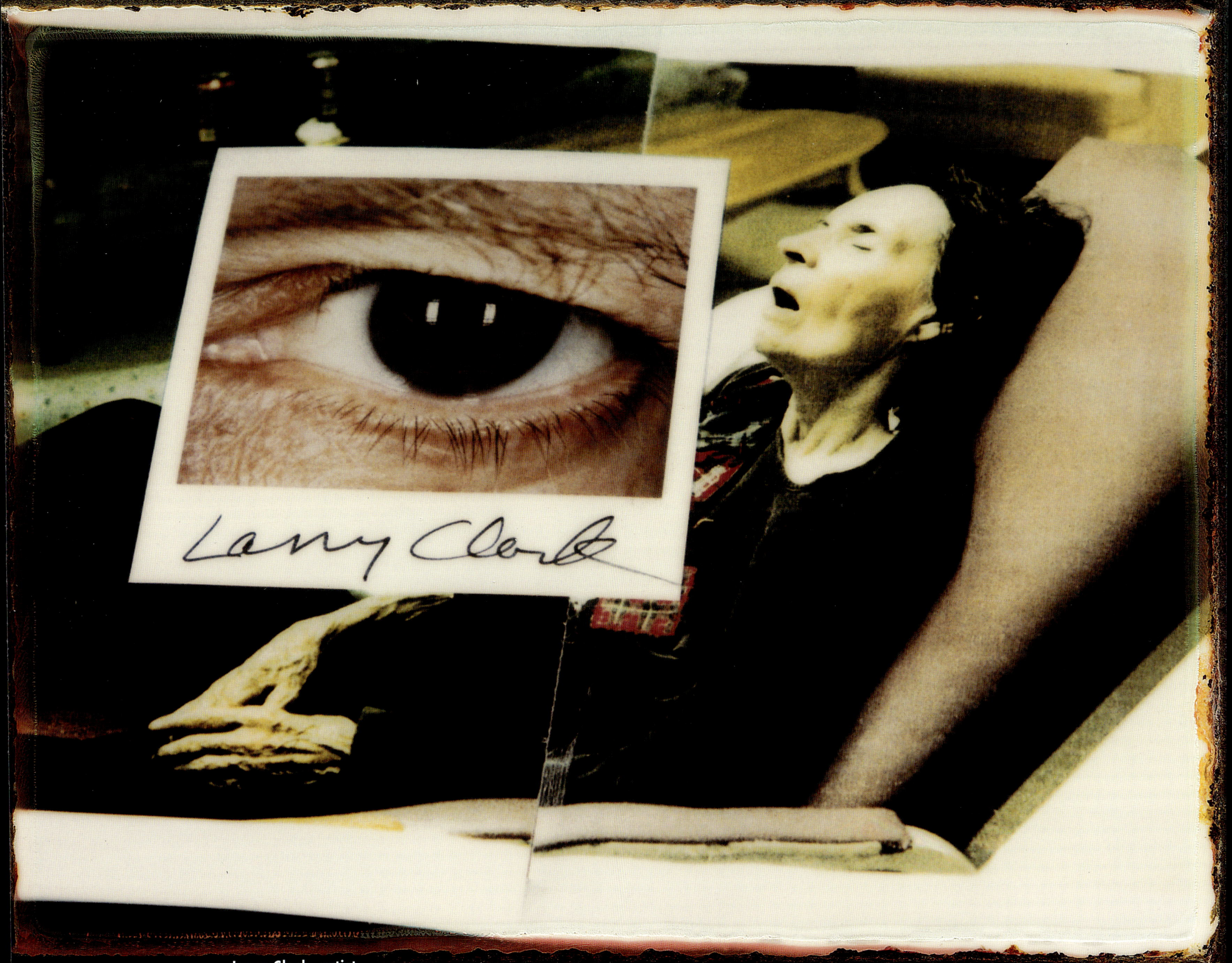

Larry Clark, artist

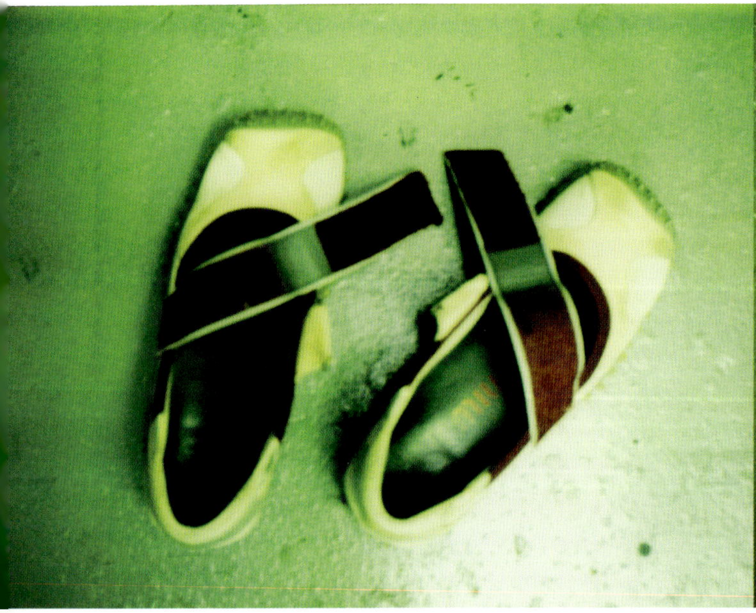
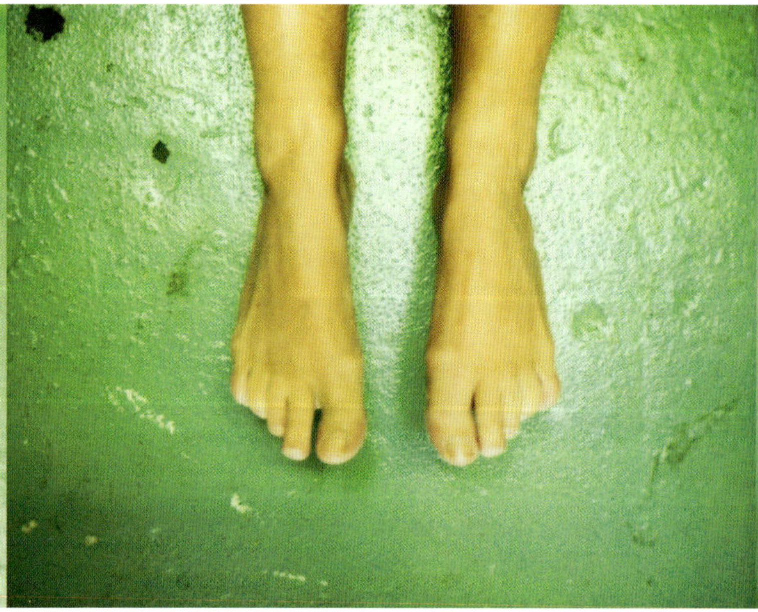
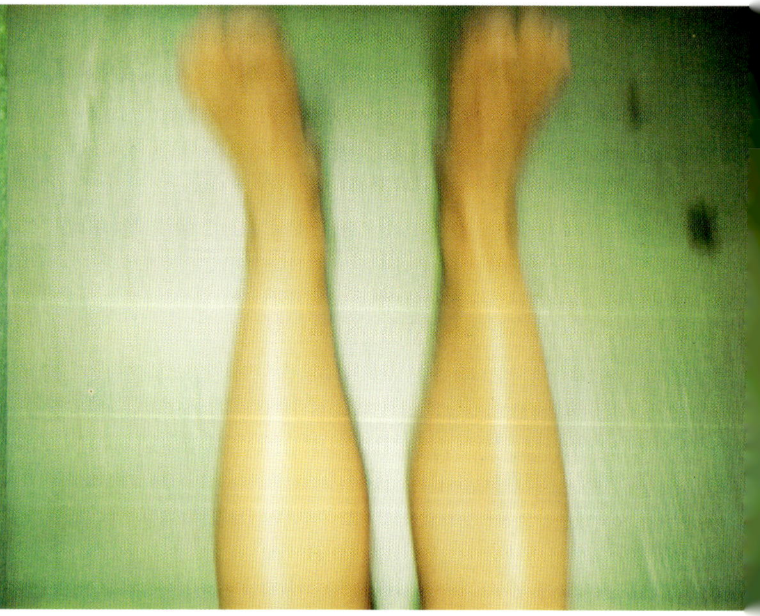

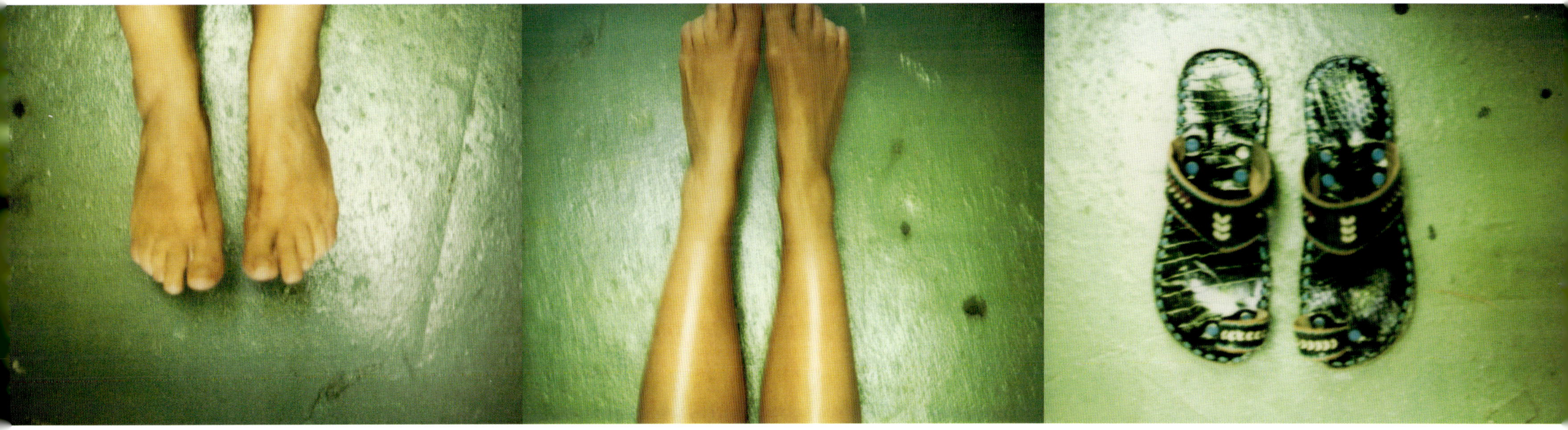

Jane & Louise Wilson, artists

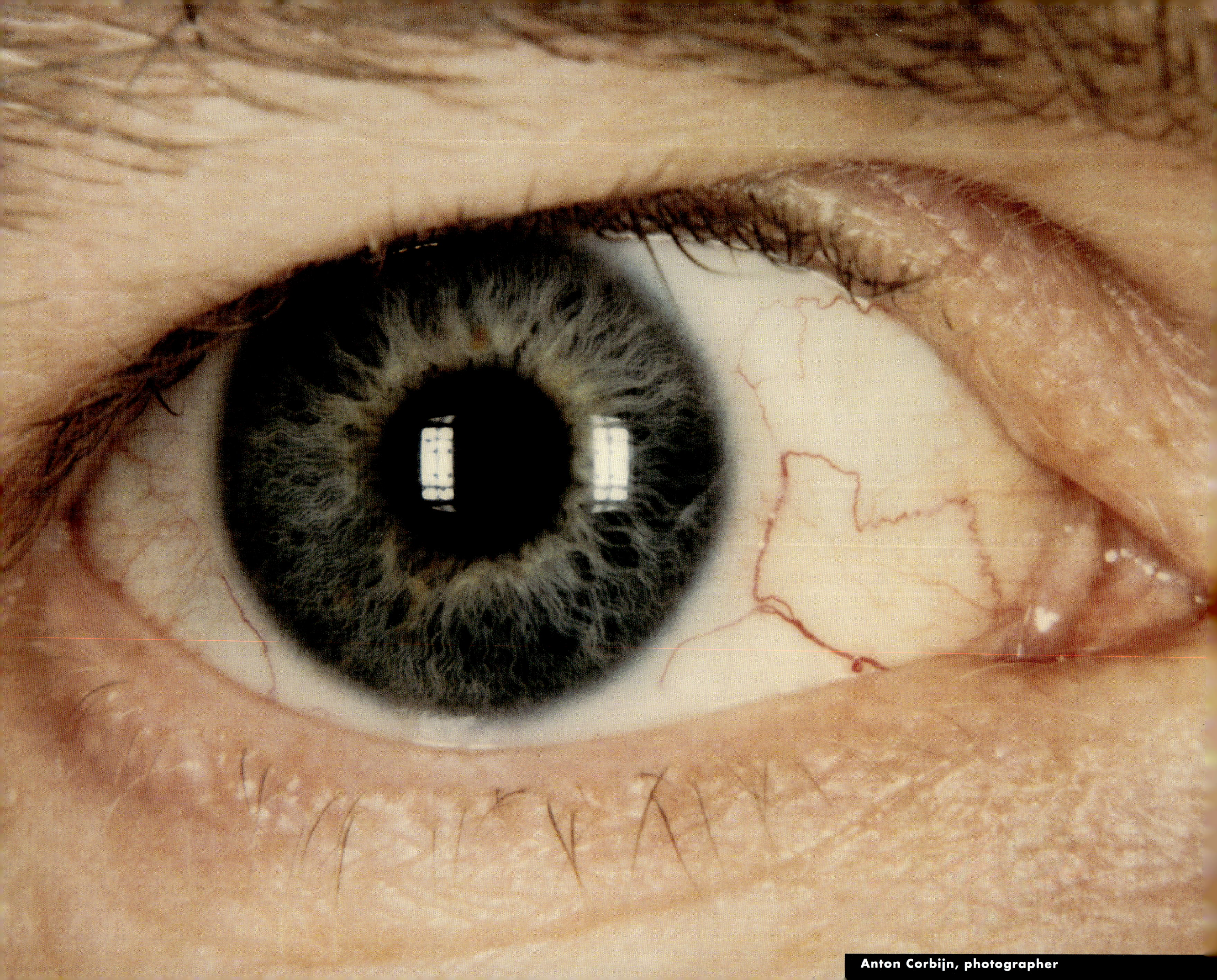

Anton Corbijn, photographer

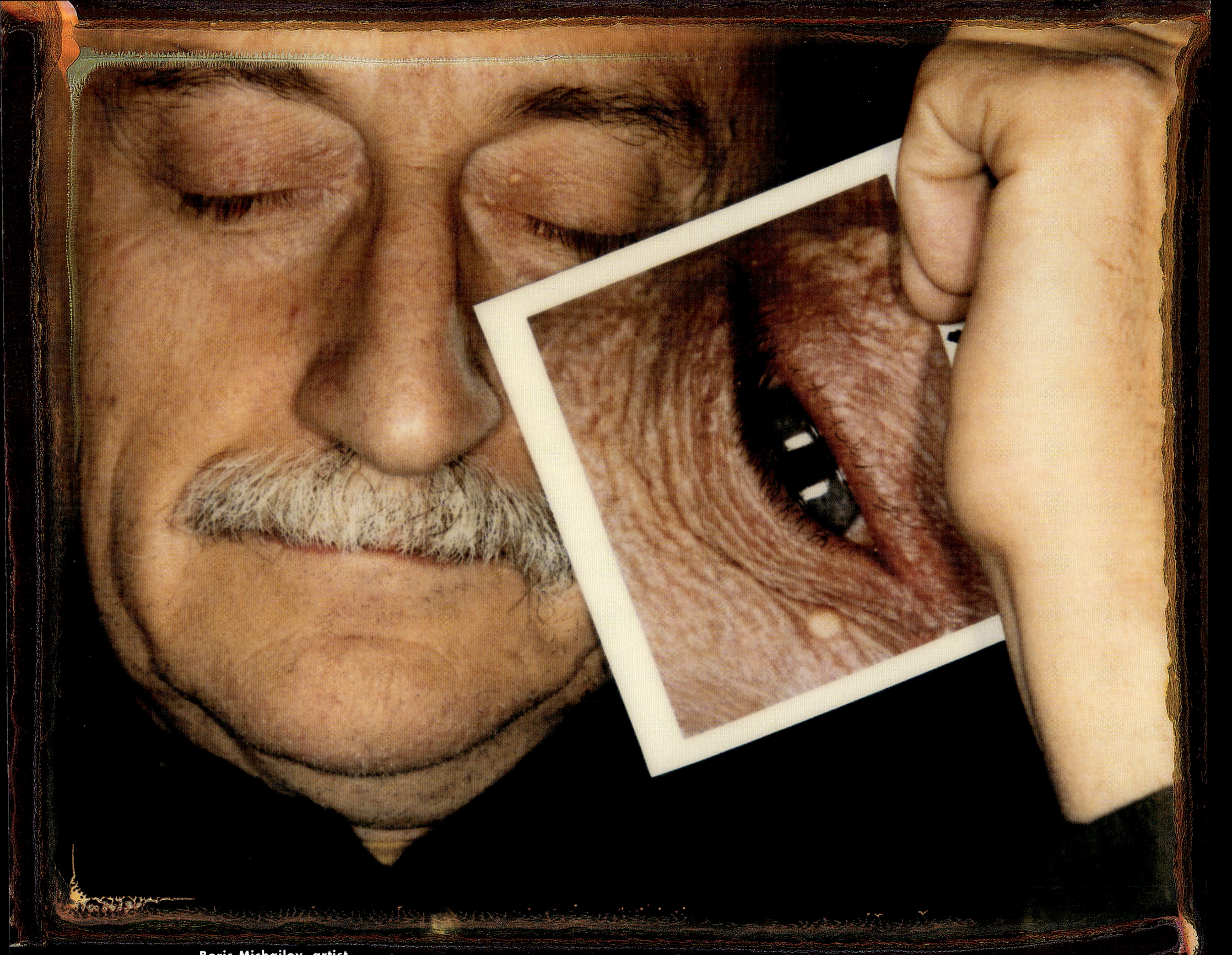

Boris Michailov, artist

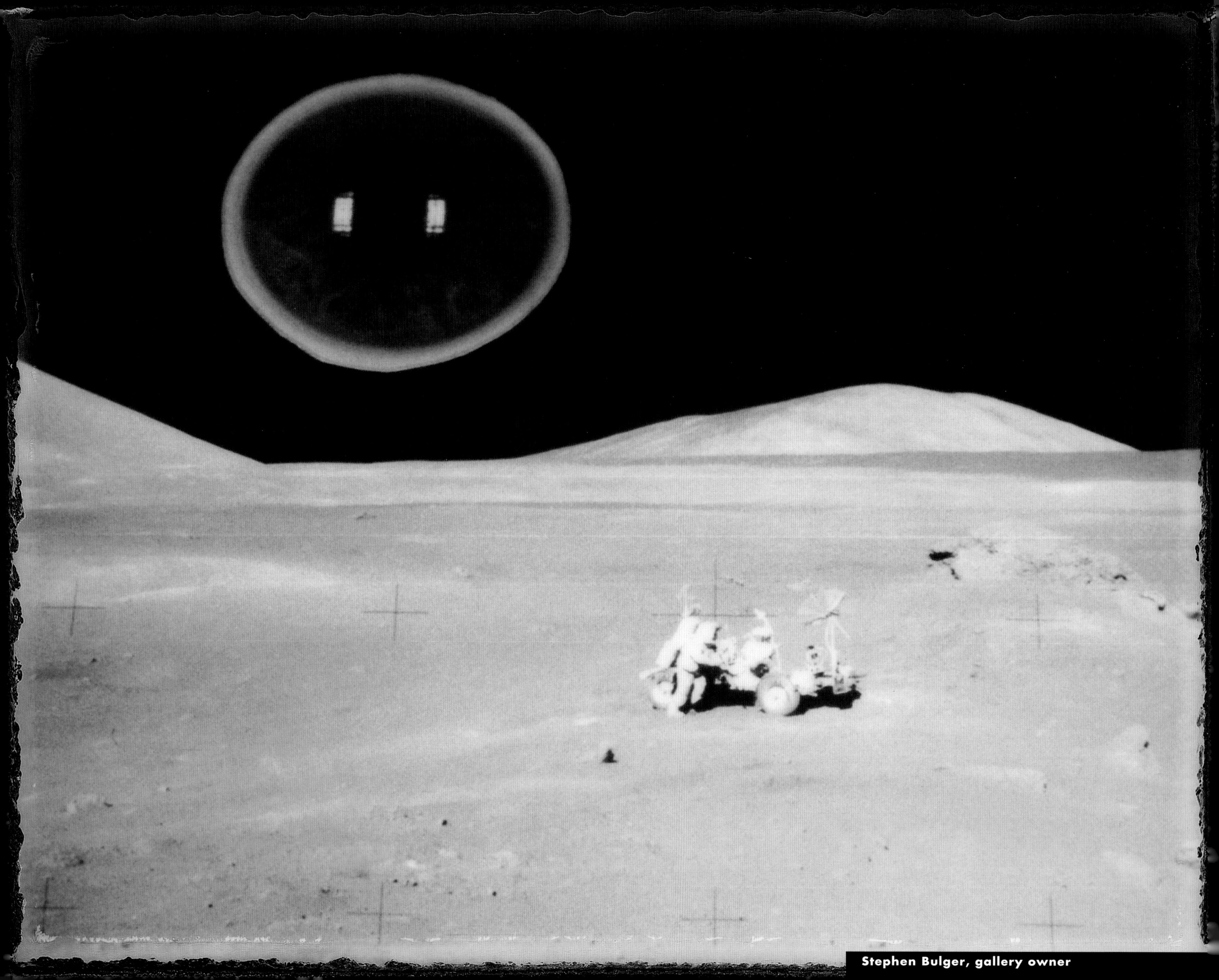

Stephen Bulger, gallery owner

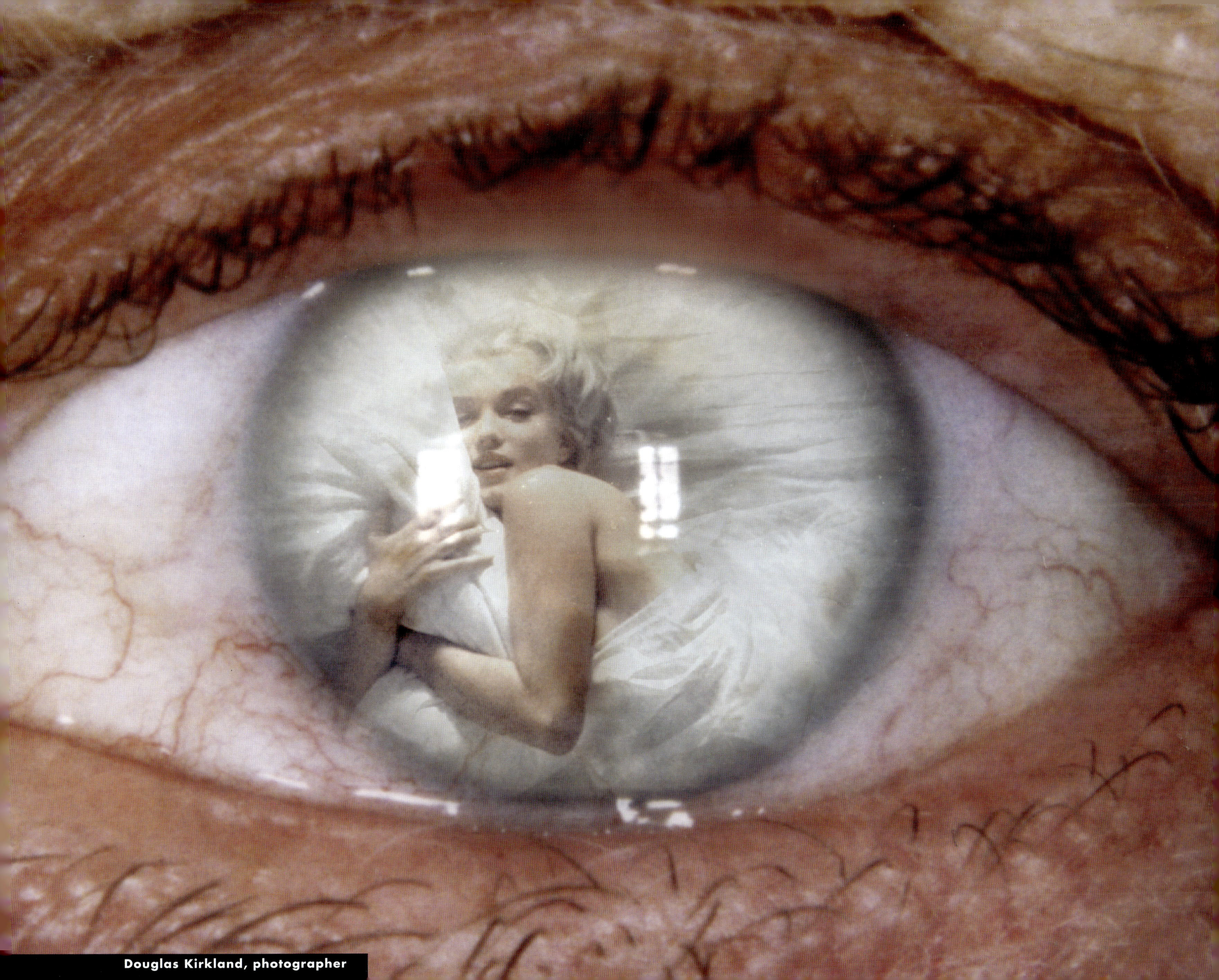
Douglas Kirkland, photographer

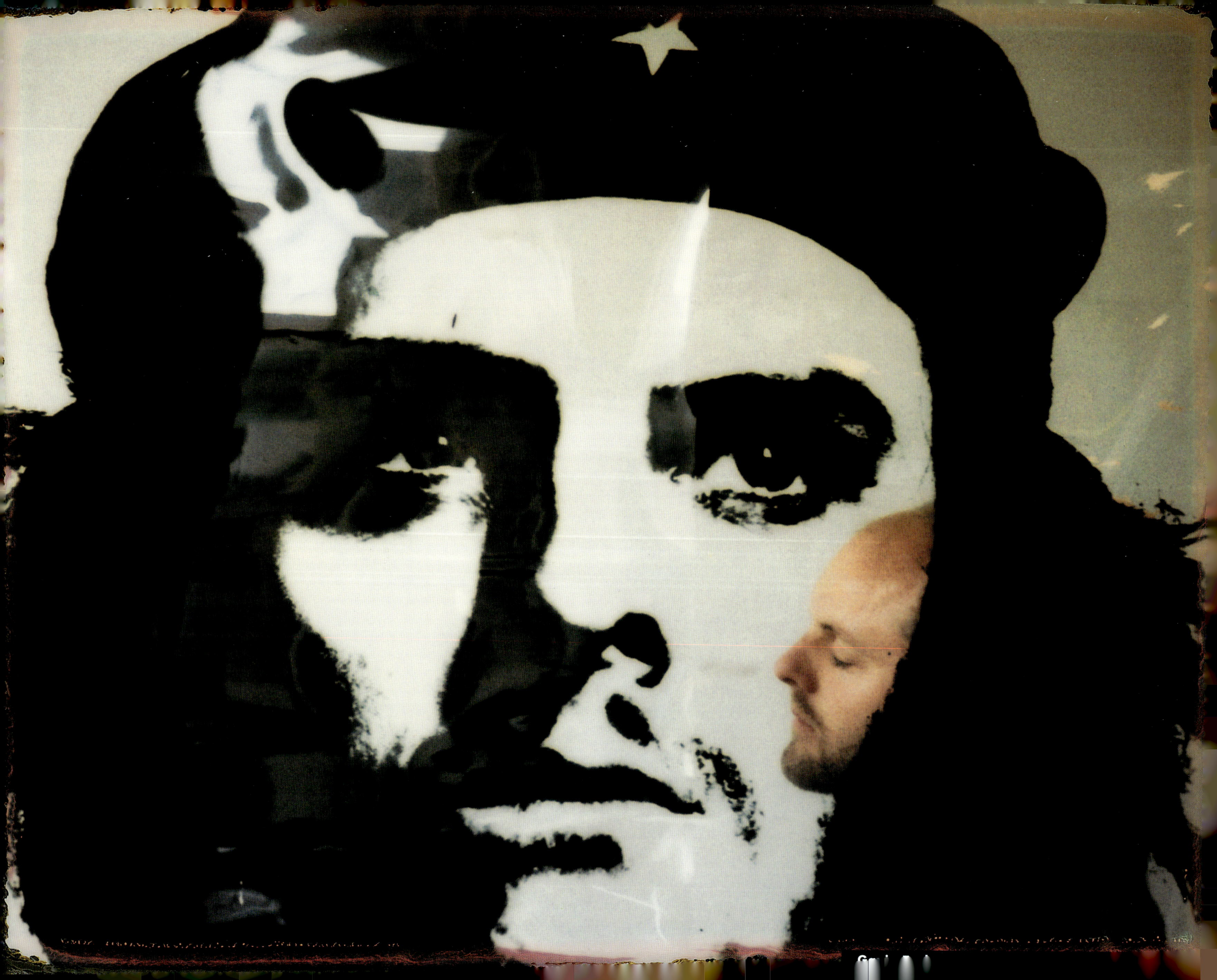

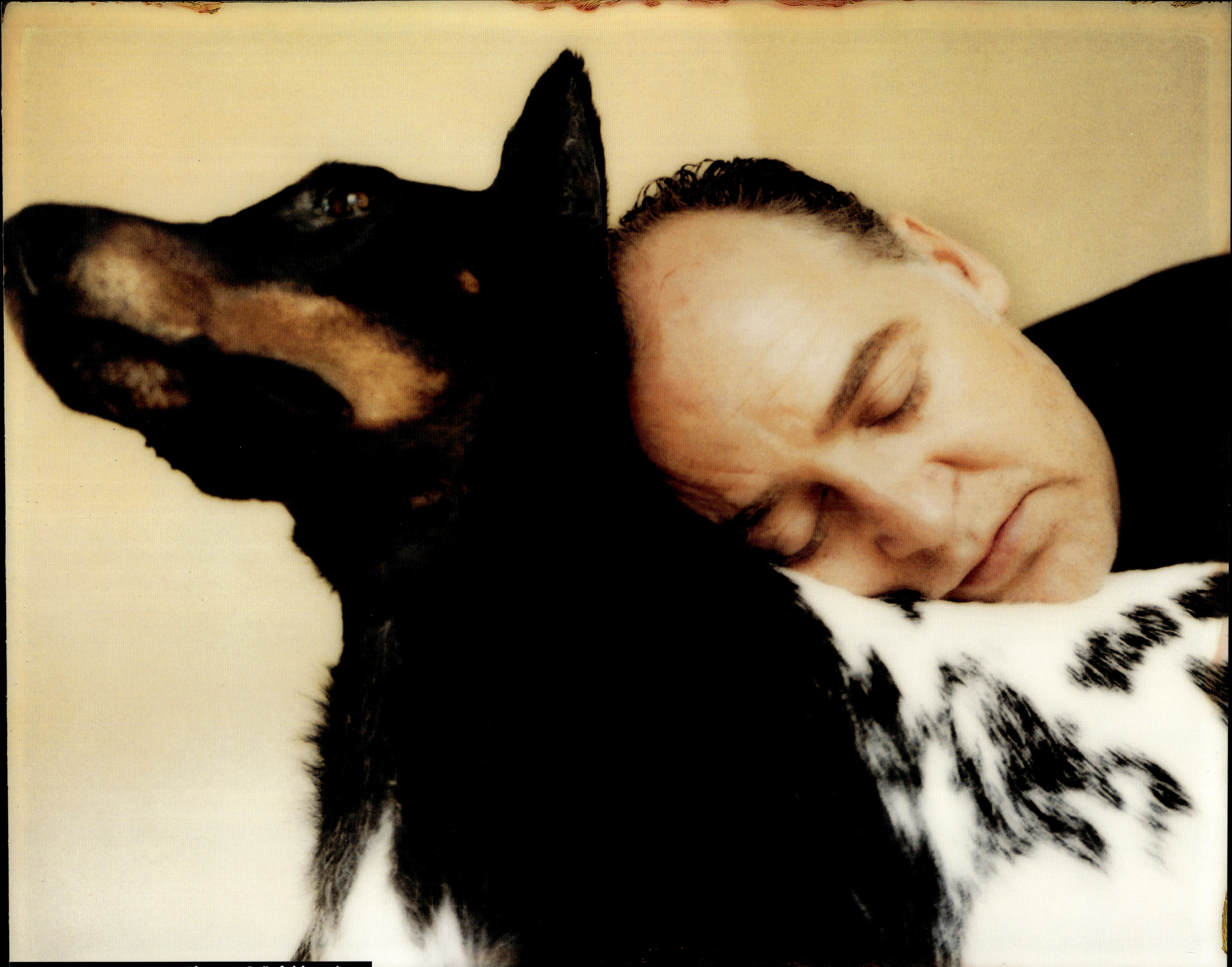

Thomas Grünfeld, artist

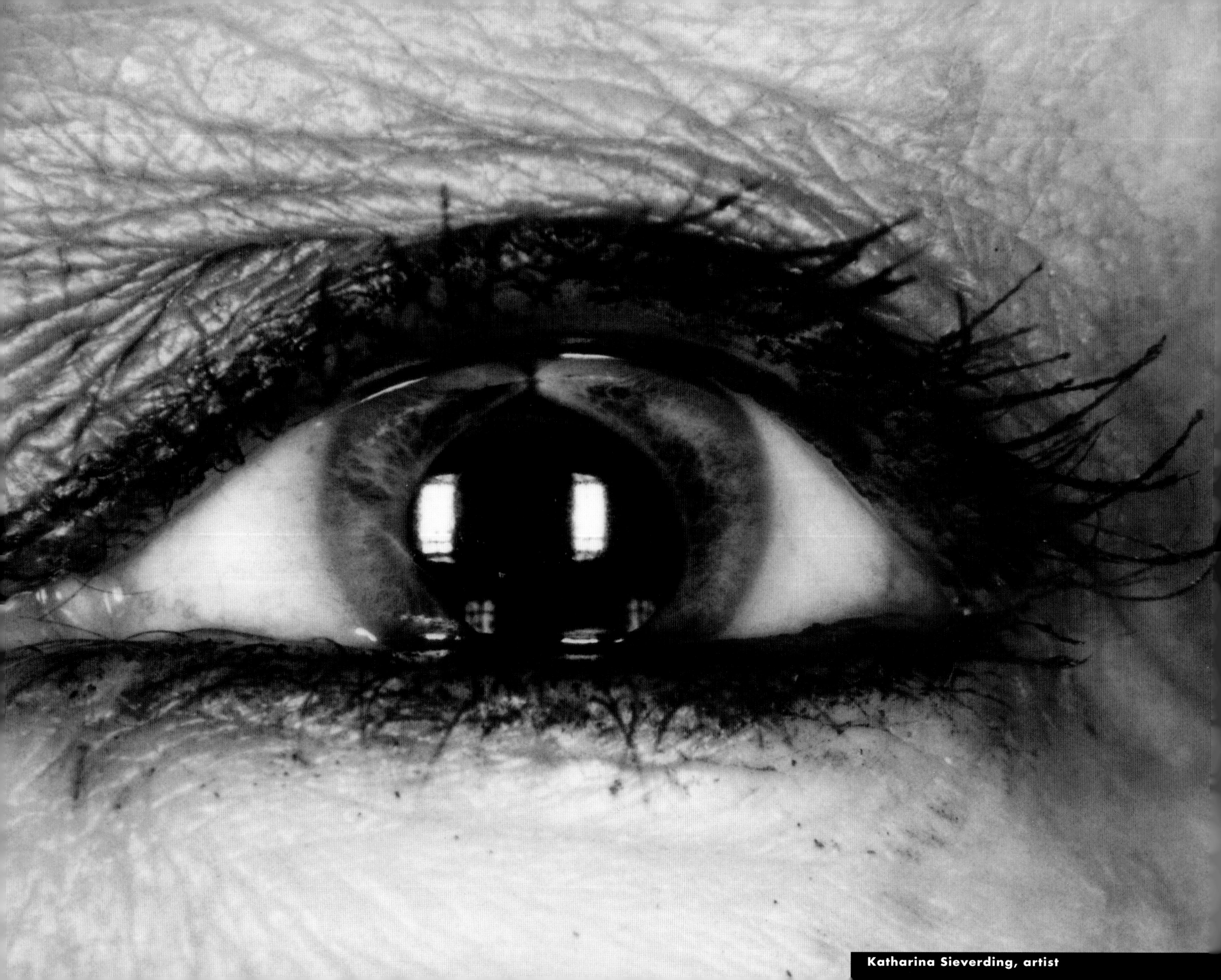

Katharina Sieverding, artist

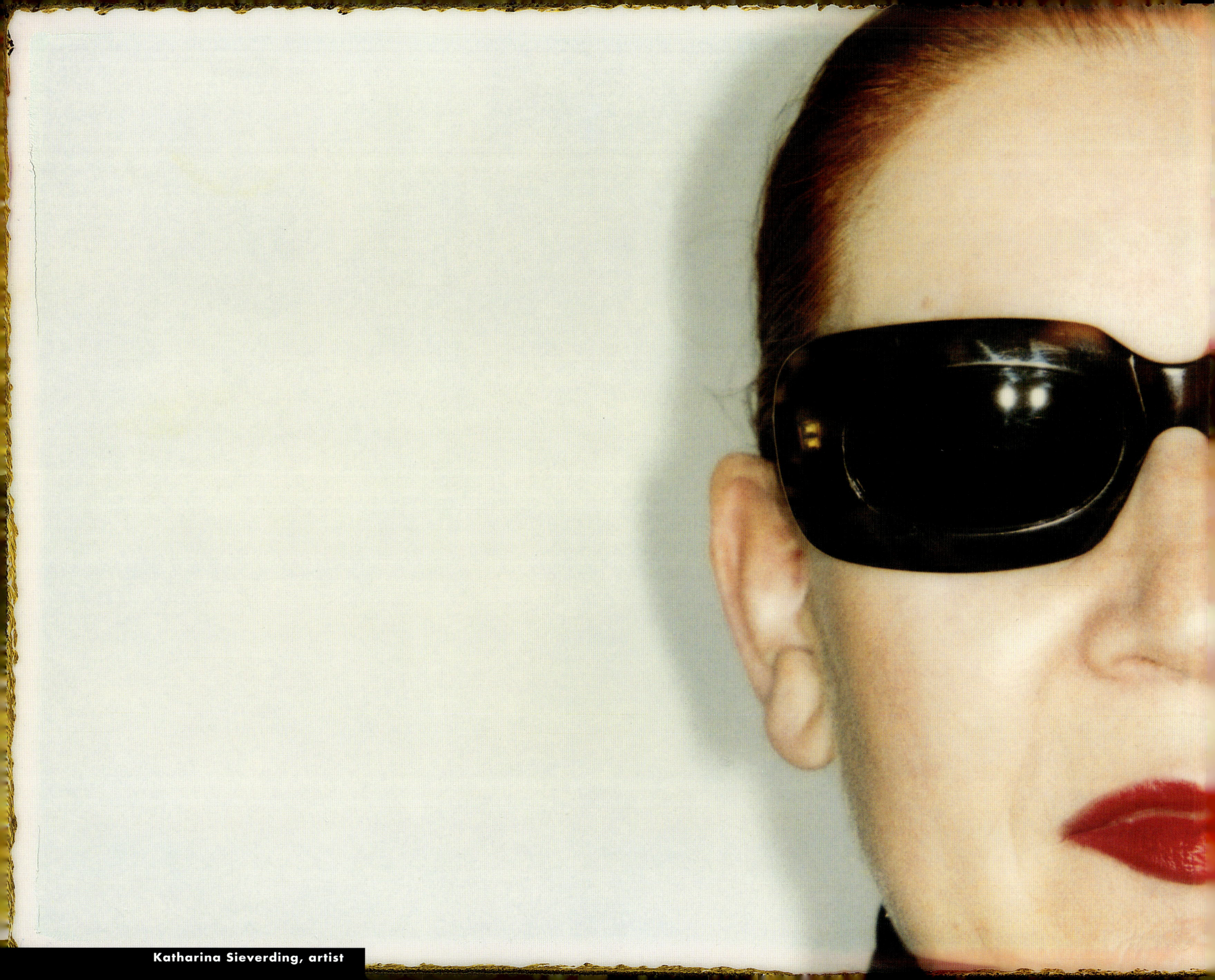

Katharina Sieverding, artist

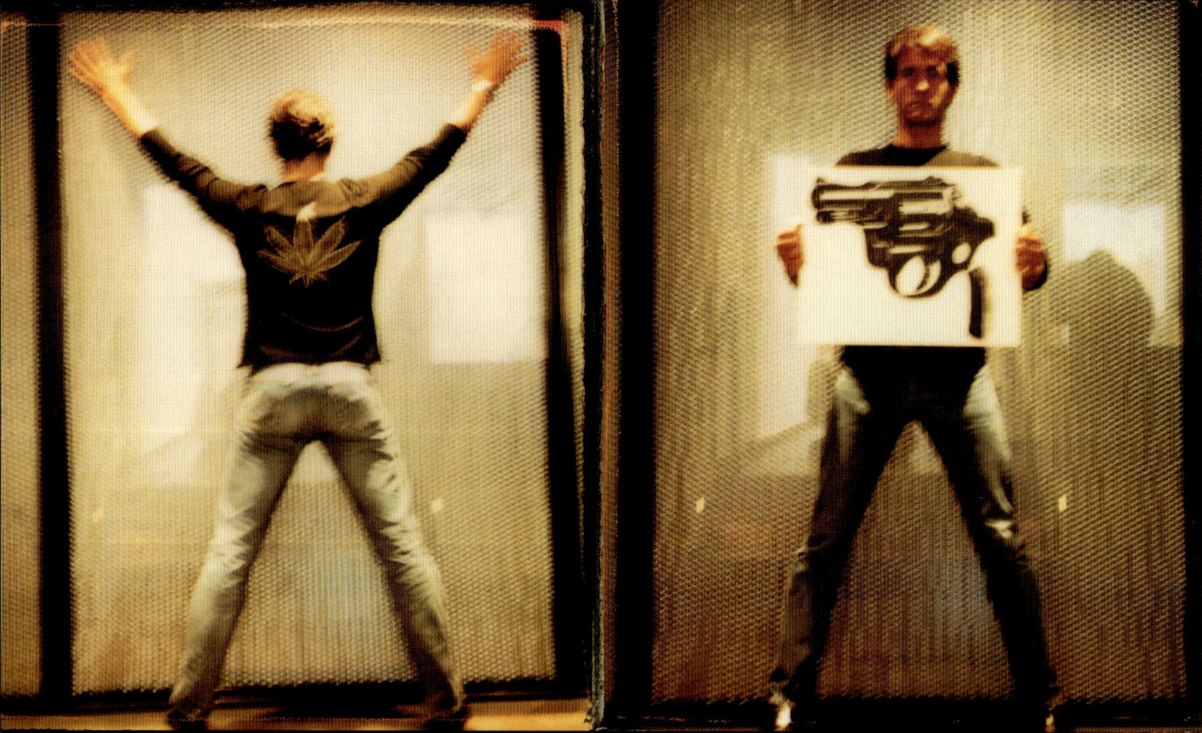

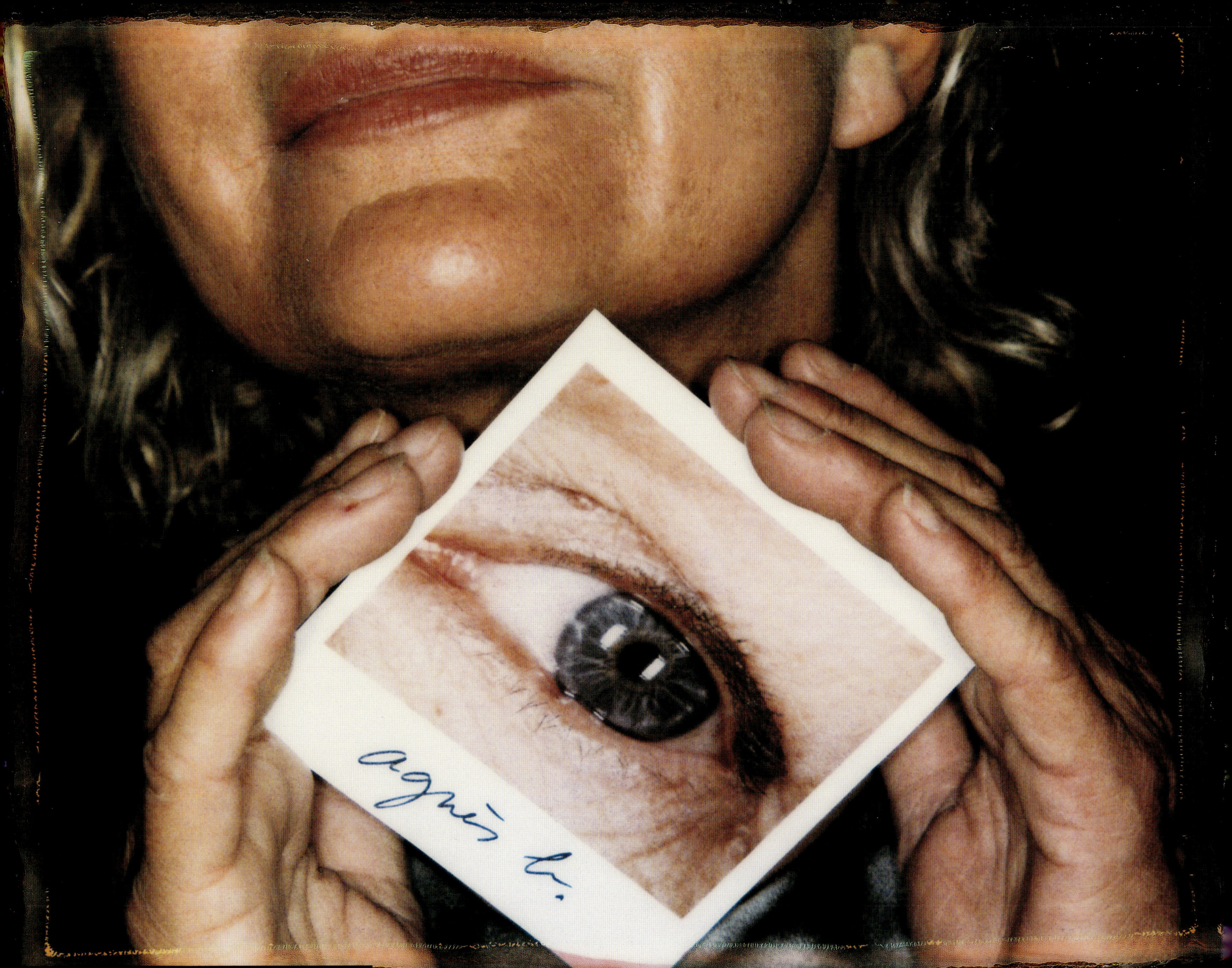

agnès b., designer

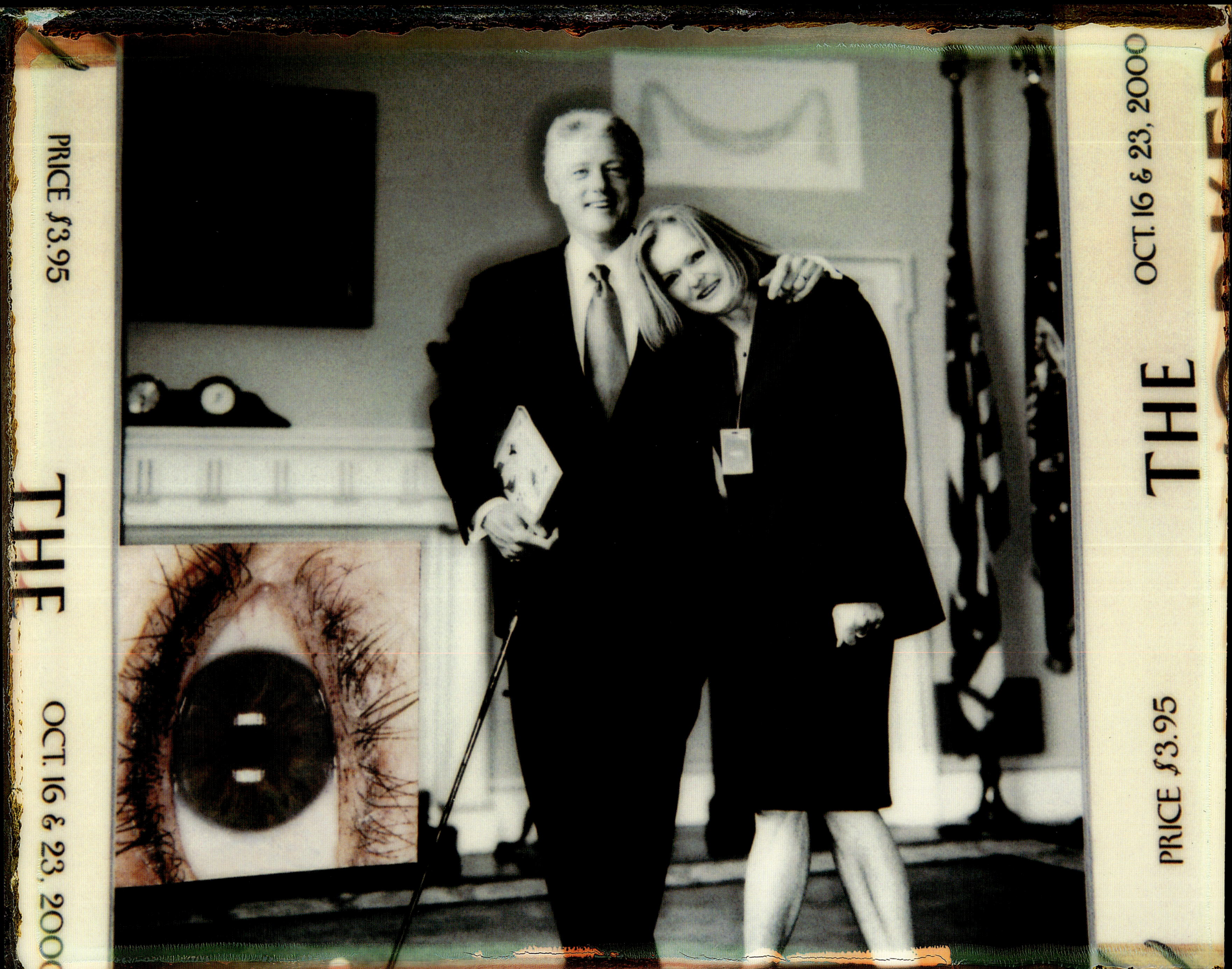

Elisabeth Biondi, director of photography

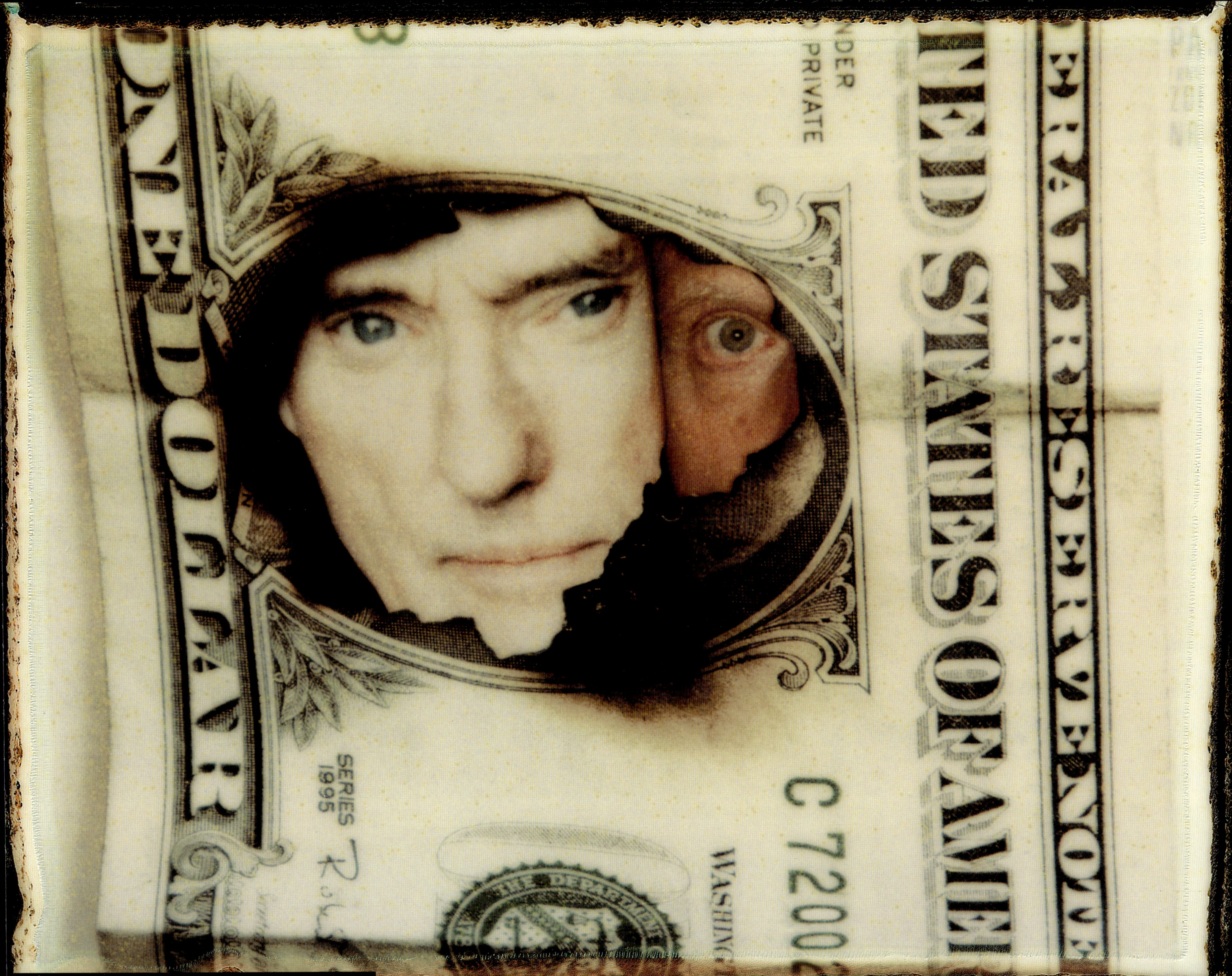

Hermann Vaske, art director

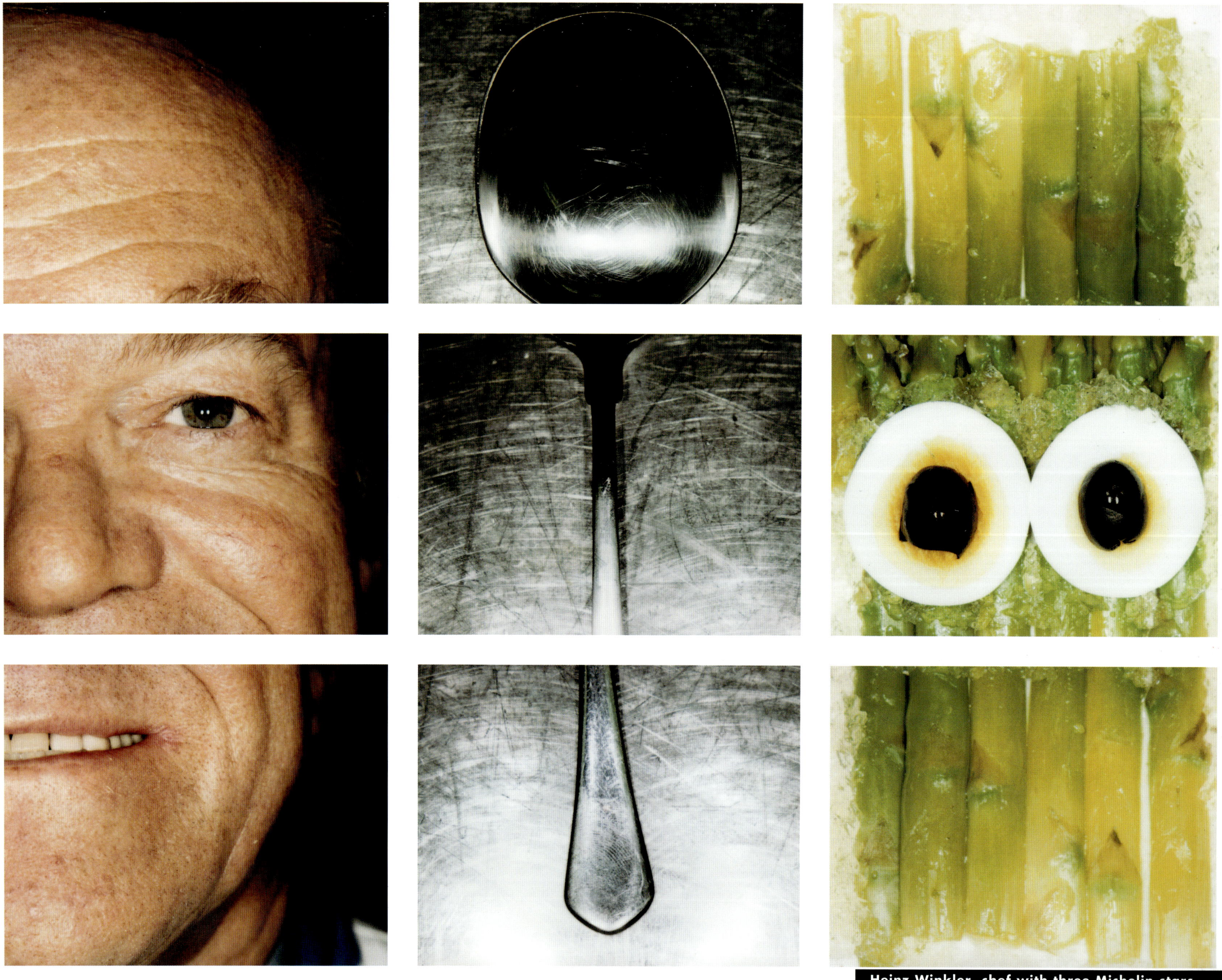

Heinz Winkler, chef with three Michelin stars

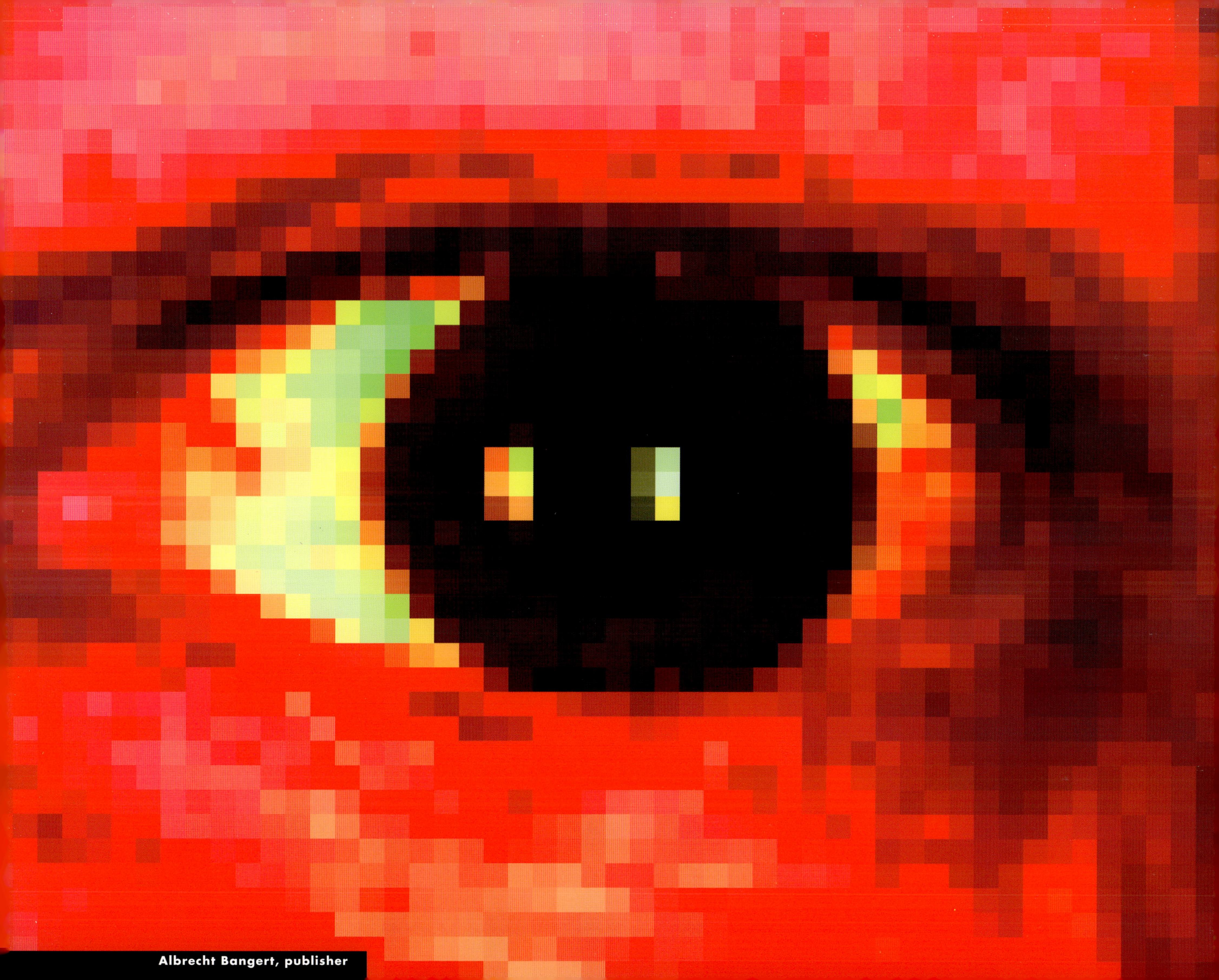

Albrecht Bangert, publisher

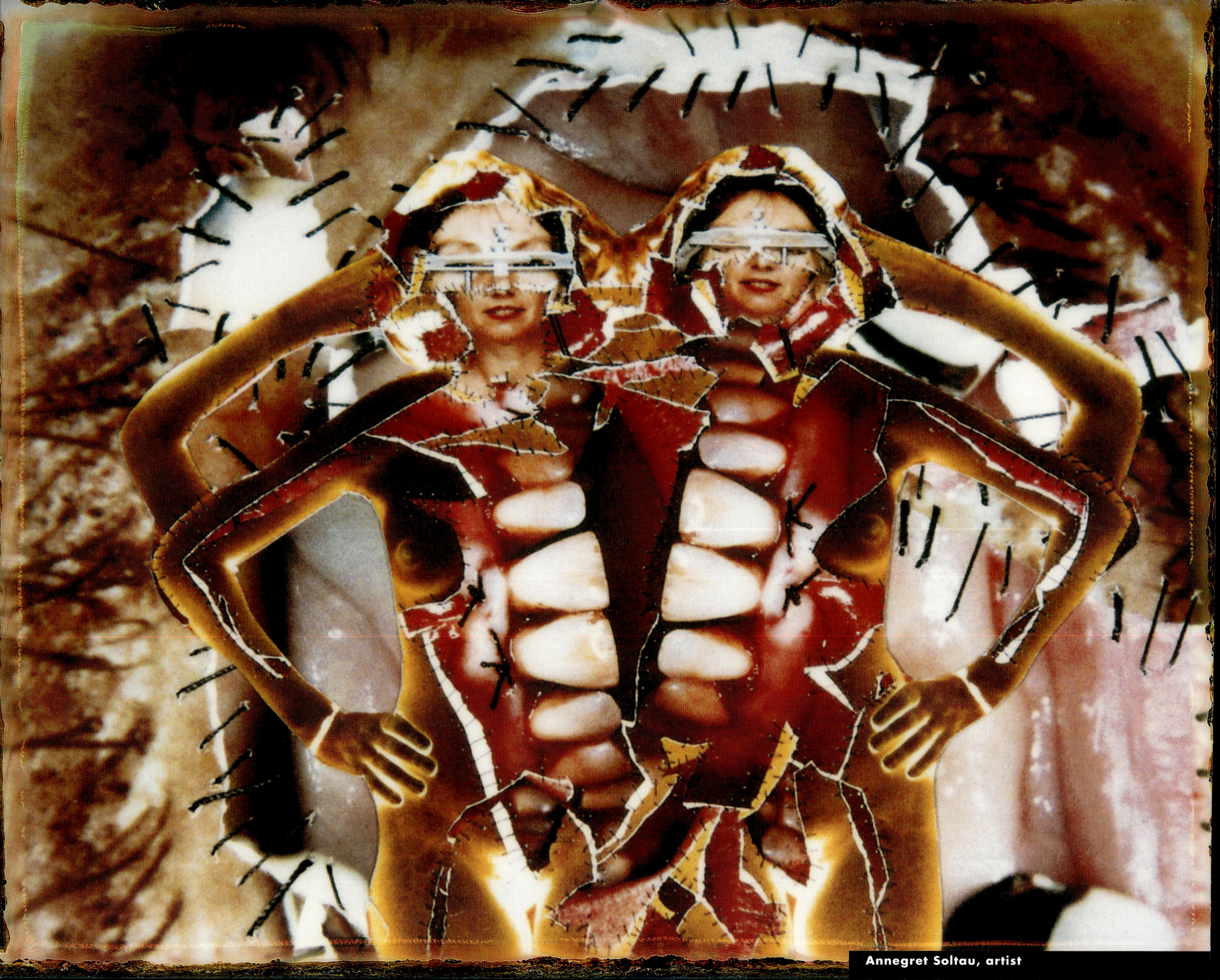

Annegret Soltau, artist

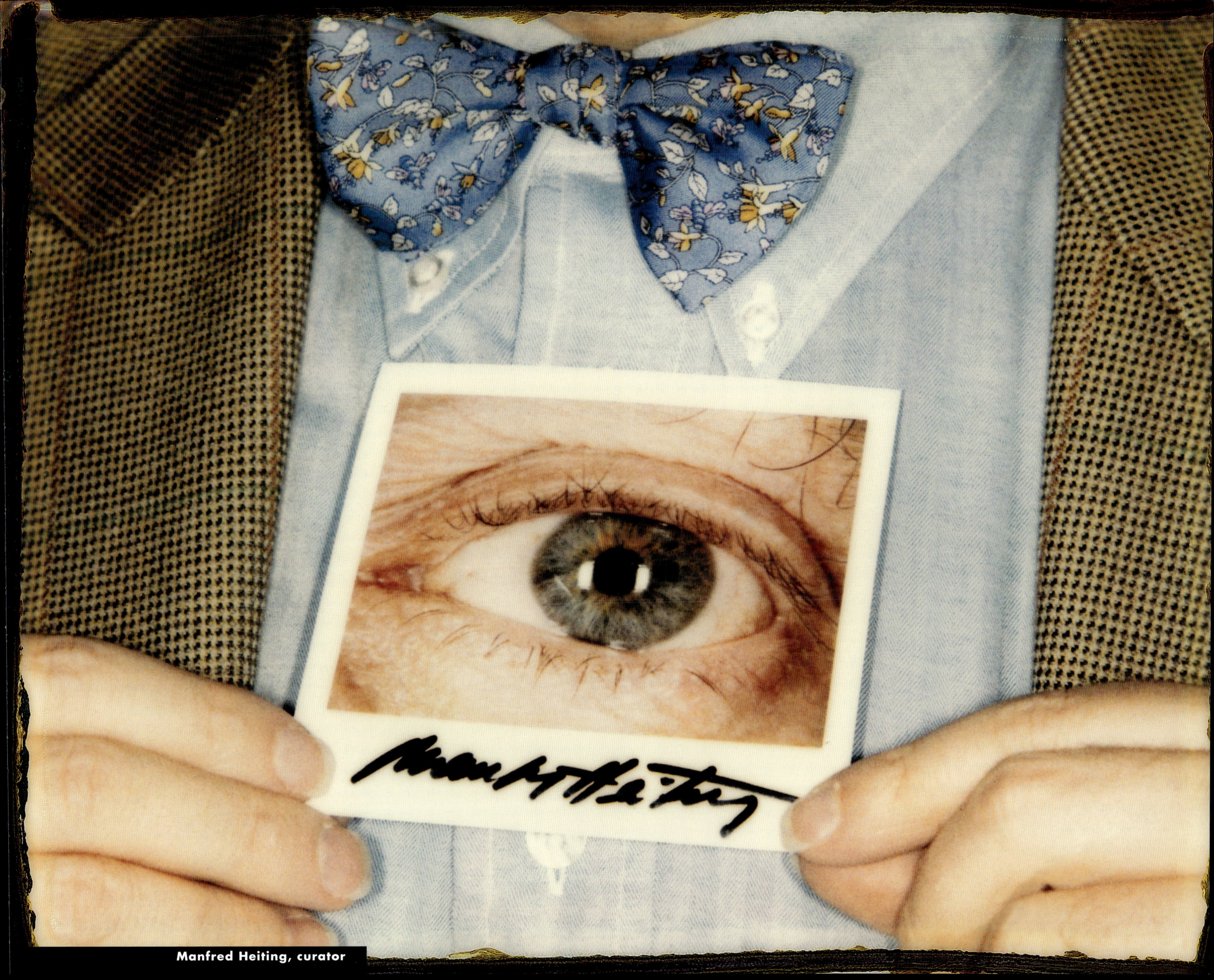

Manfred Heiting, curator

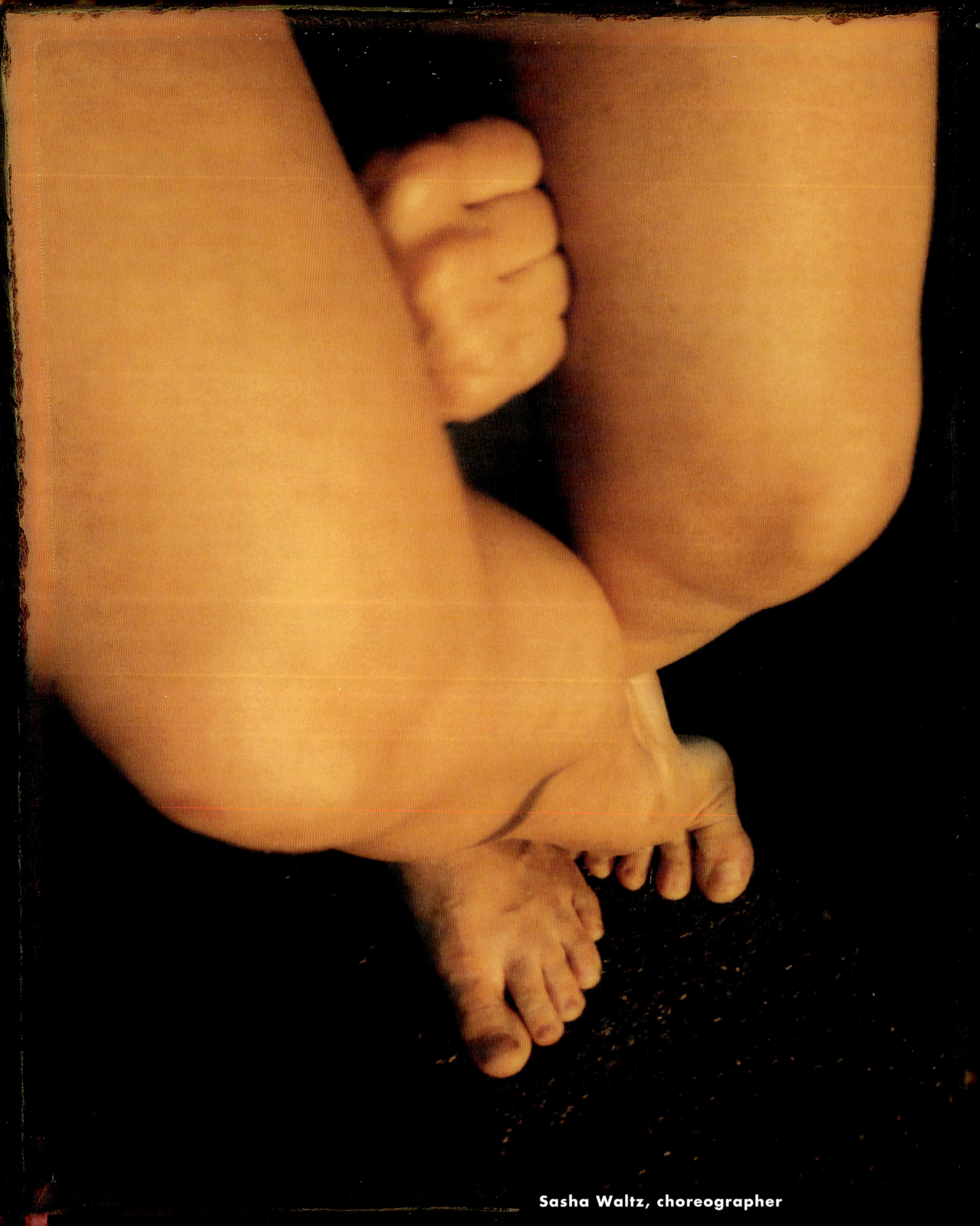

Sasha Waltz, choreographer

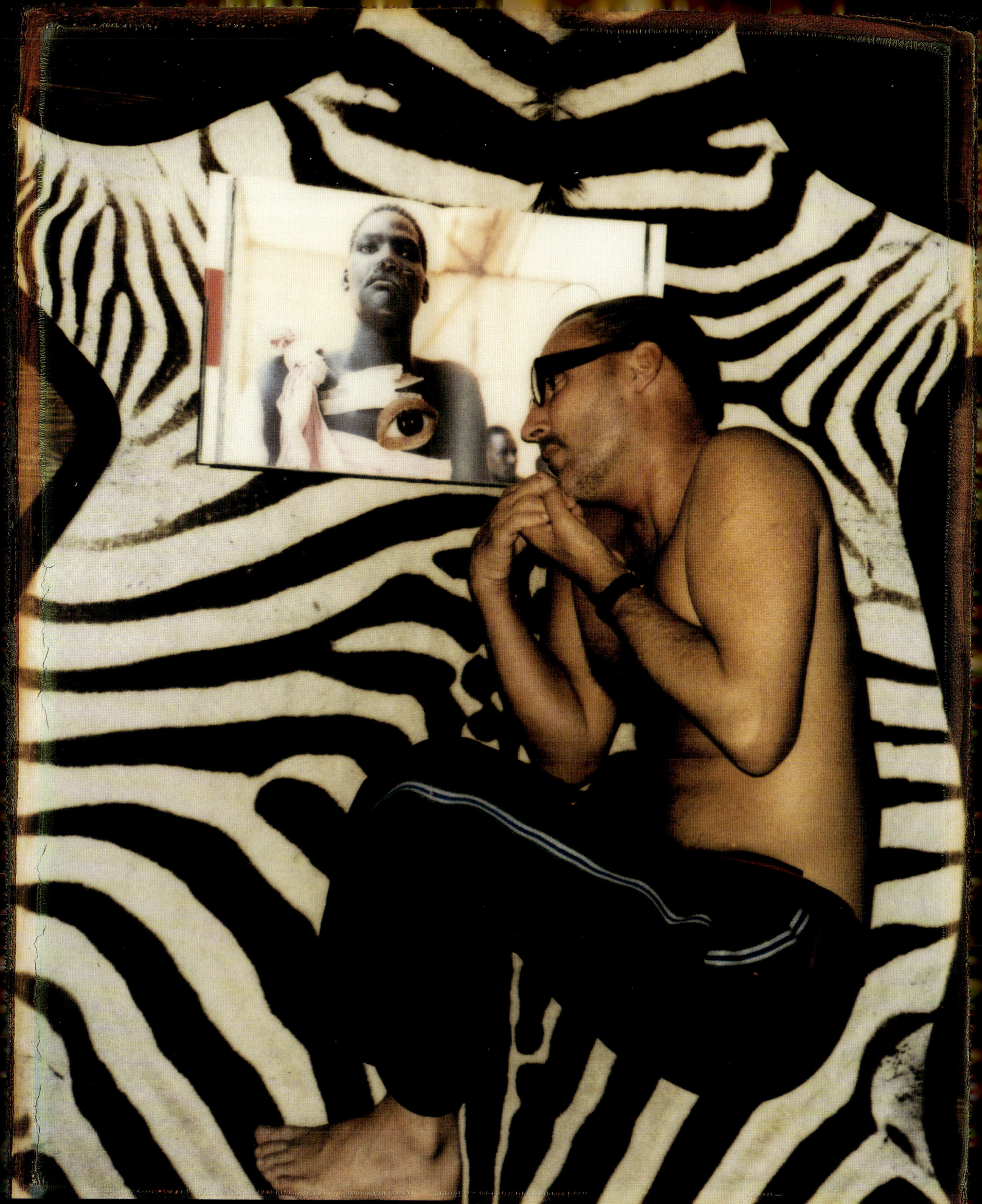

Michel Comte, photographer

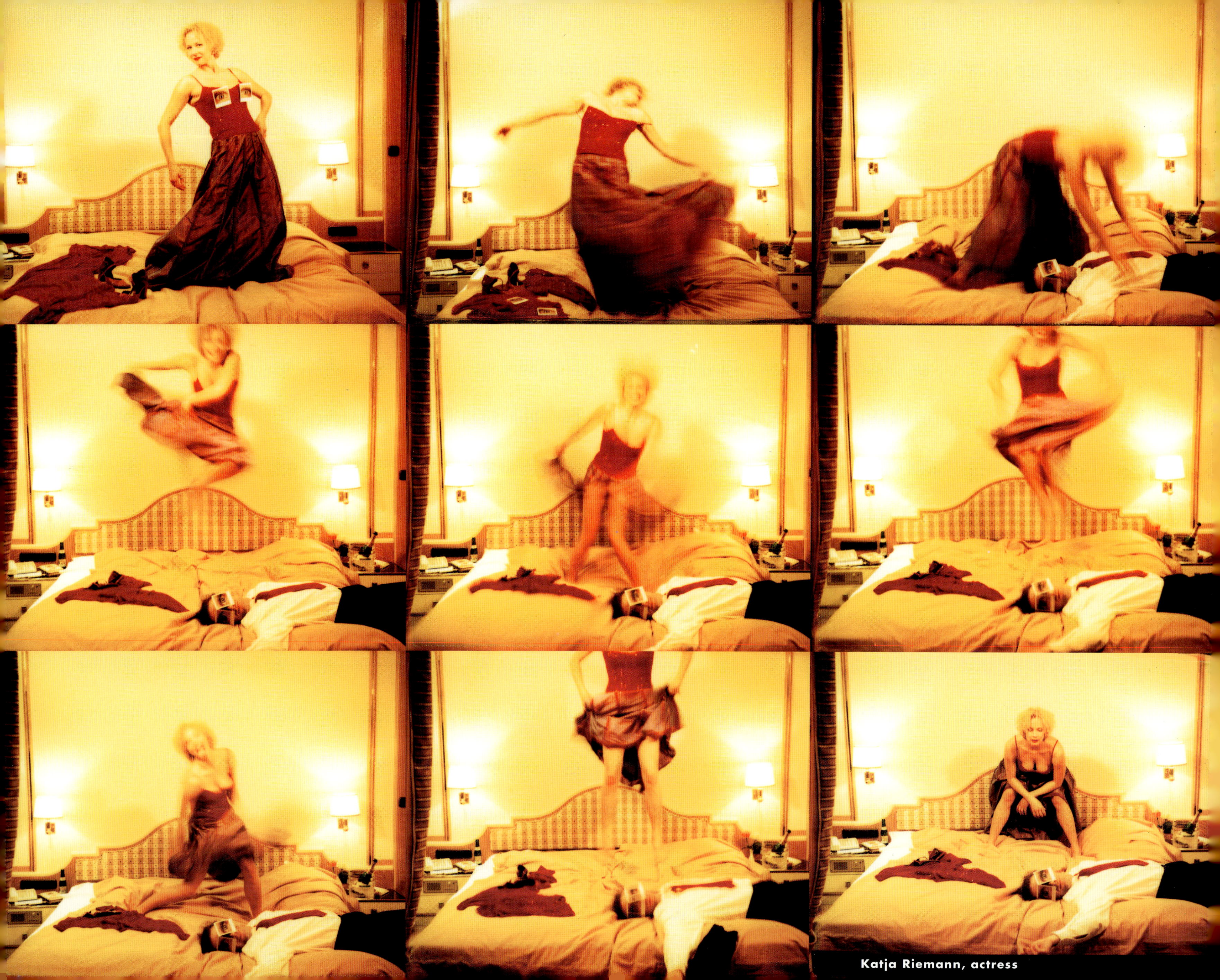

Katja Riemann, actress

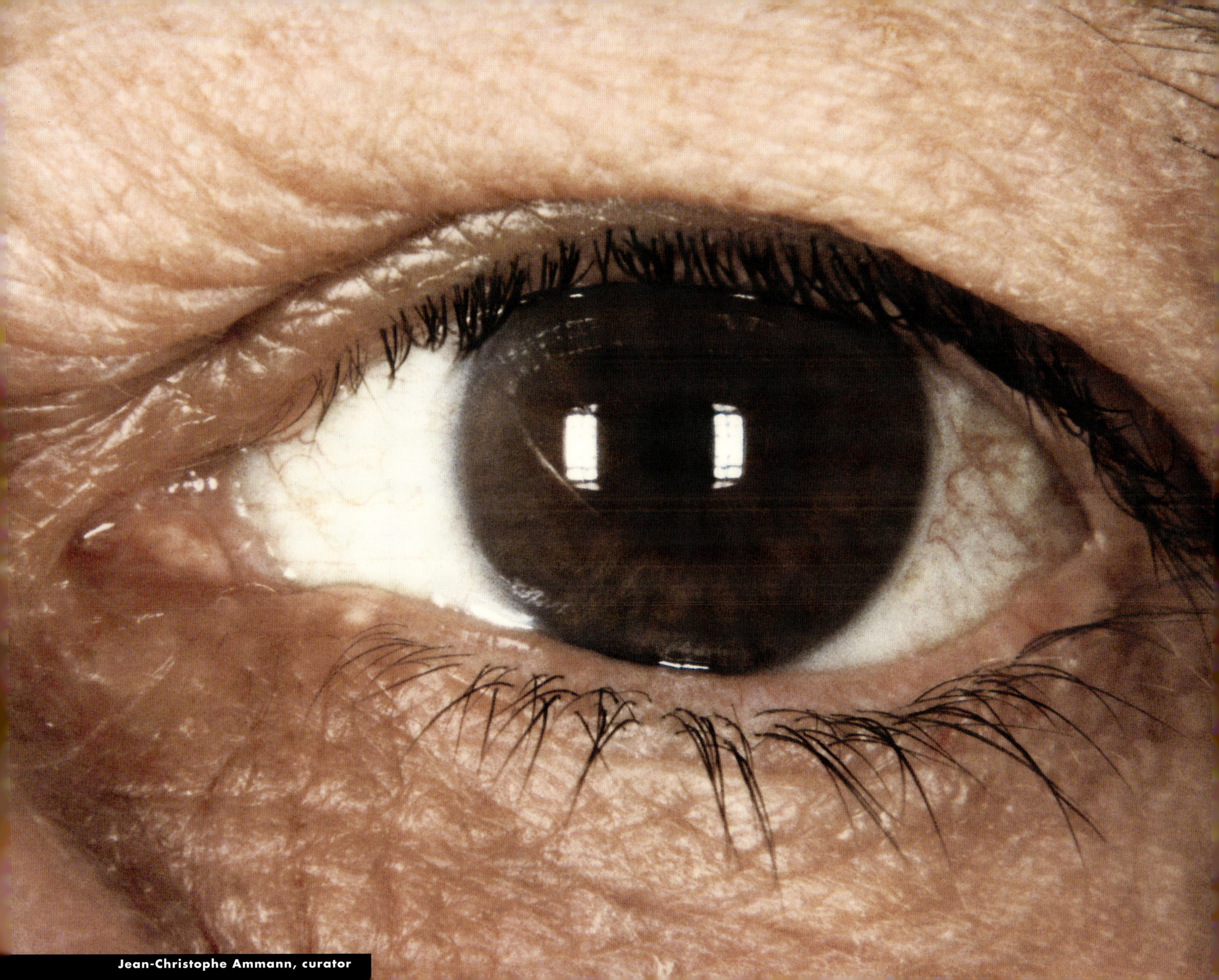

Jean-Christophe Ammann, curator

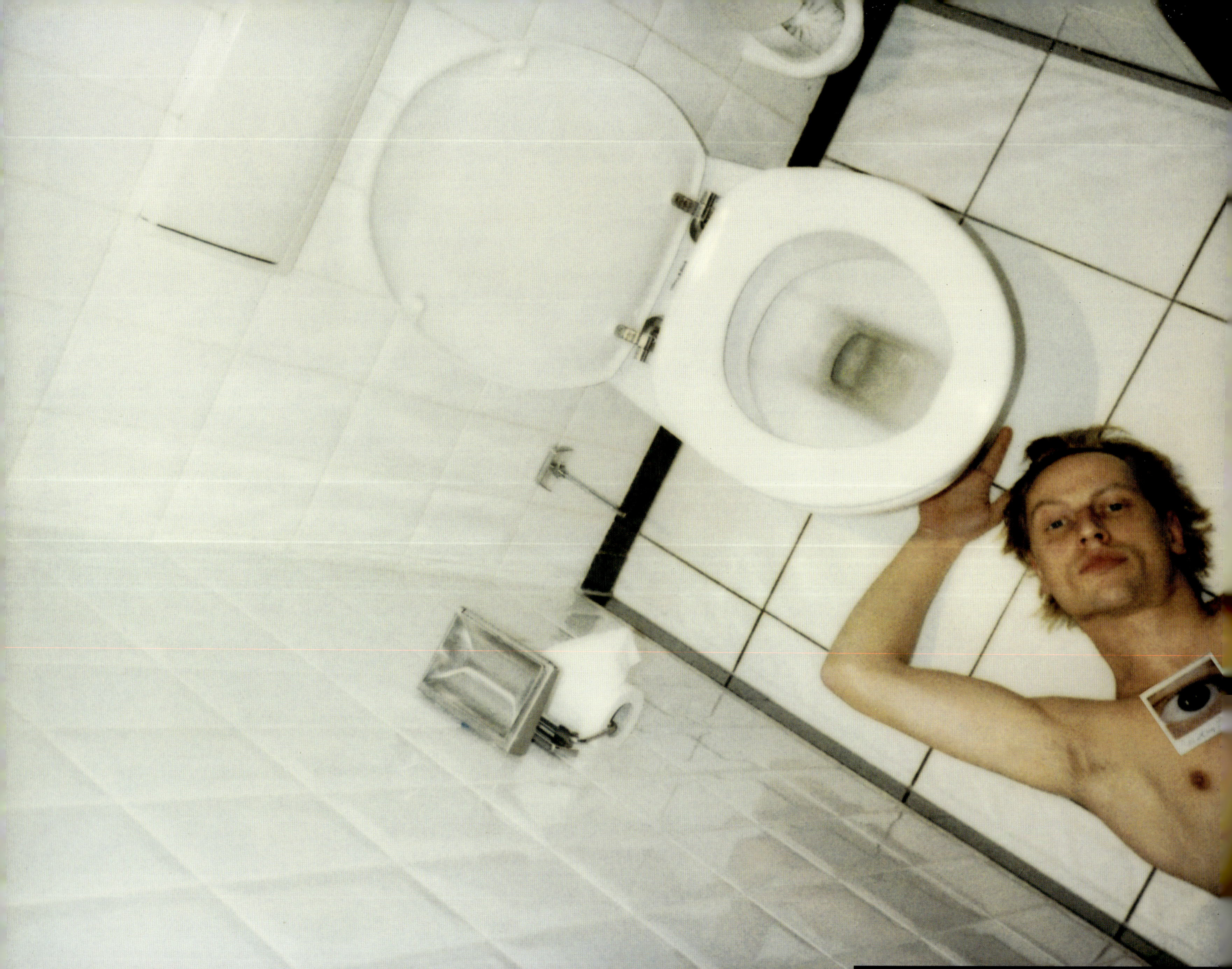

Victor Schefé, actor

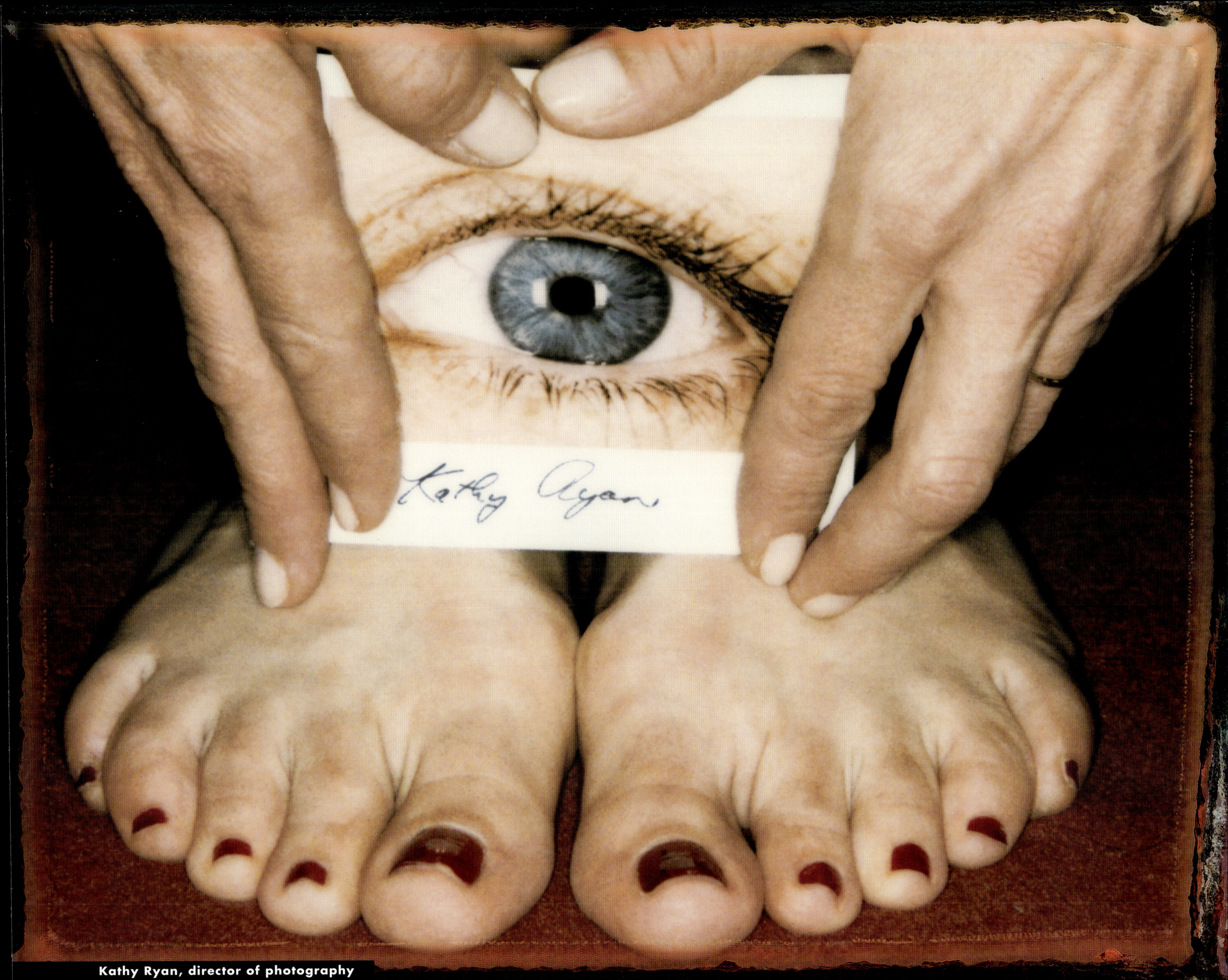

Kathy Ryan, director of photography

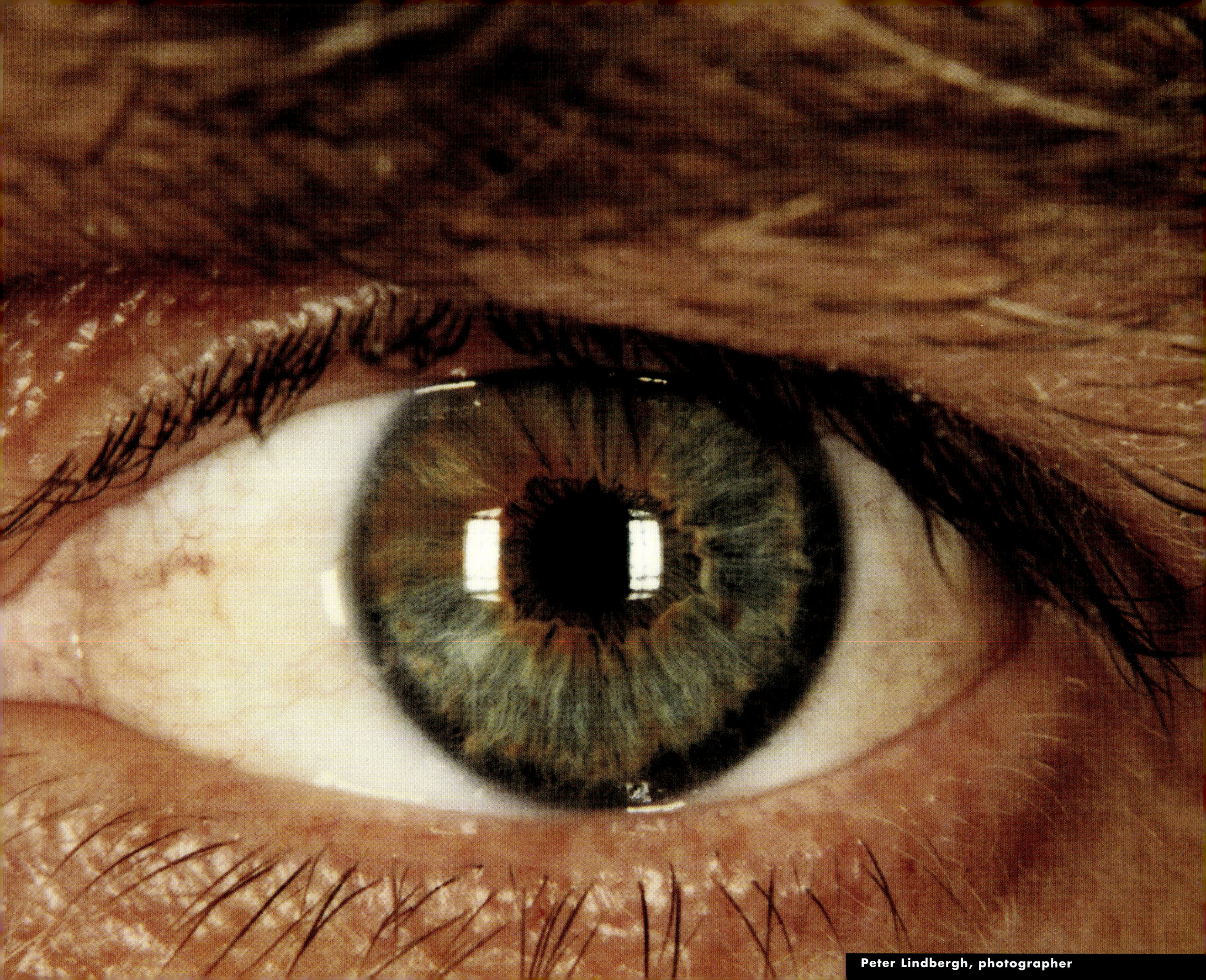

Peter Lindbergh, photographer

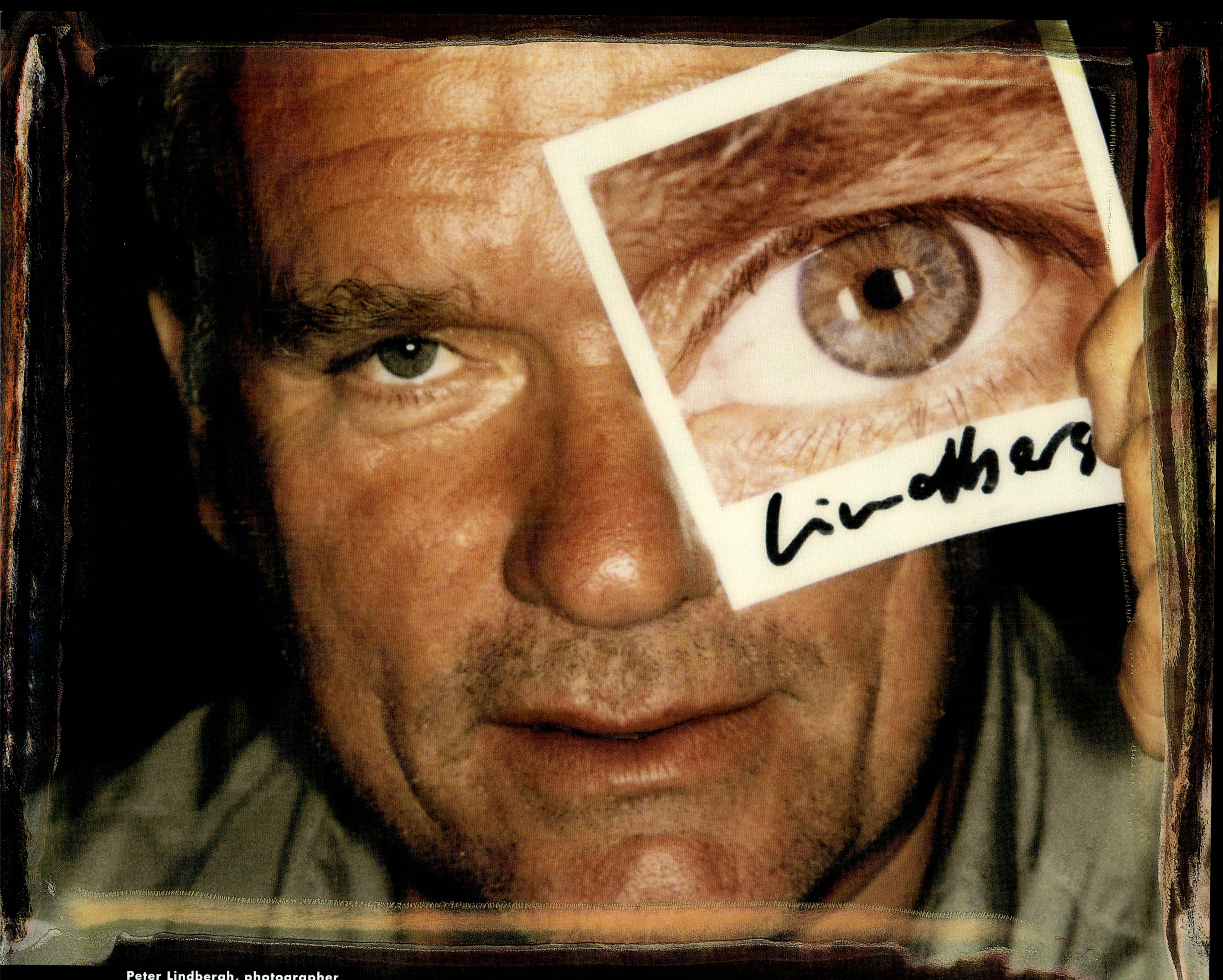

Peter Lindbergh, photographer

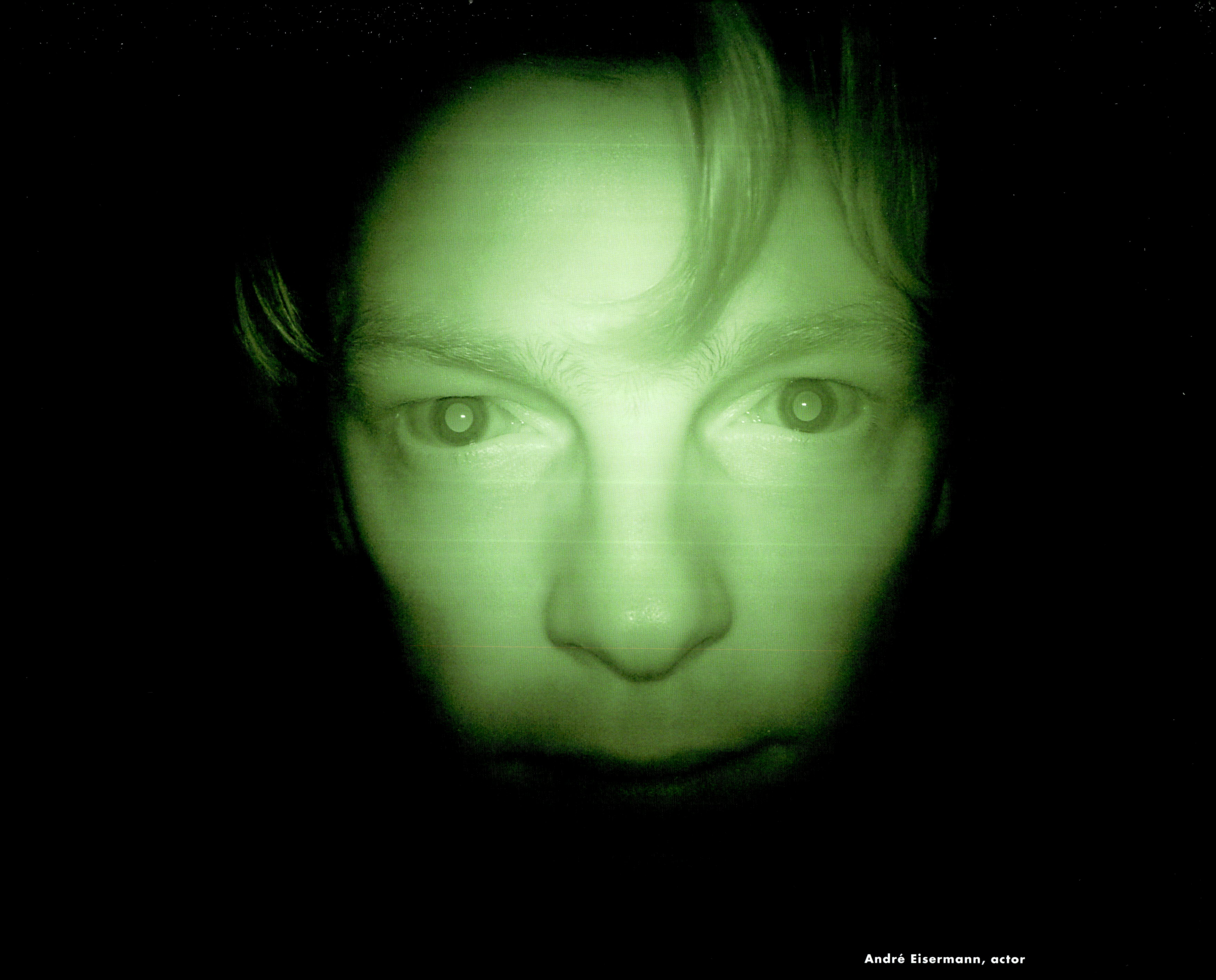

André Eisermann, actor

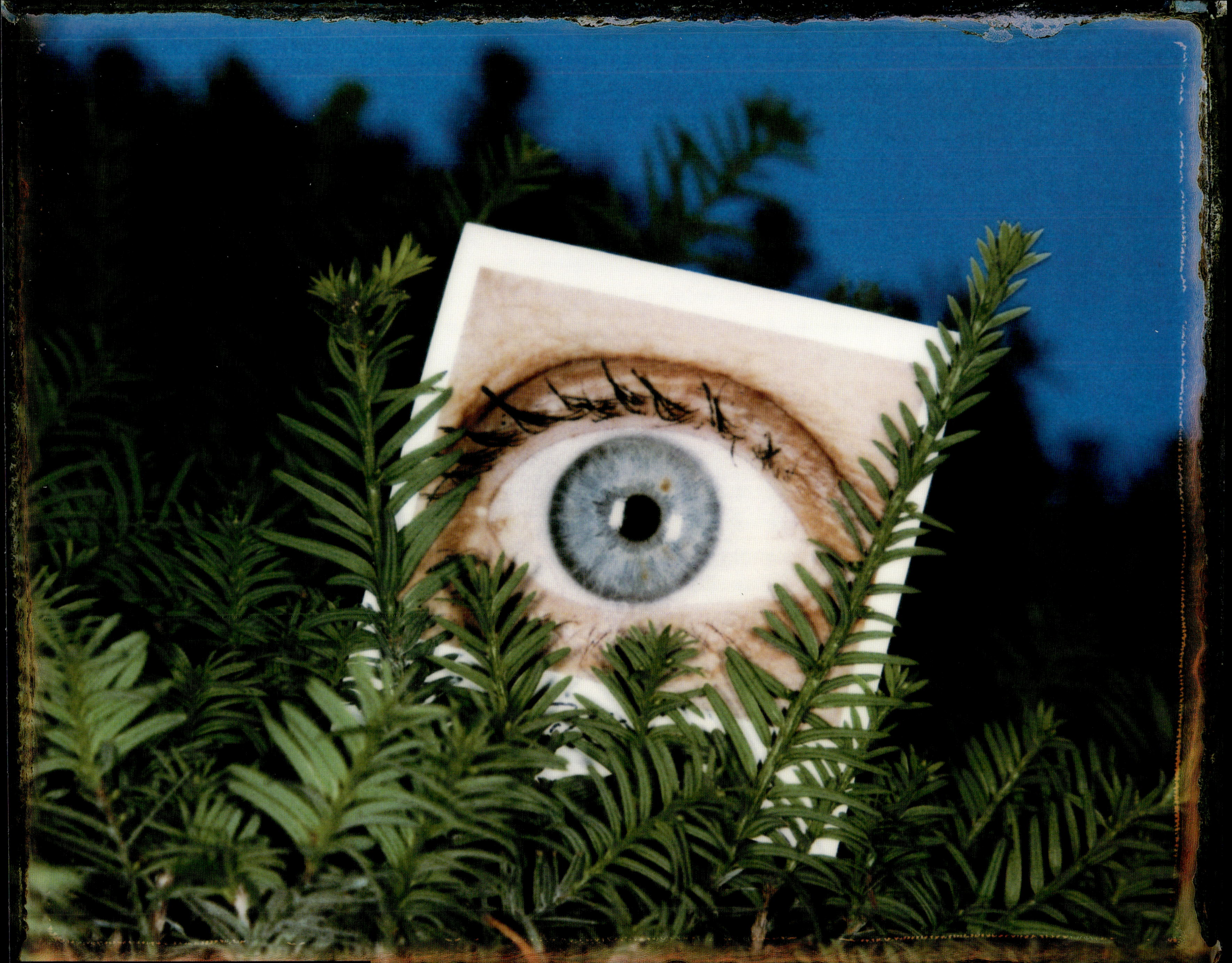

Ruth Eichhorn, director of photography

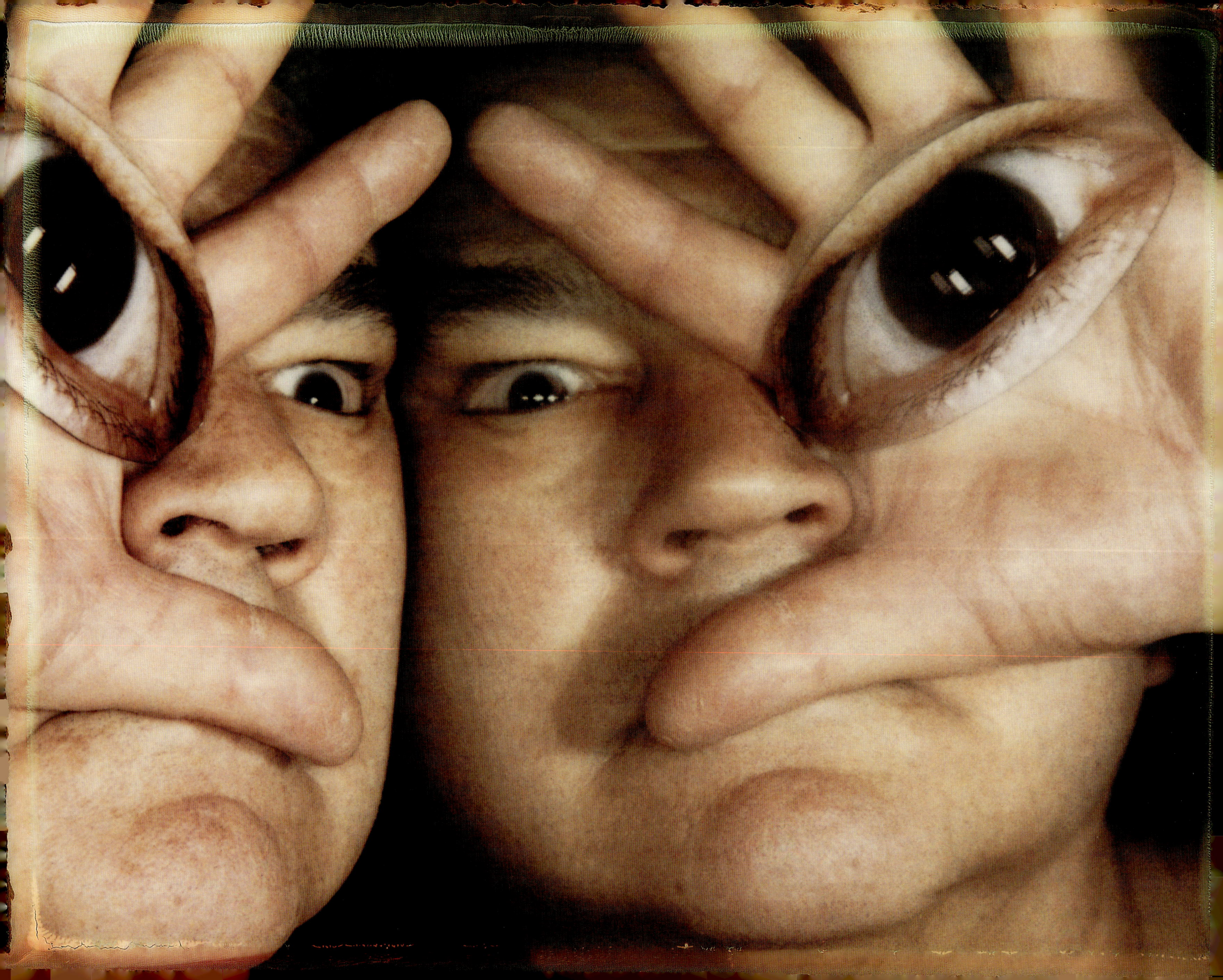

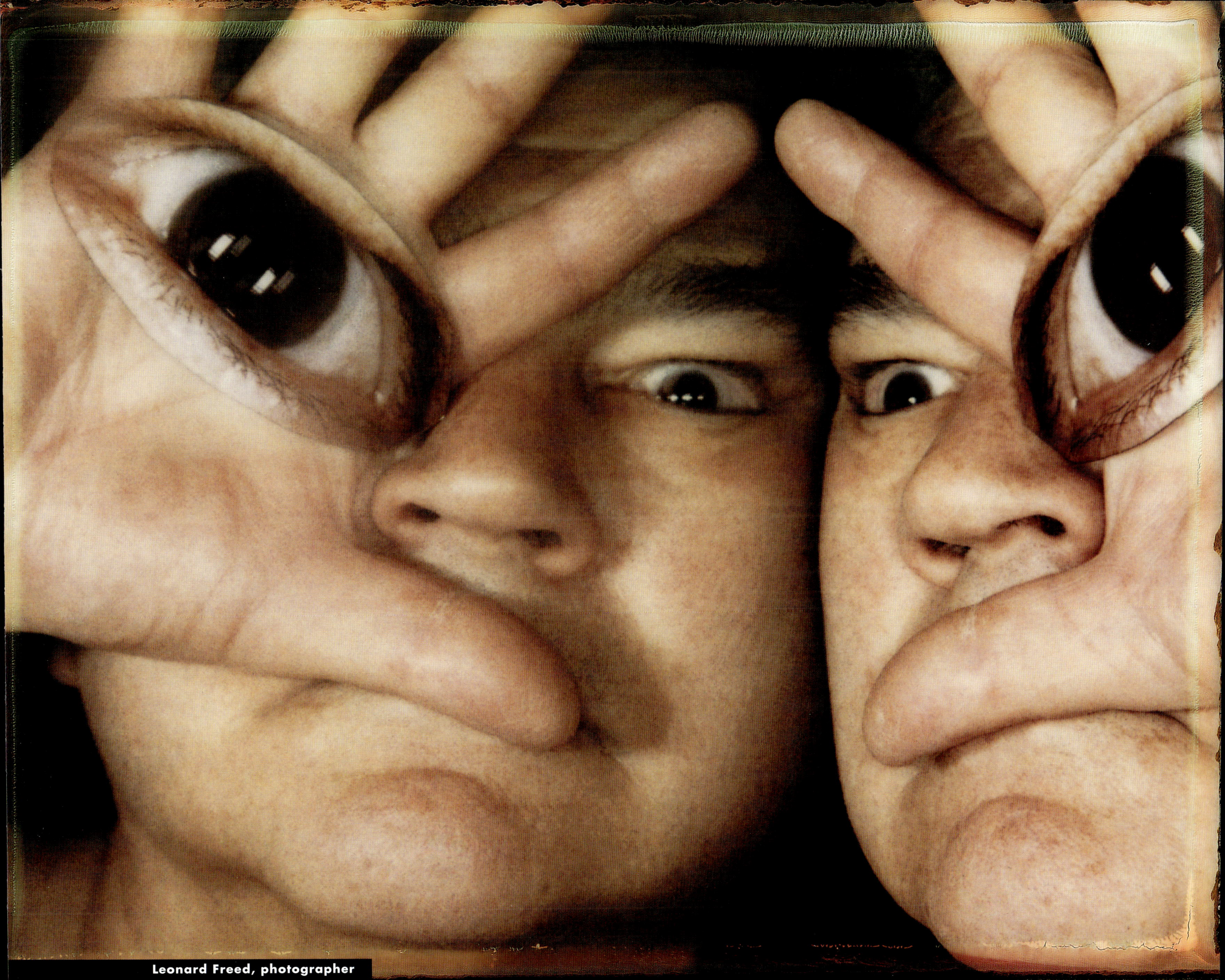

Leonard Freed, photographer

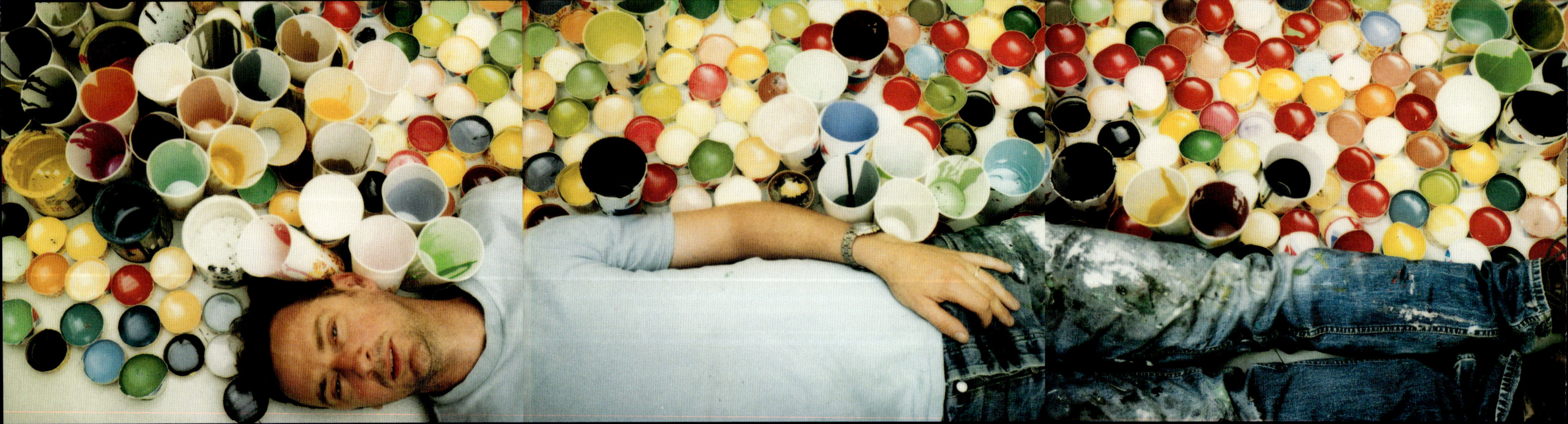

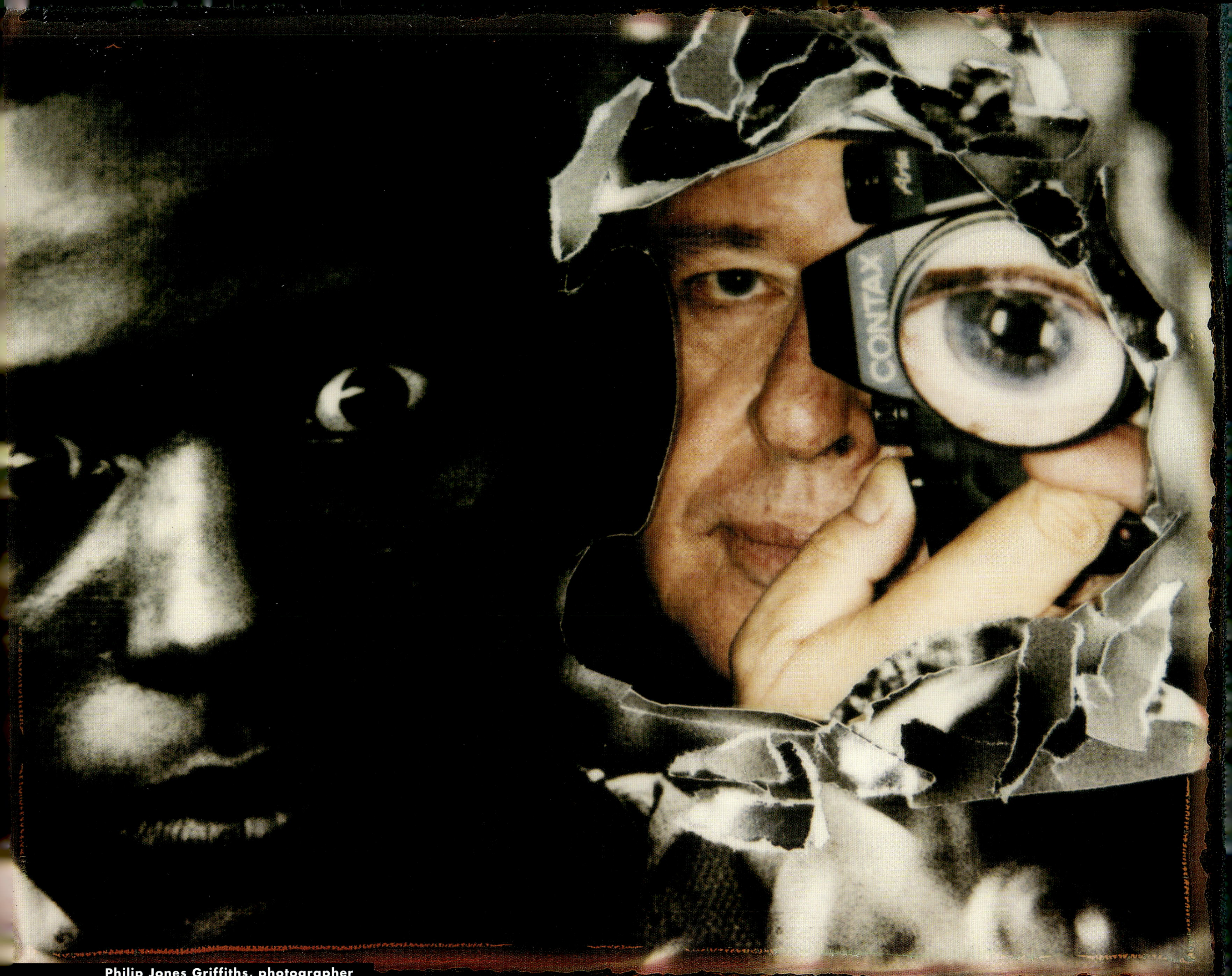

Philip Jones Griffiths, photographer

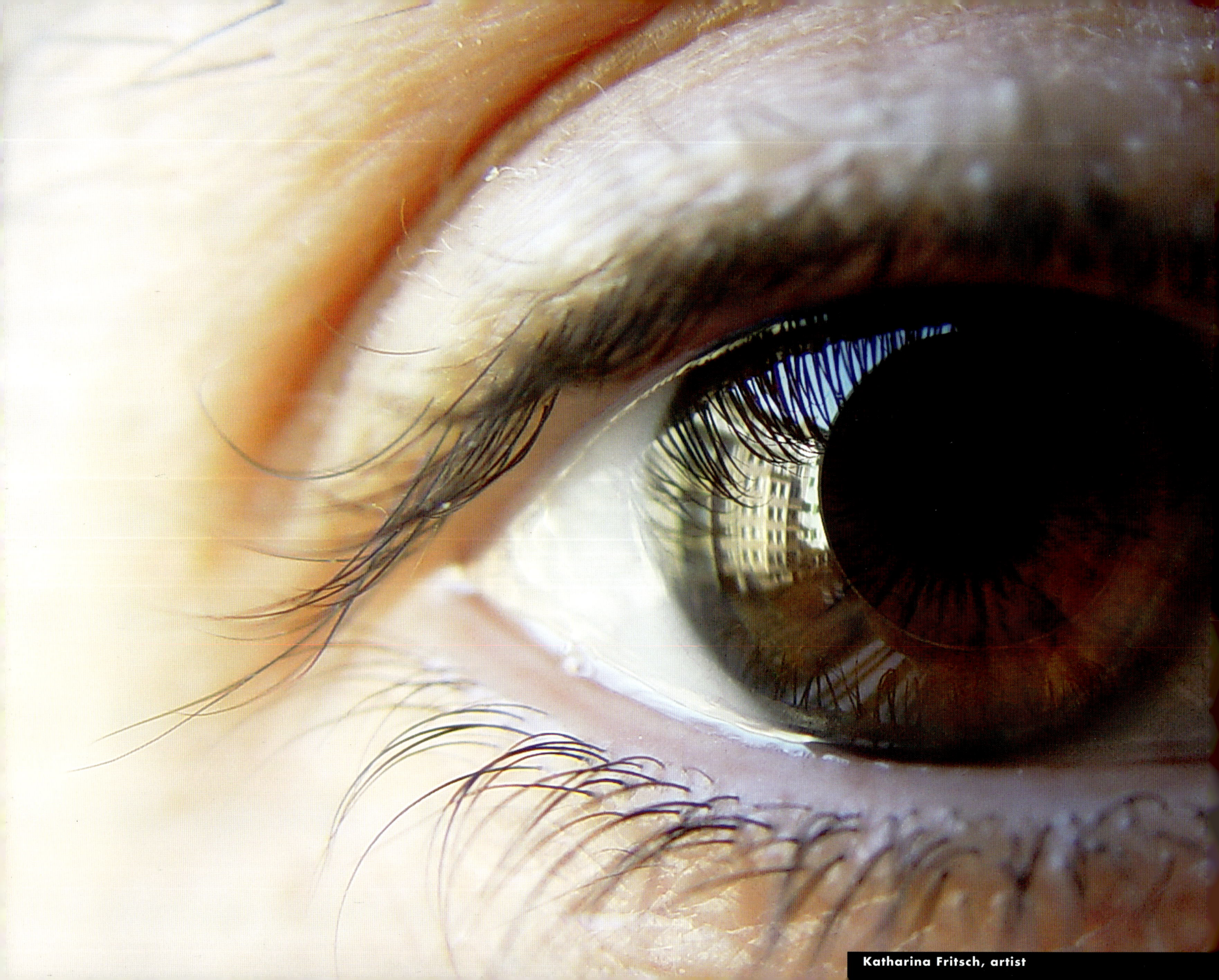

Katharina Fritsch, artist

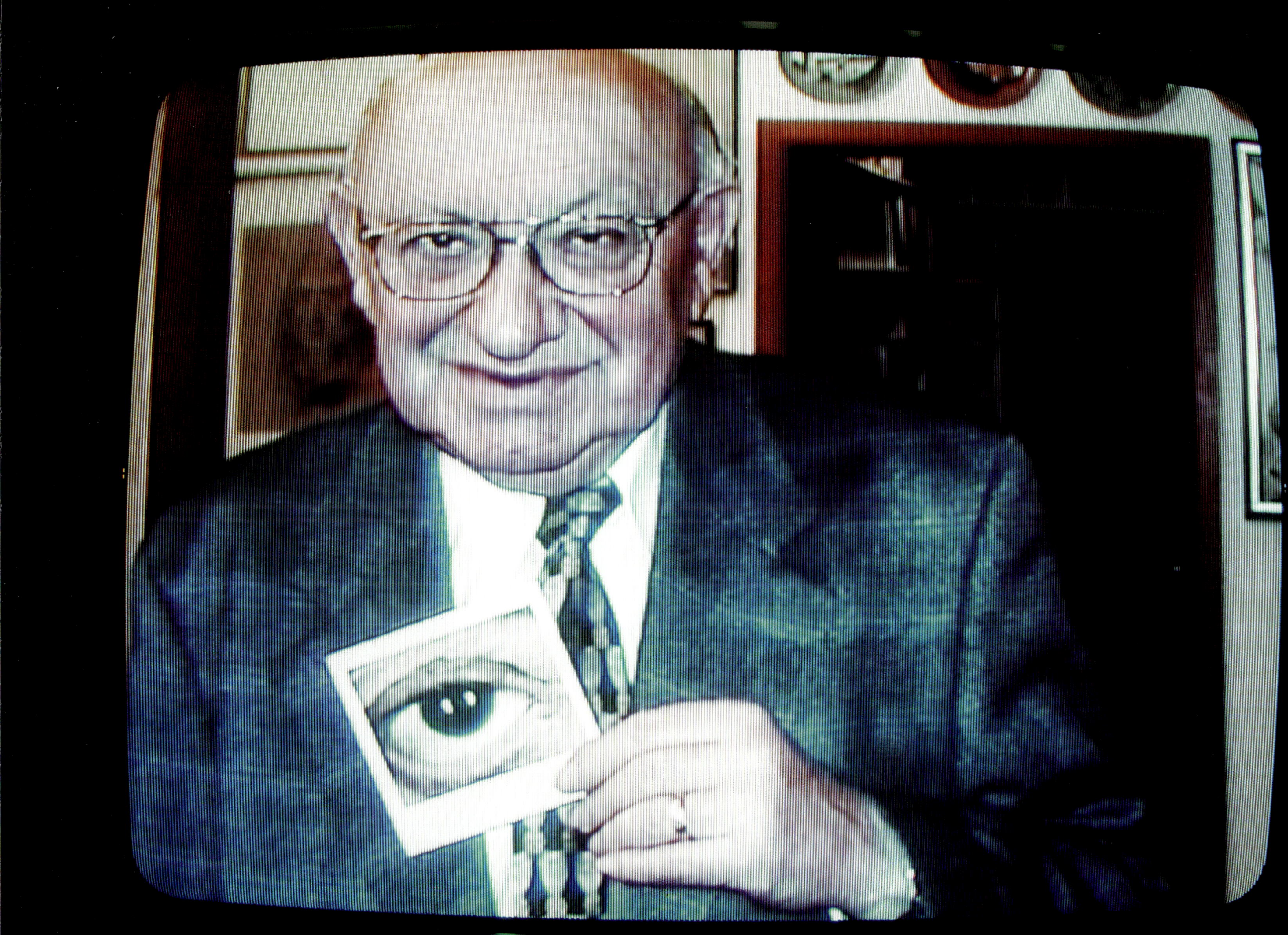

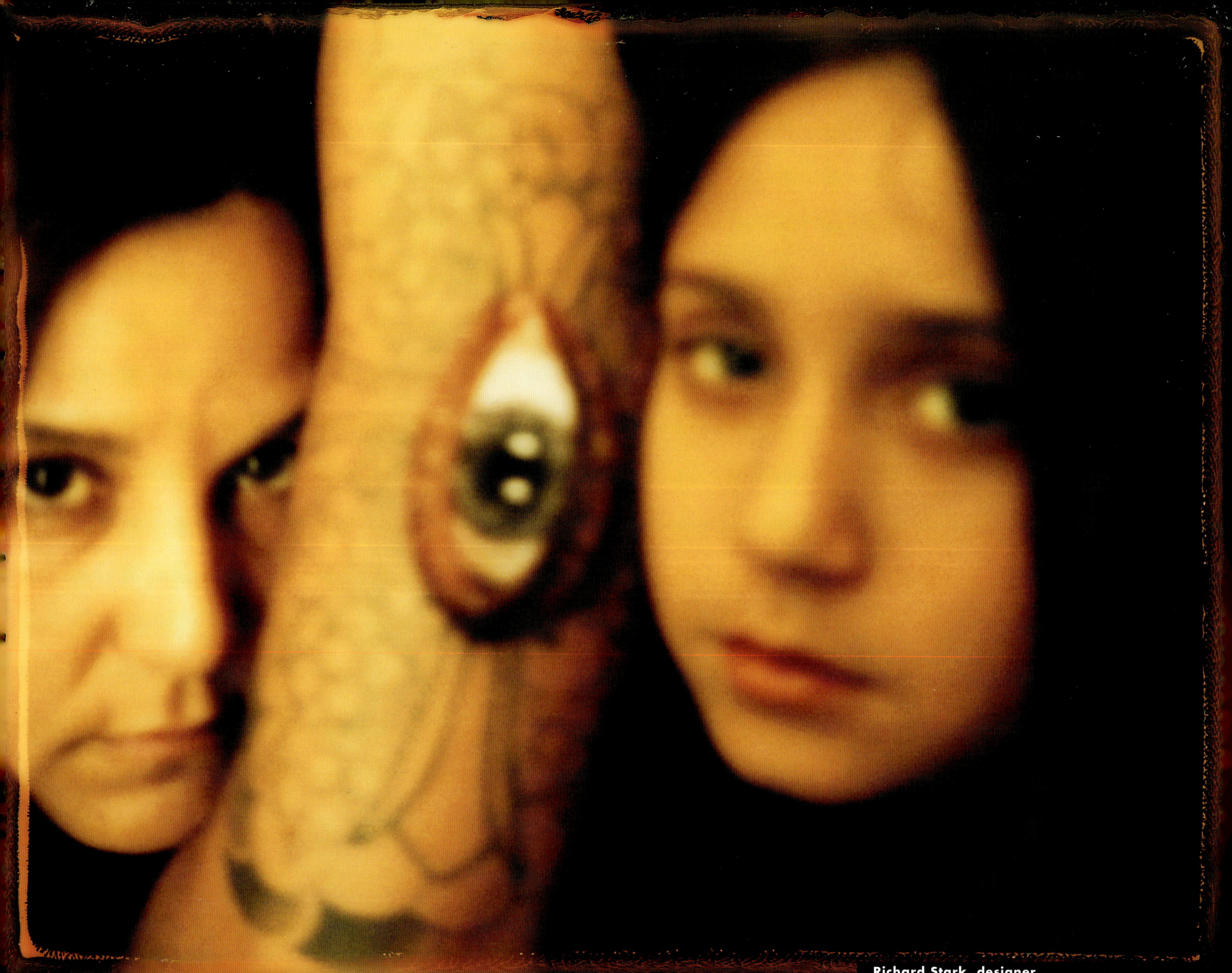
Richard Stark, designer

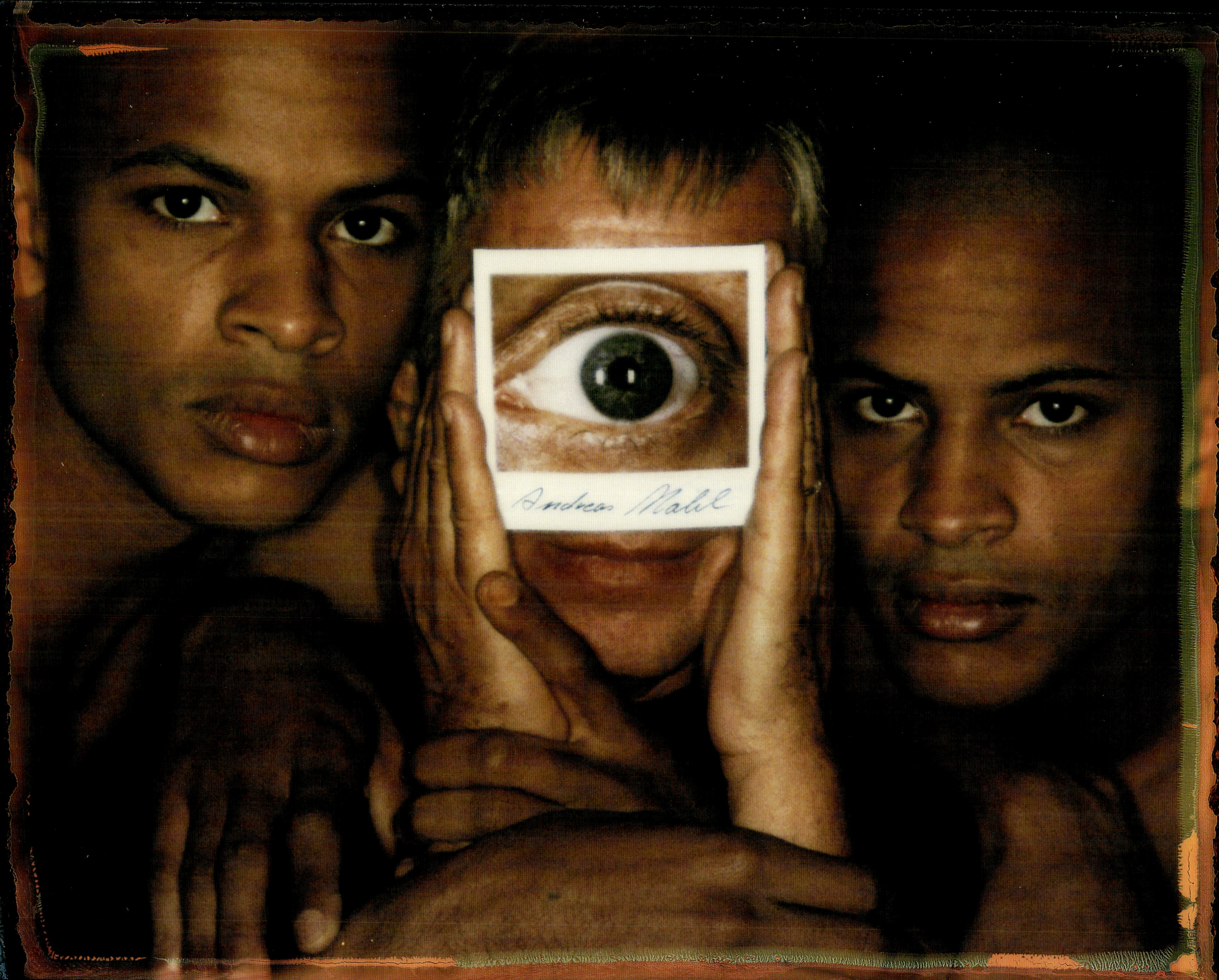

Andreas Mahl, photographer

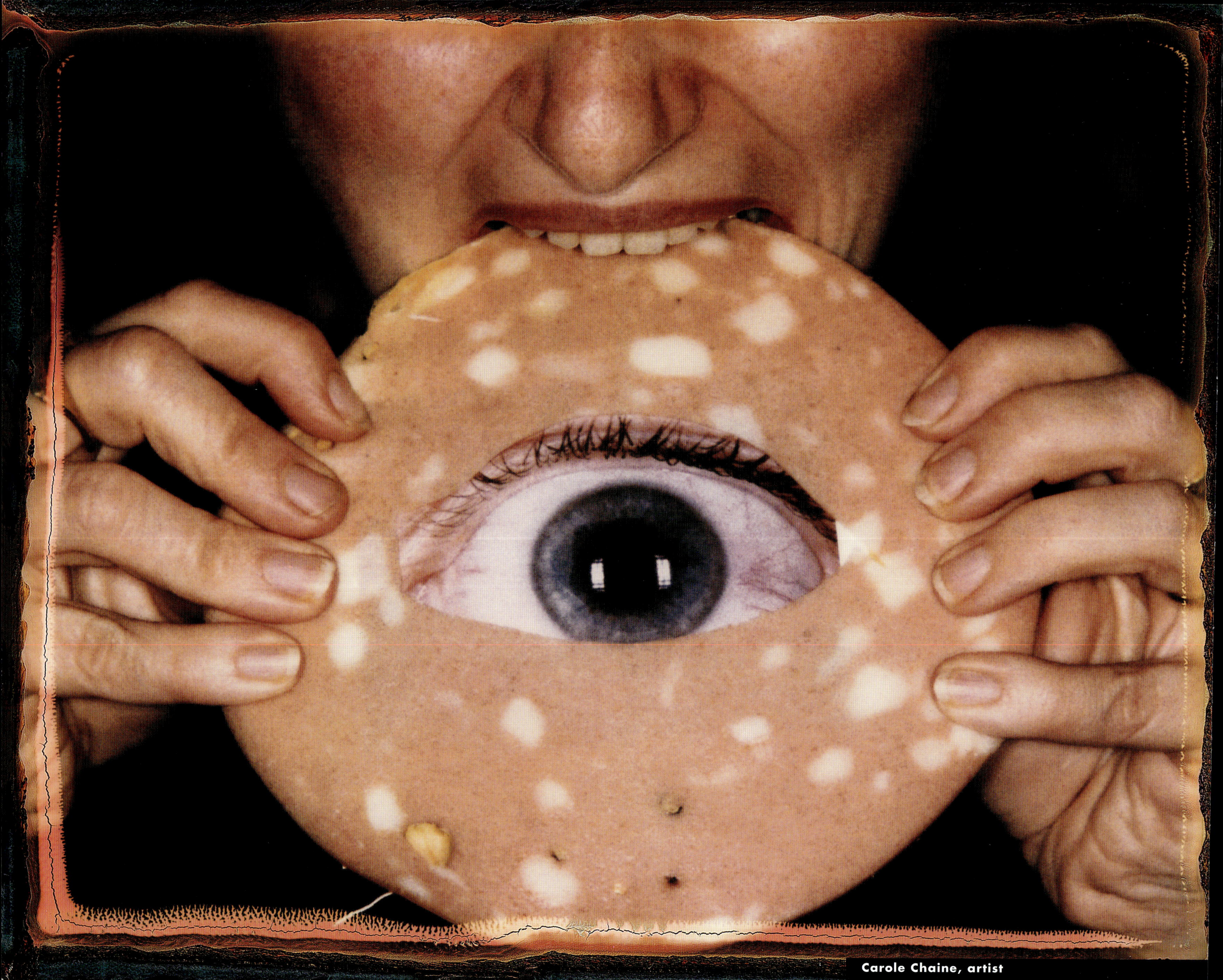

Carole Chaine, artist

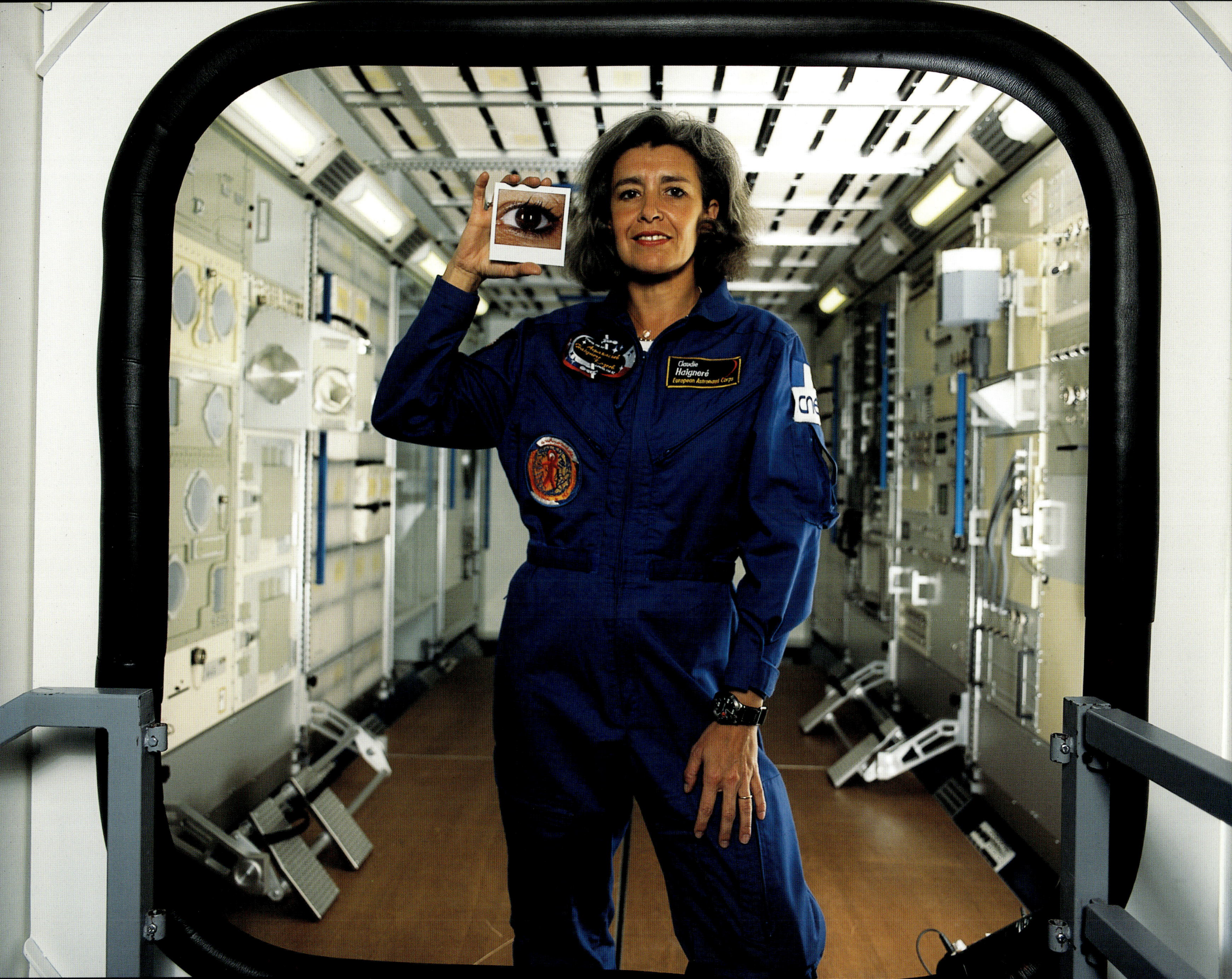

Claudie Haigneré, astronaut

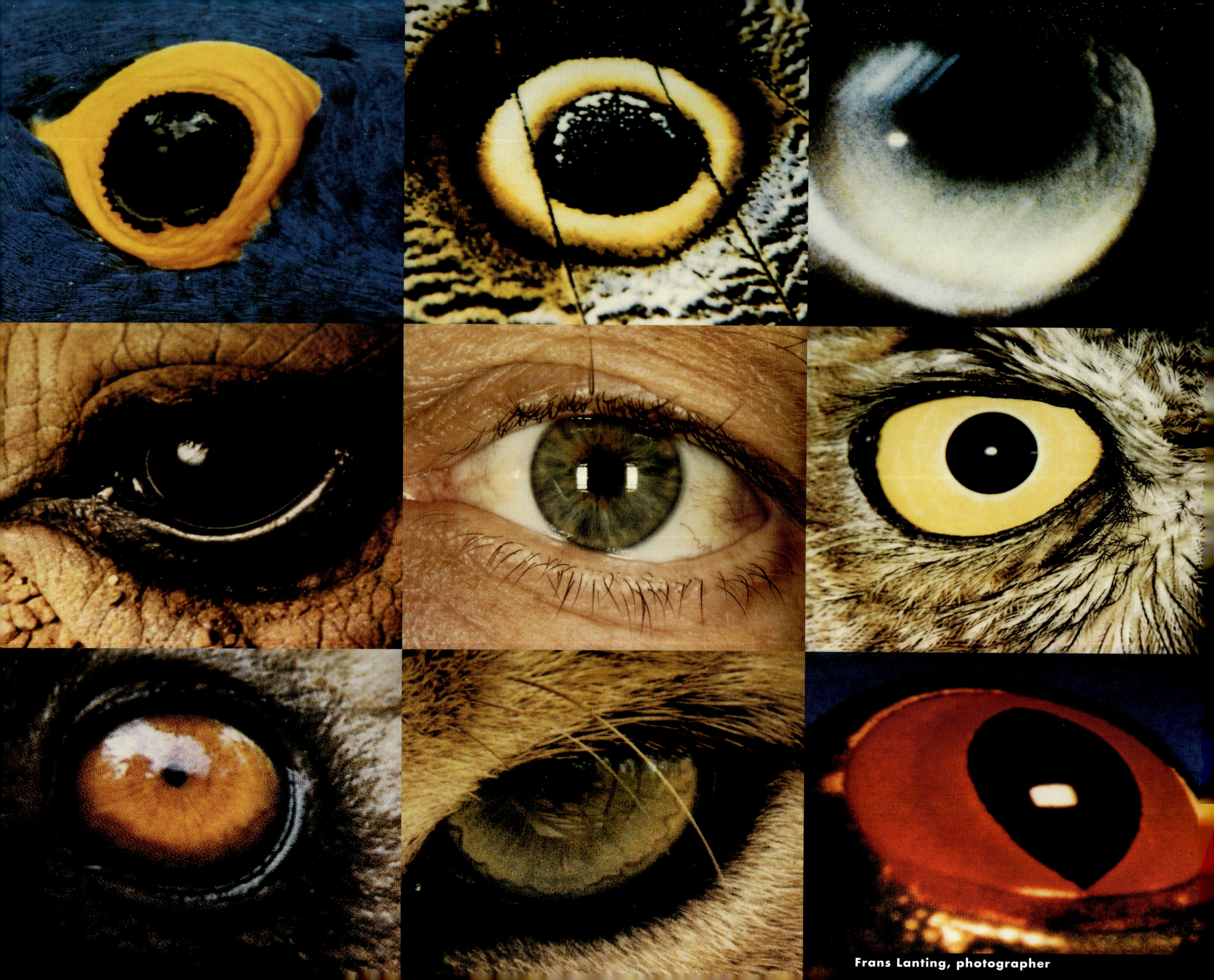

Frans Lanting, photographer

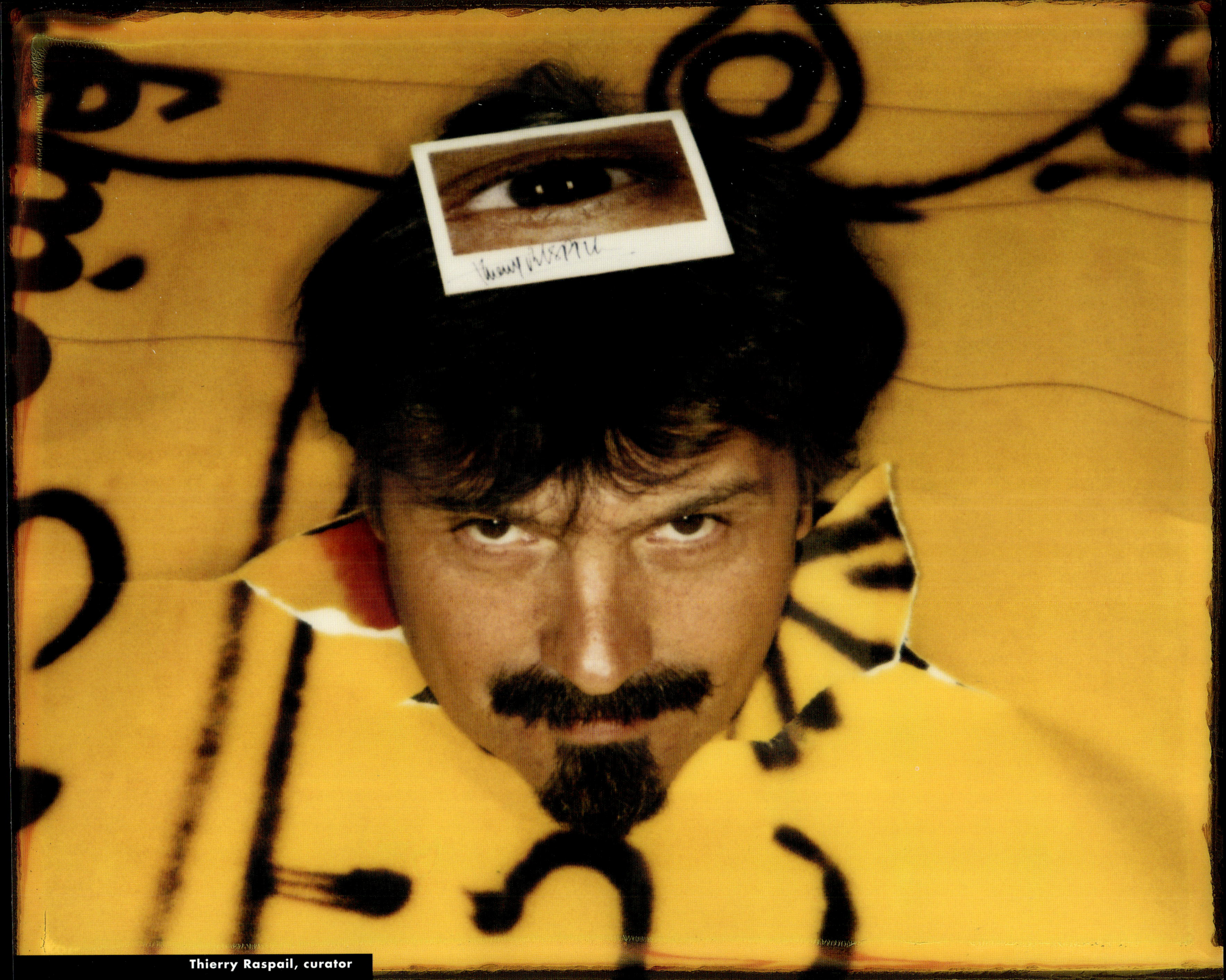

Thierry Raspail, curator

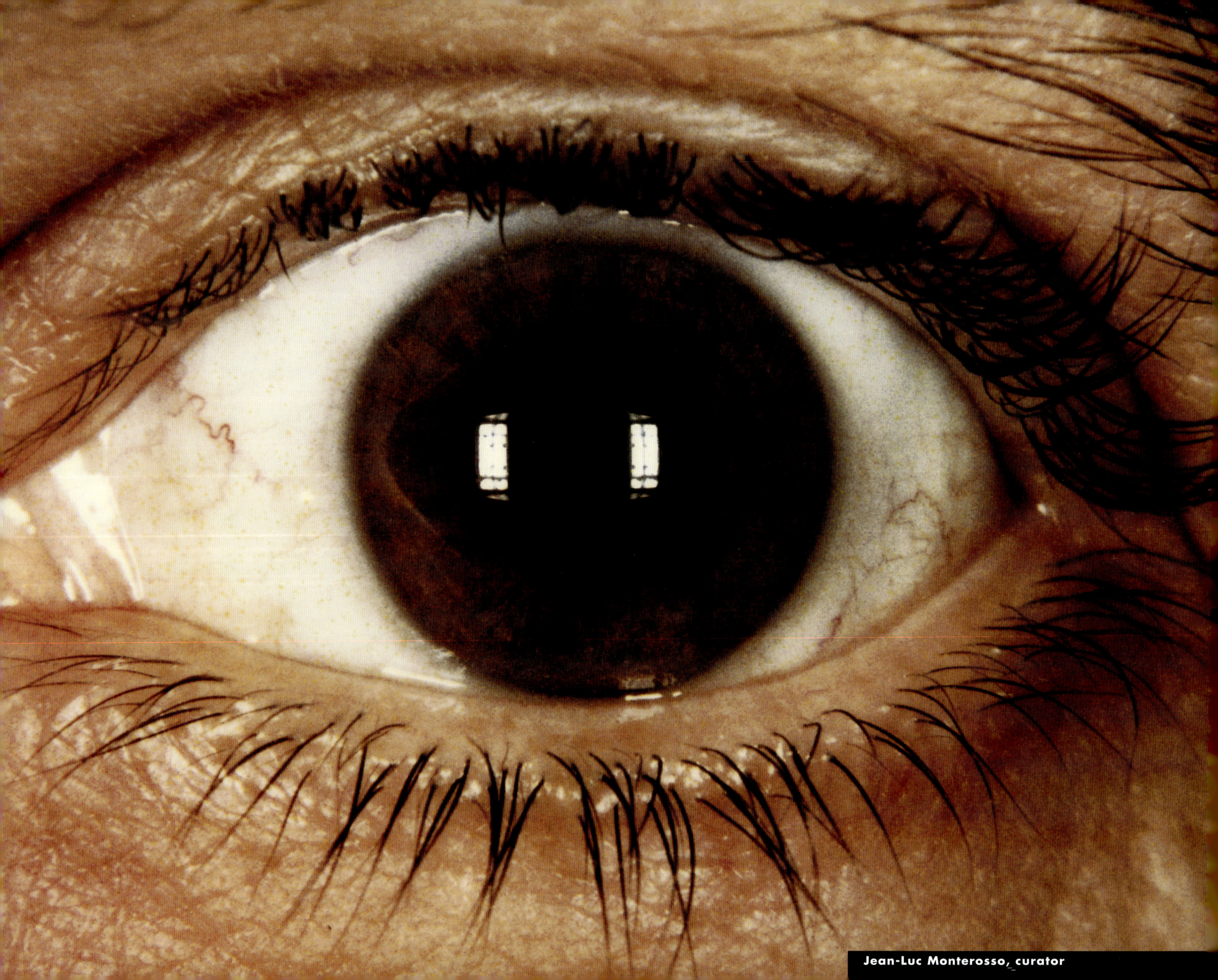

Jean-Luc Monterosso, curator

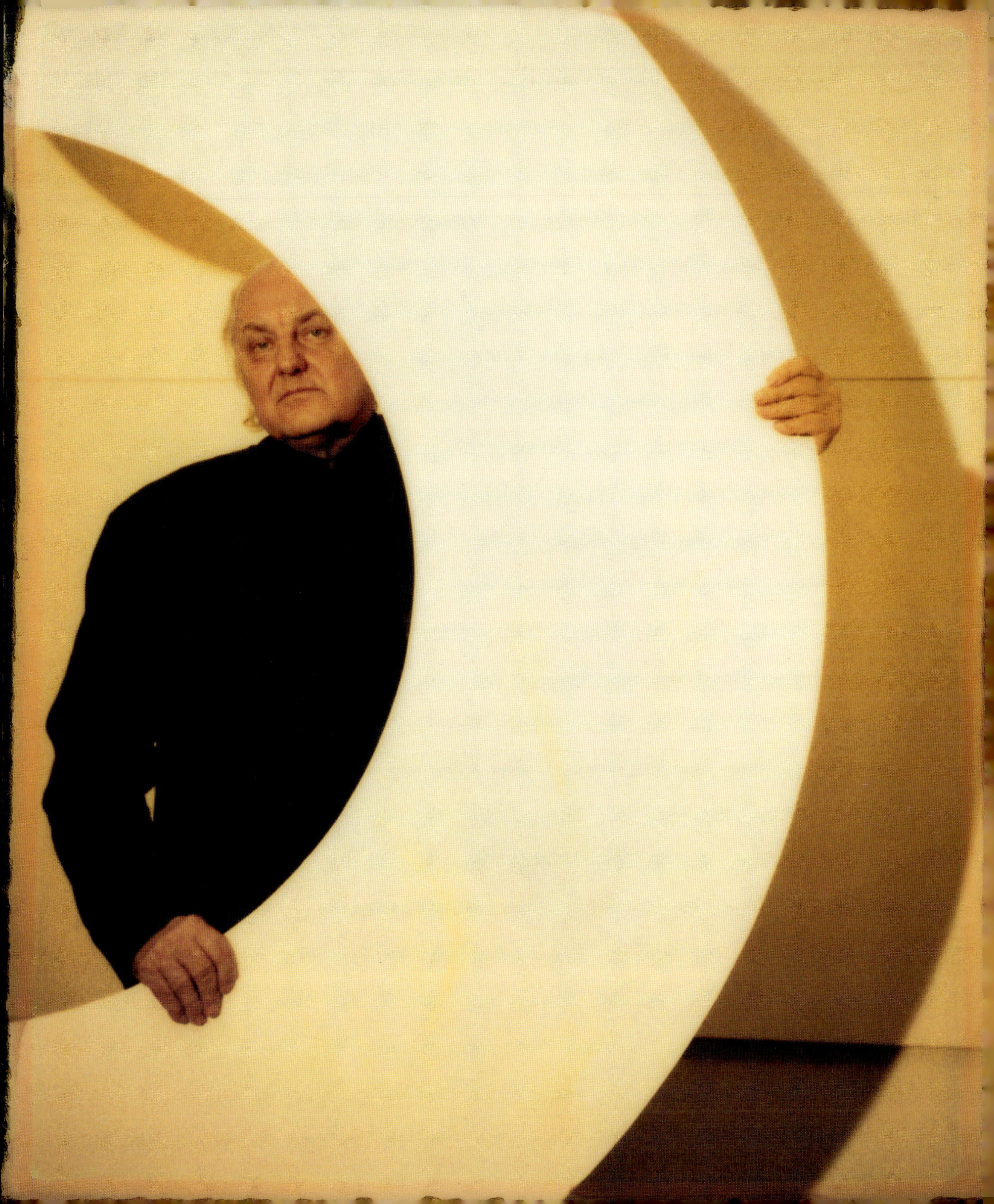

Hans Hollein, architect

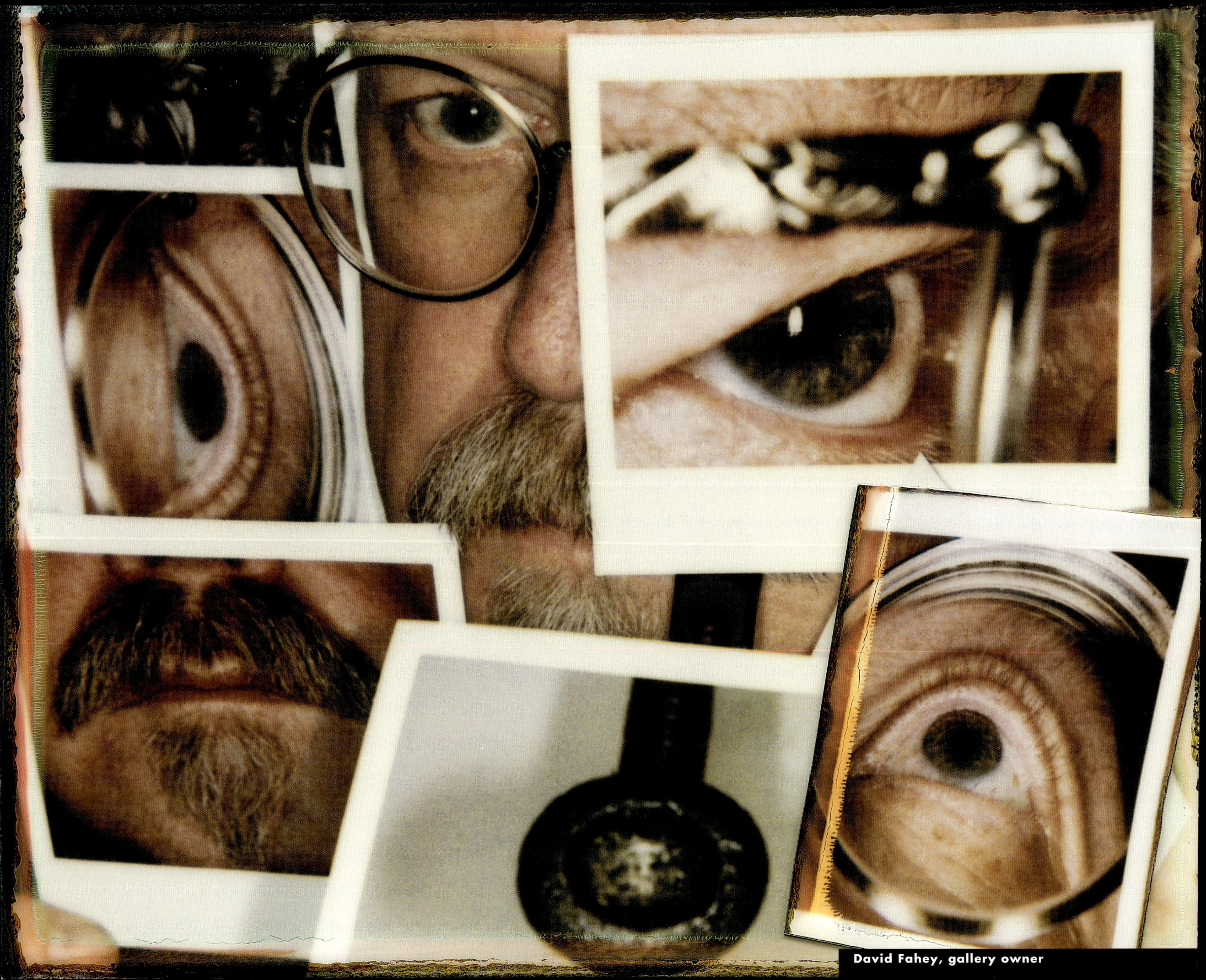

David Fahey, gallery owner

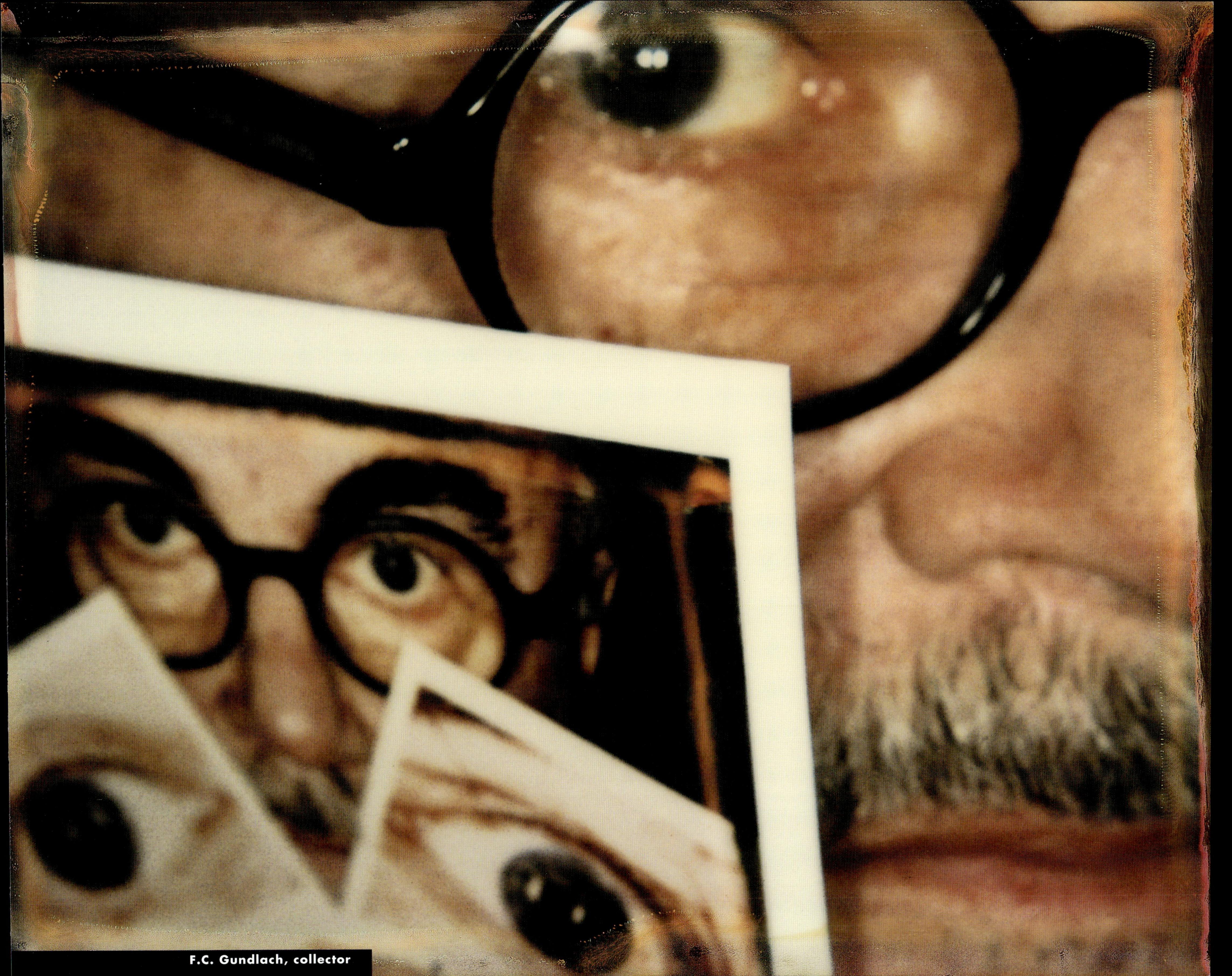

F.C. Gundlach, collector

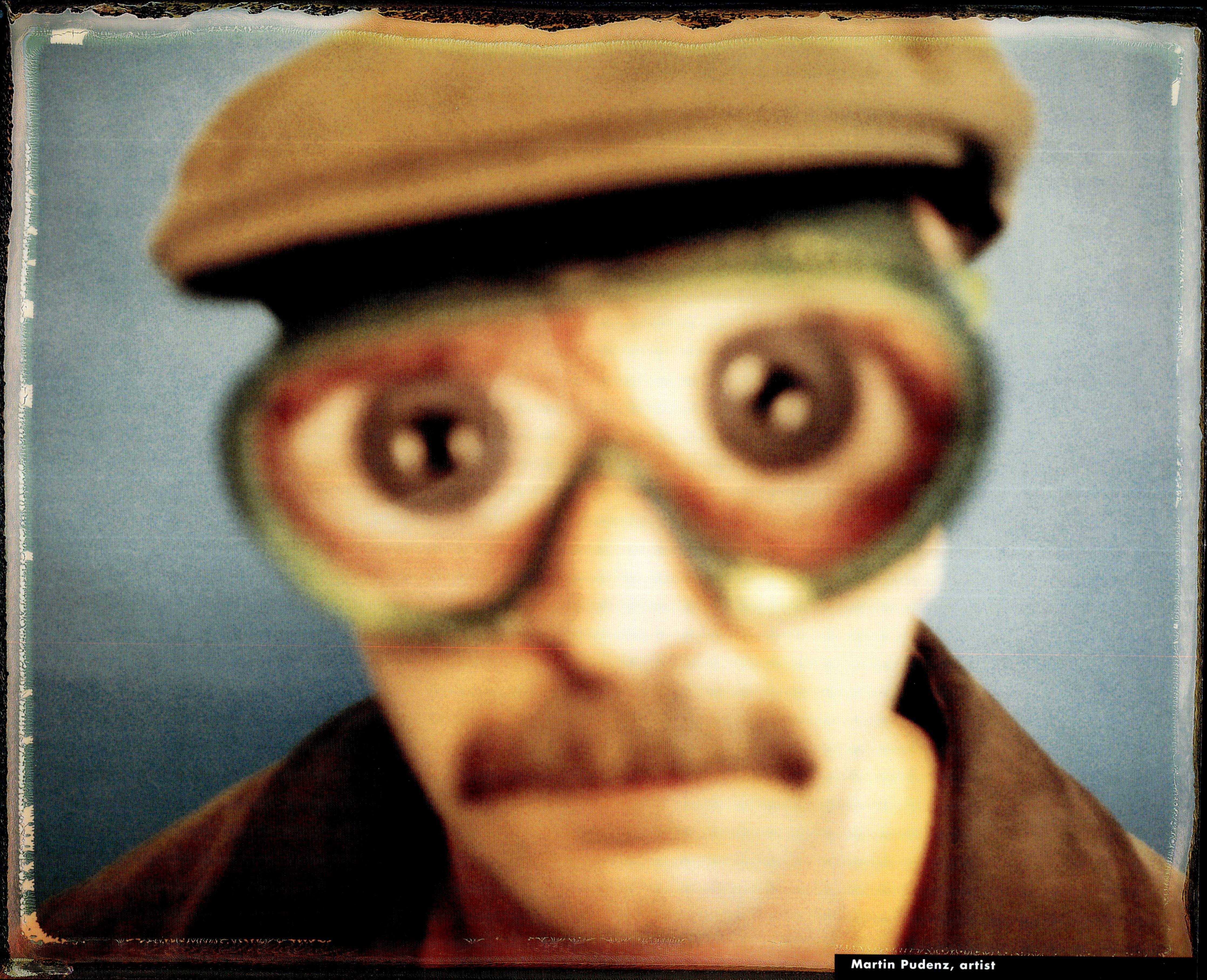

Martin Pudenz, artist

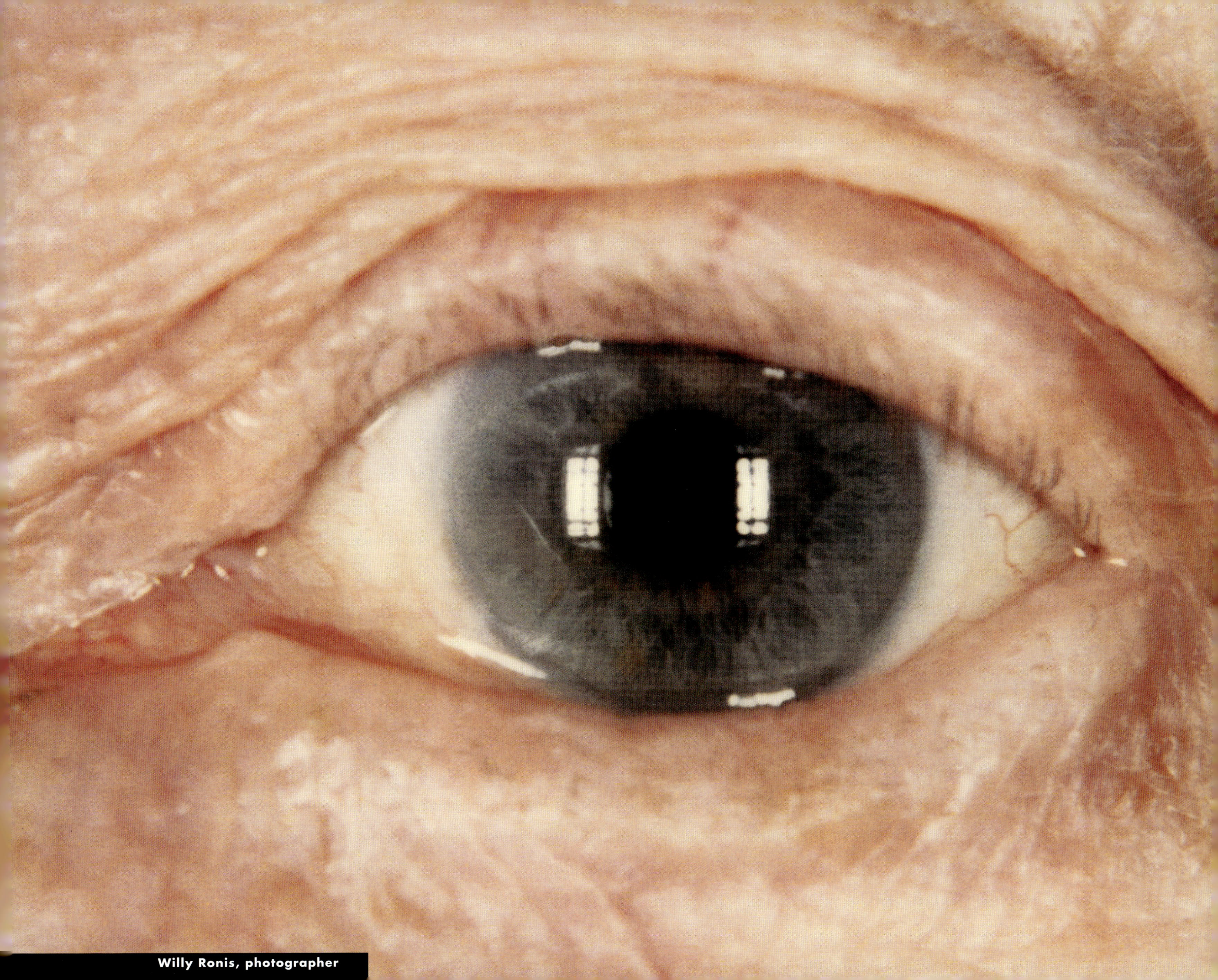

Willy Ronis, photographer

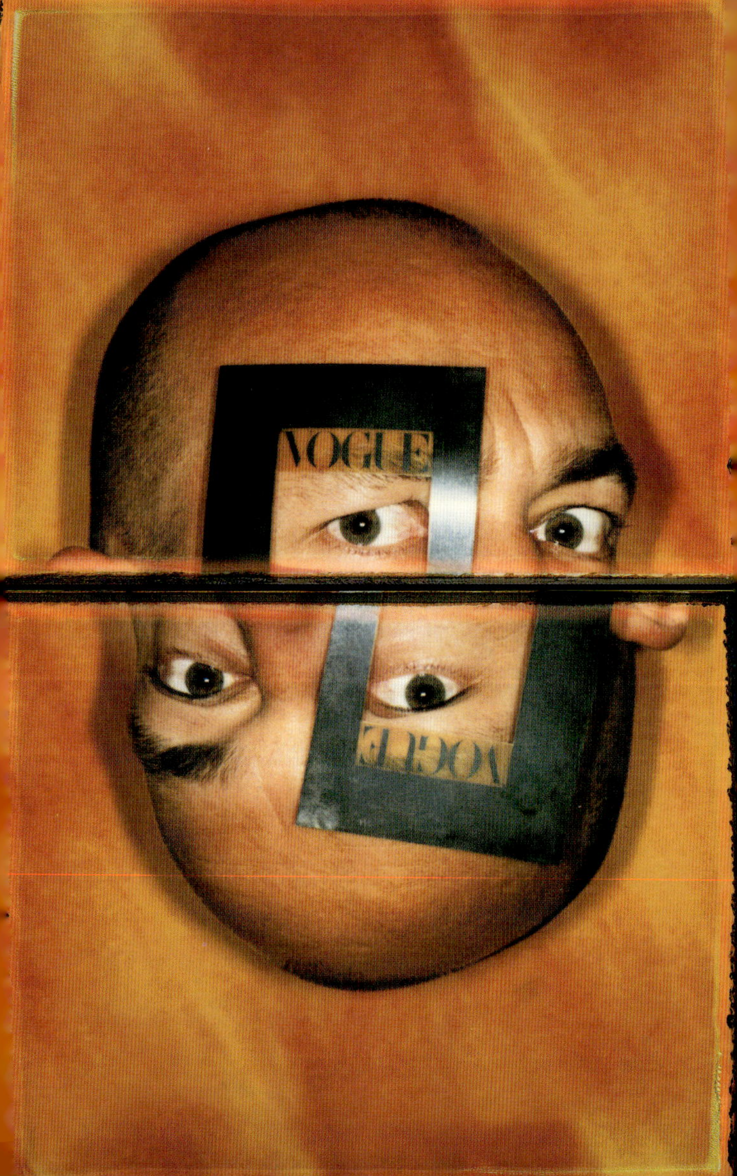
VOGUE

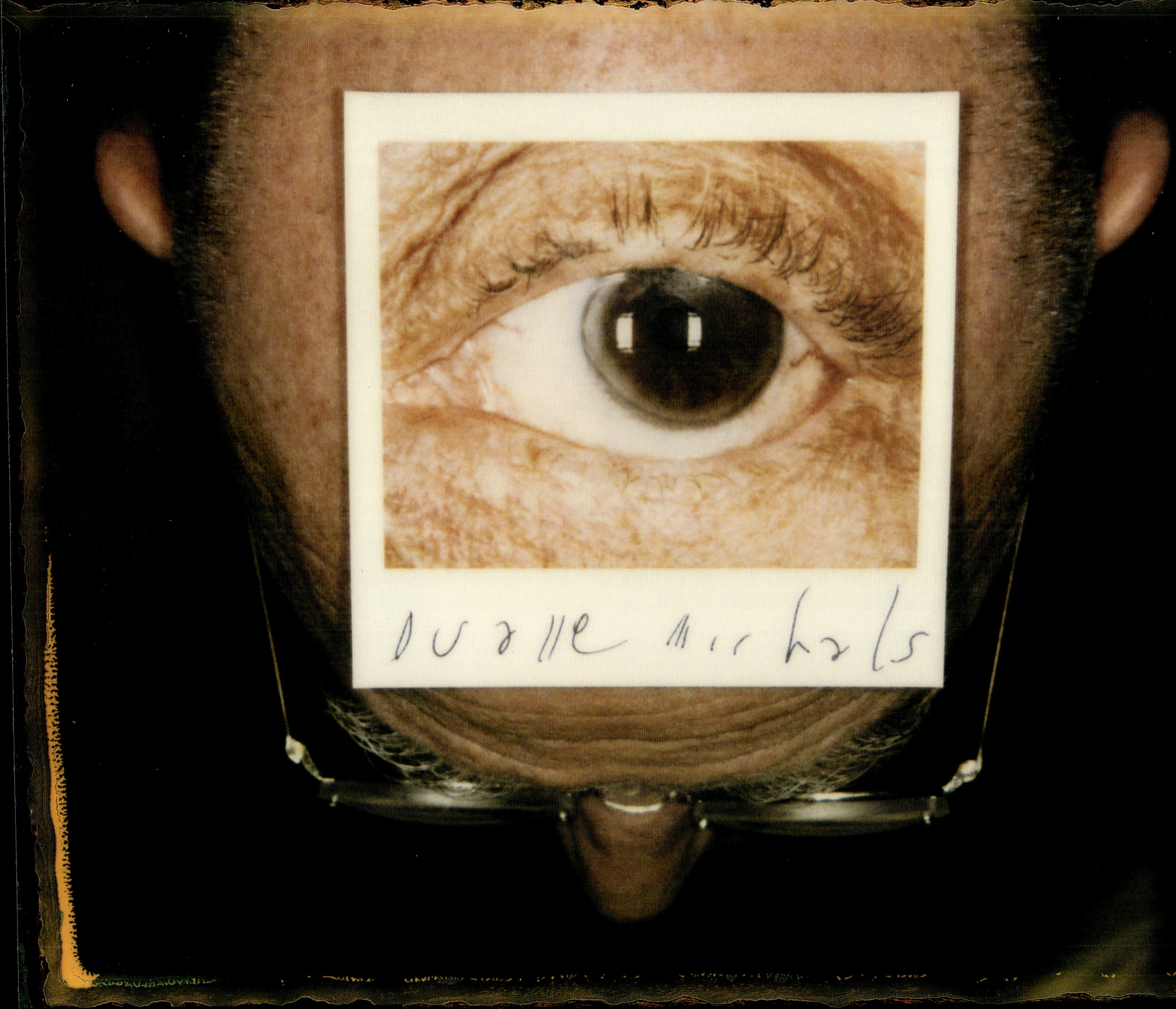

Duane Michals, artist

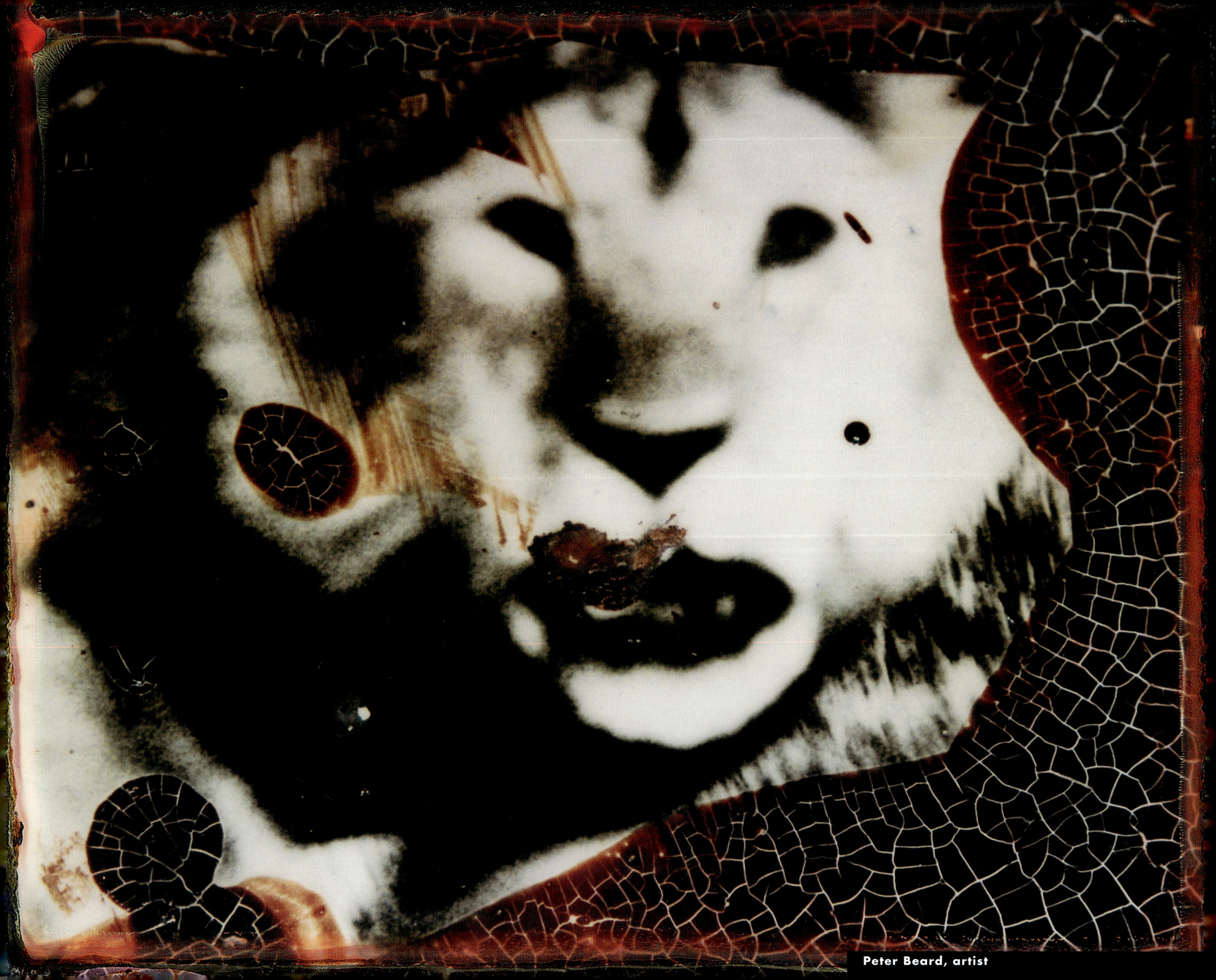

Peter Beard, artist

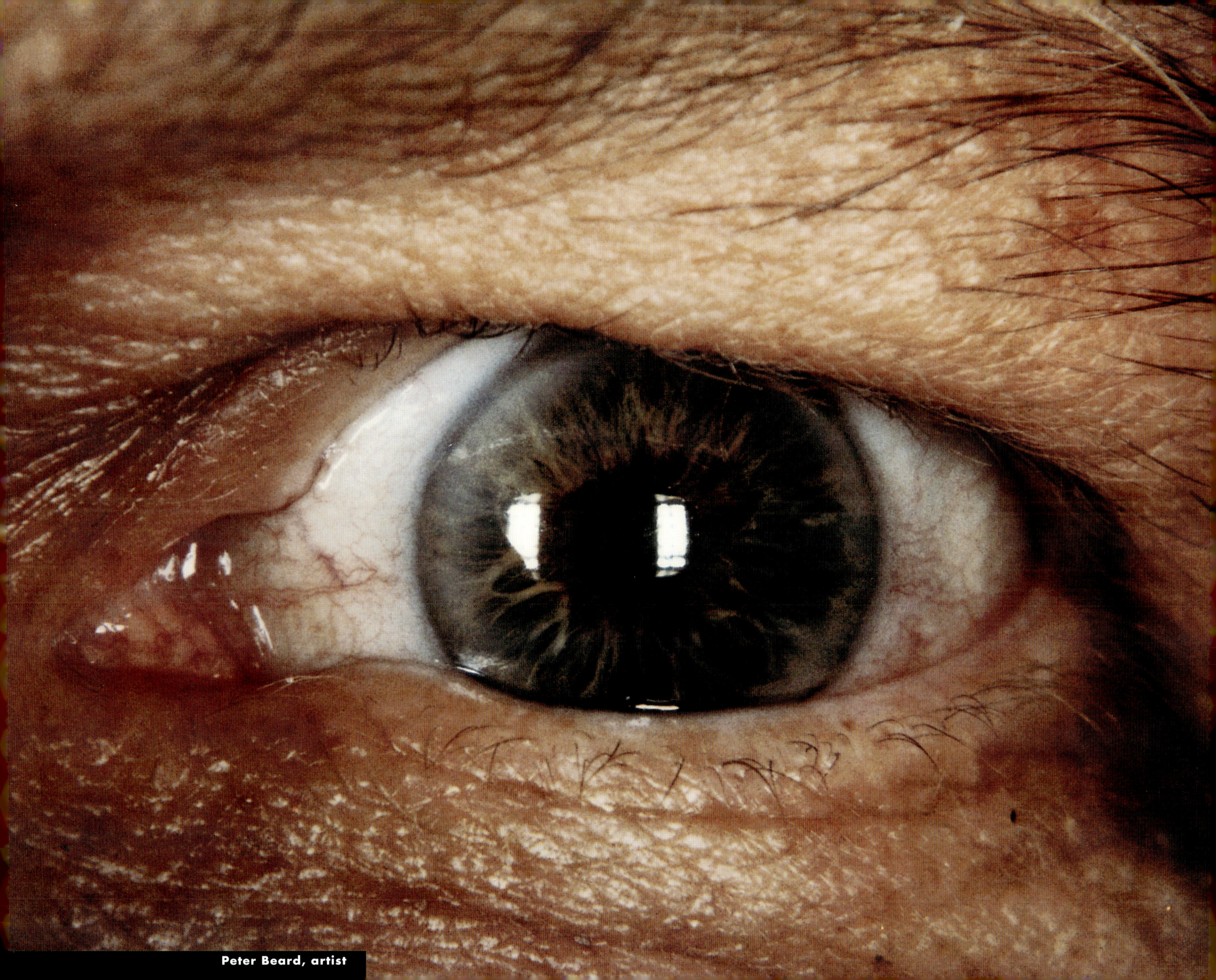

Peter Beard, artist

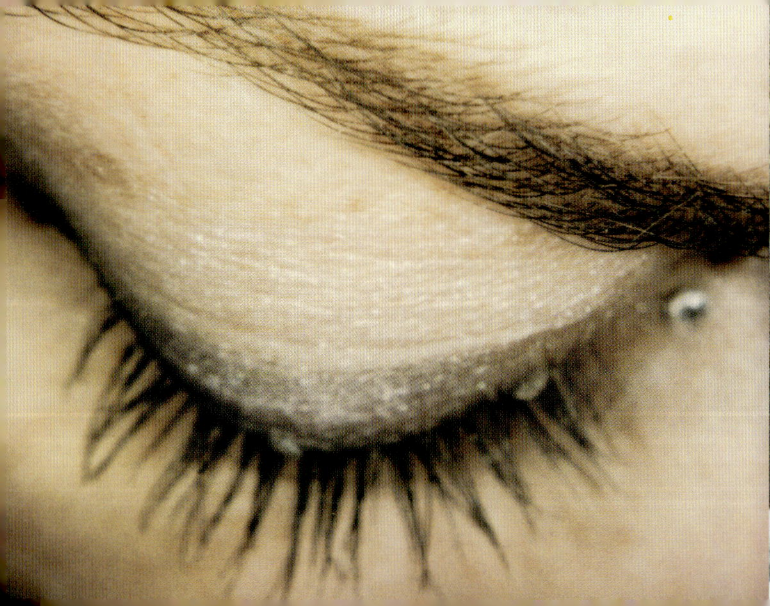
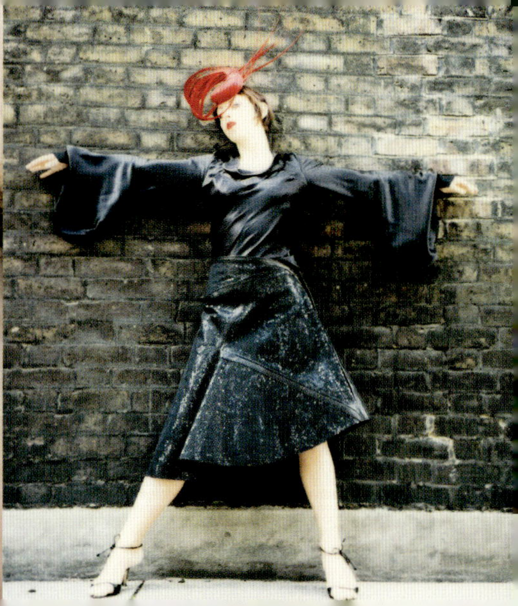

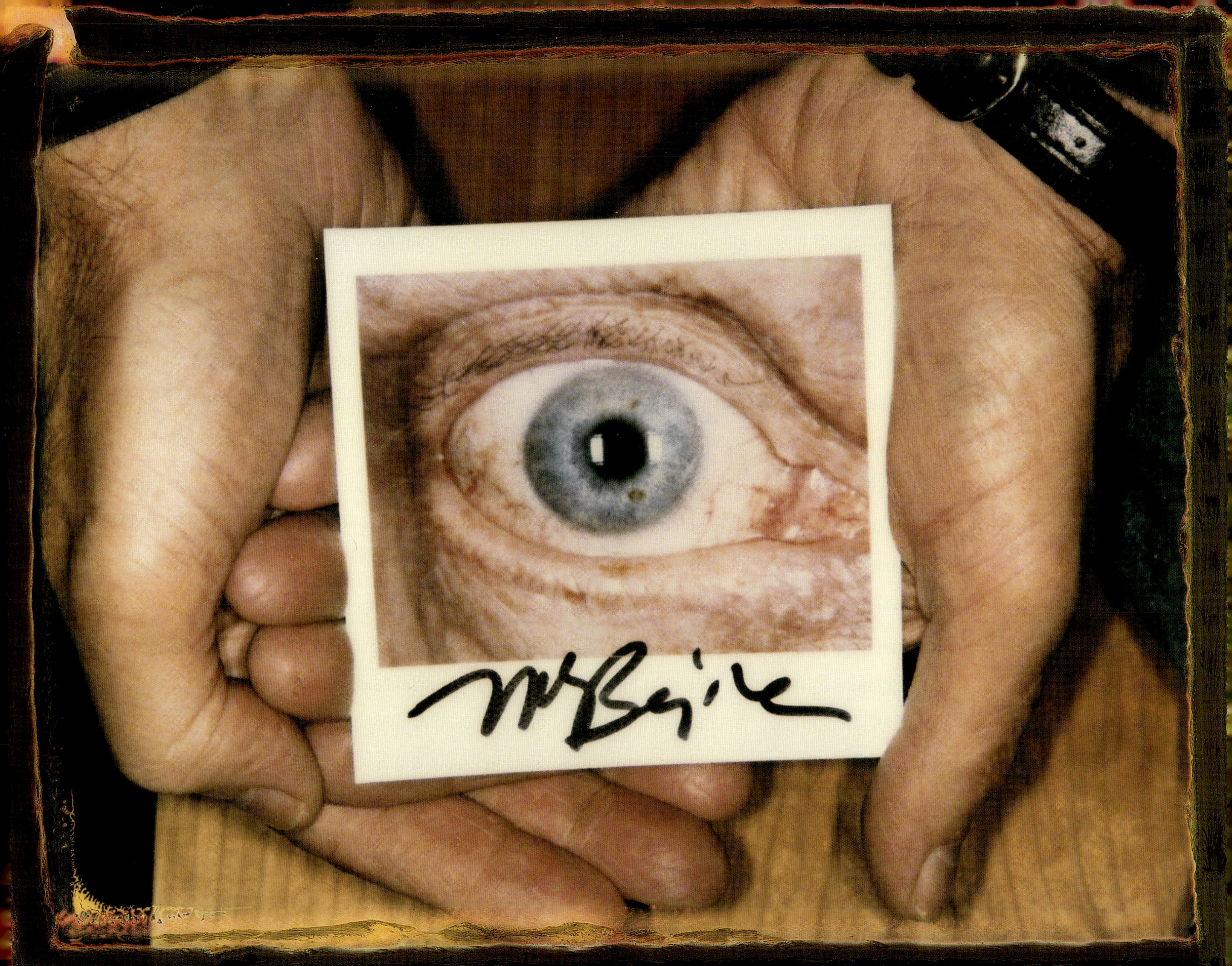

Will McBride, photographer

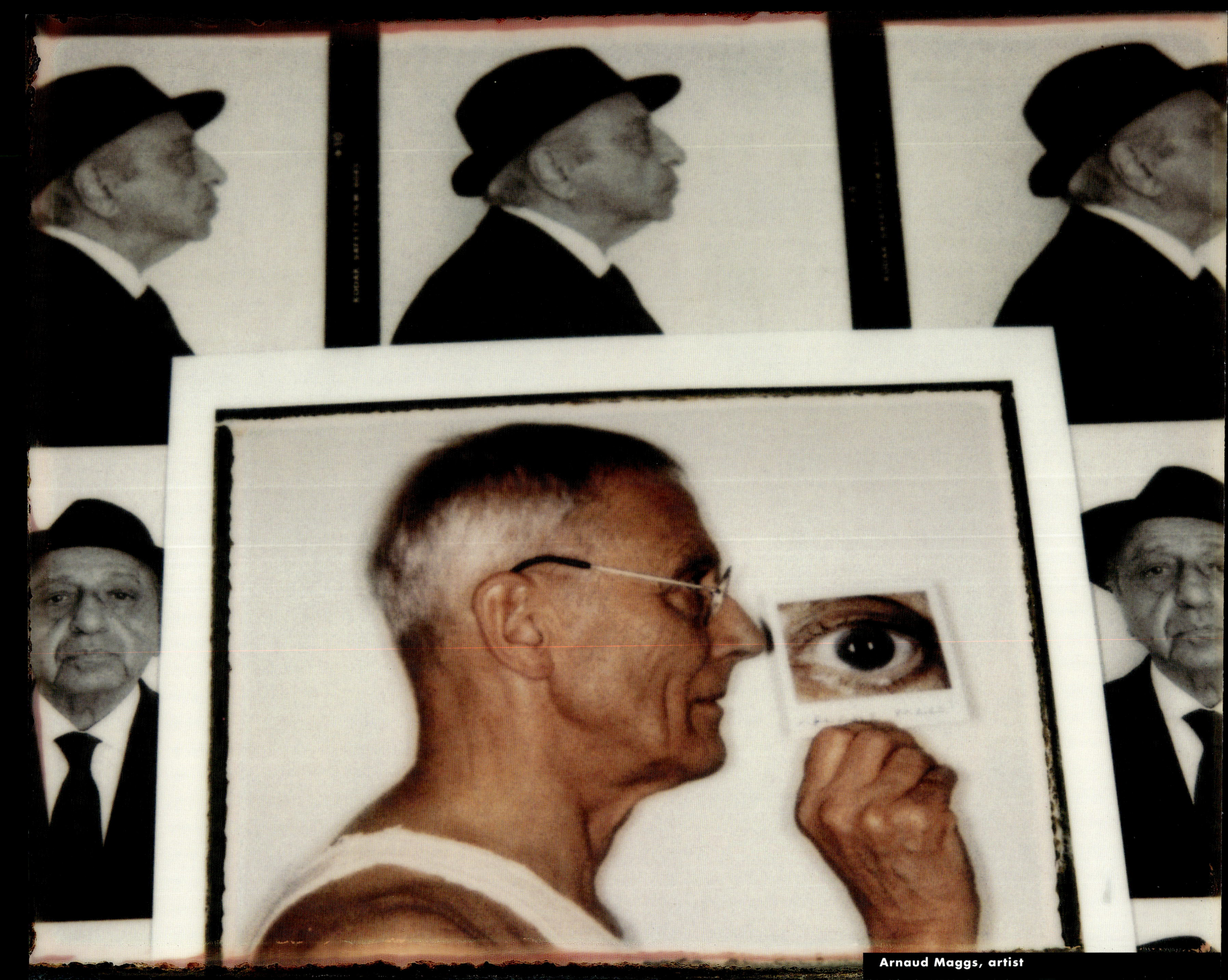

Arnaud Maggs, artist

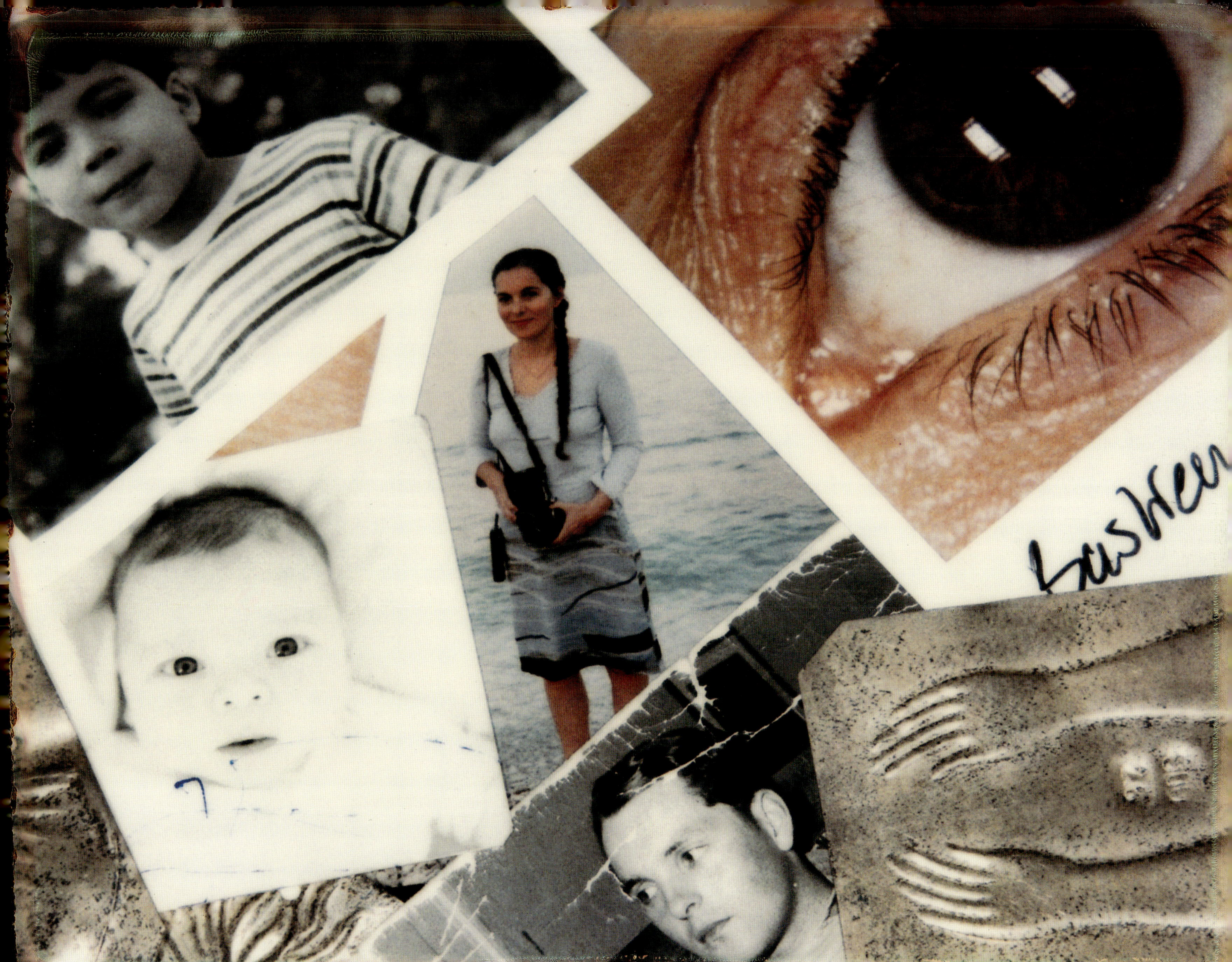

Bastienne Schmidt, photographer

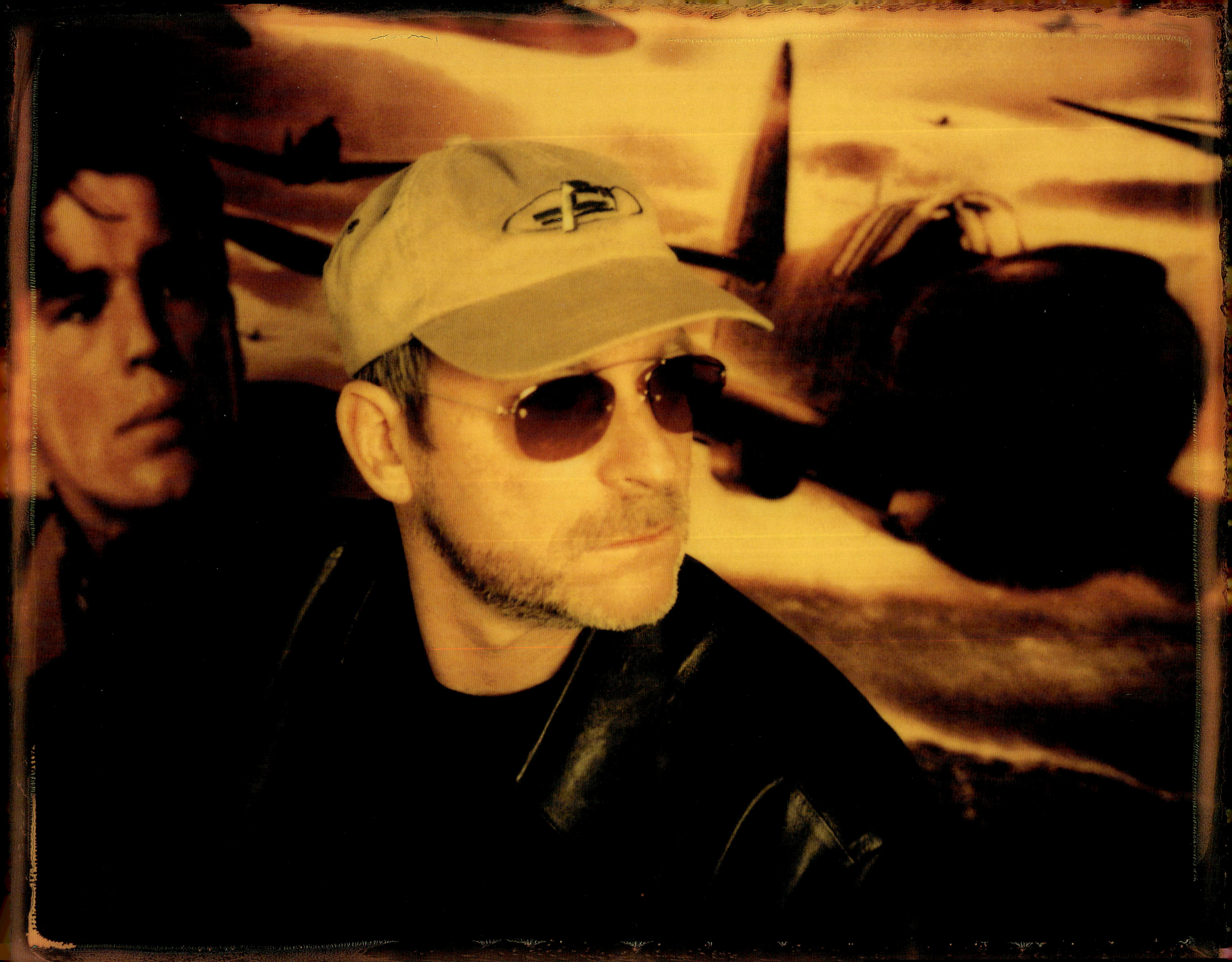

Greg Gorman, photographer

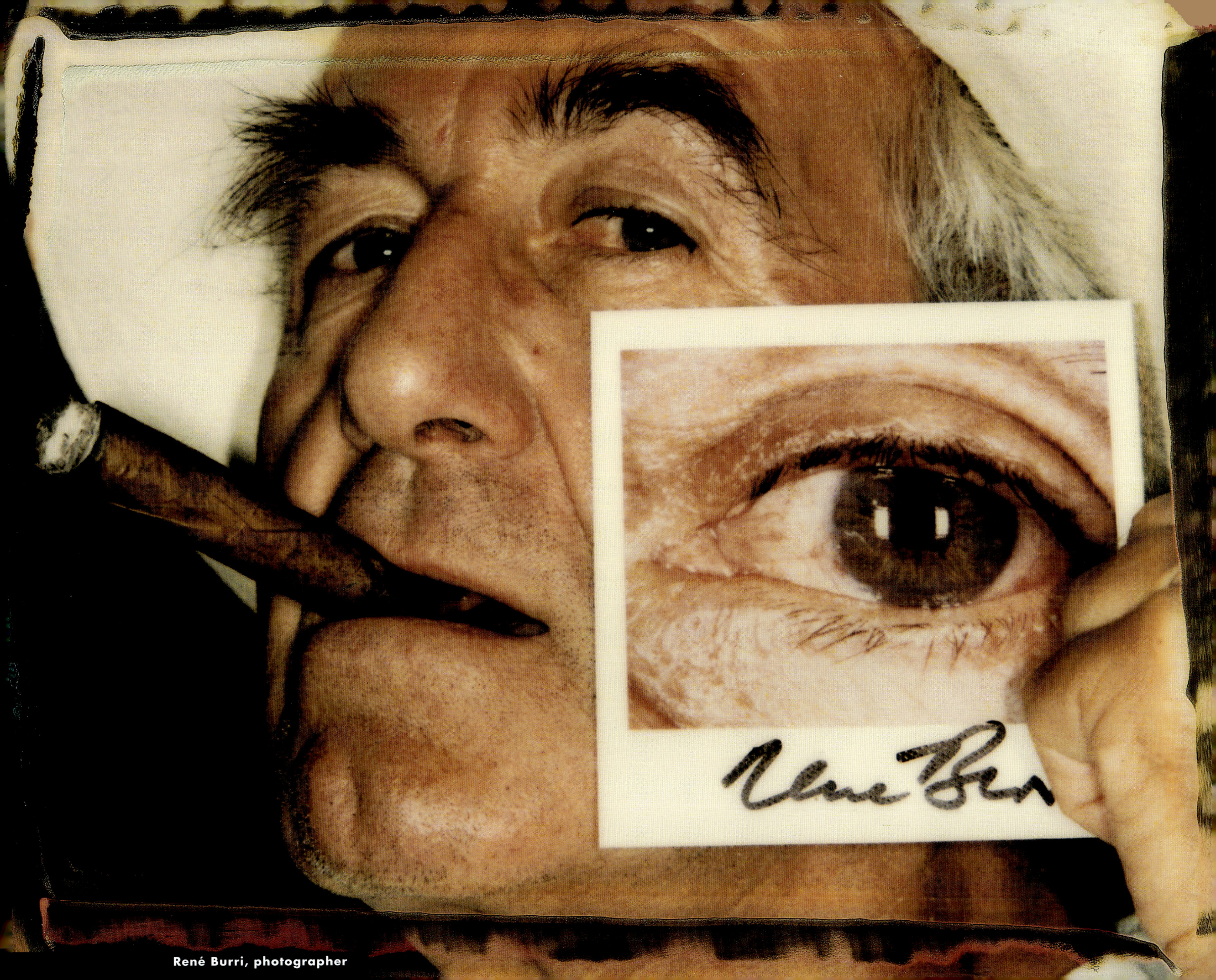

René Burri, photographer

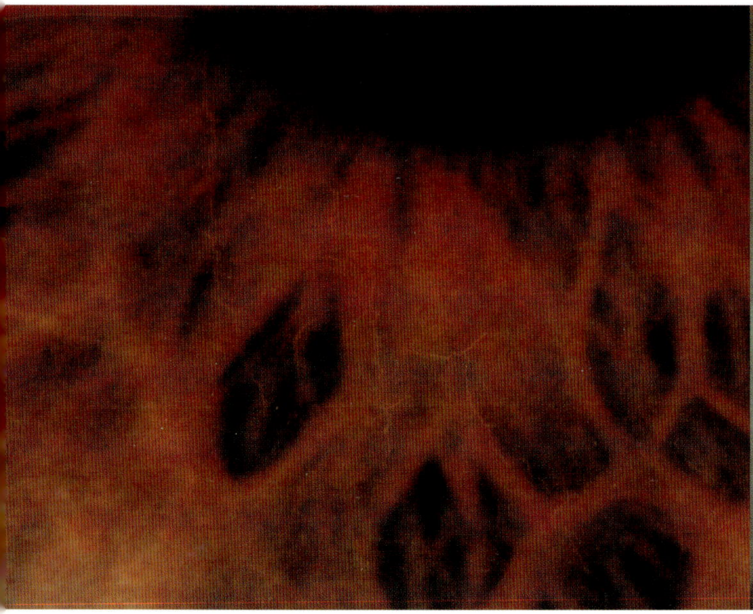
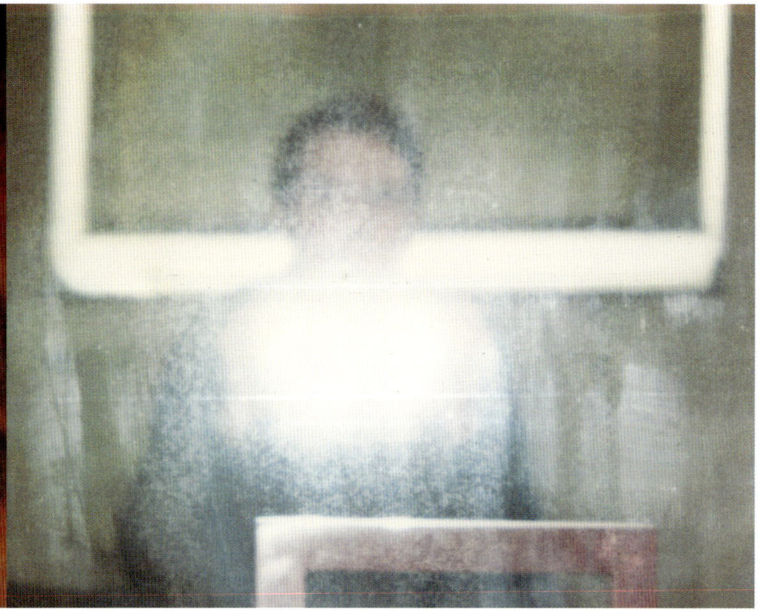

Thomas Schütte, artist

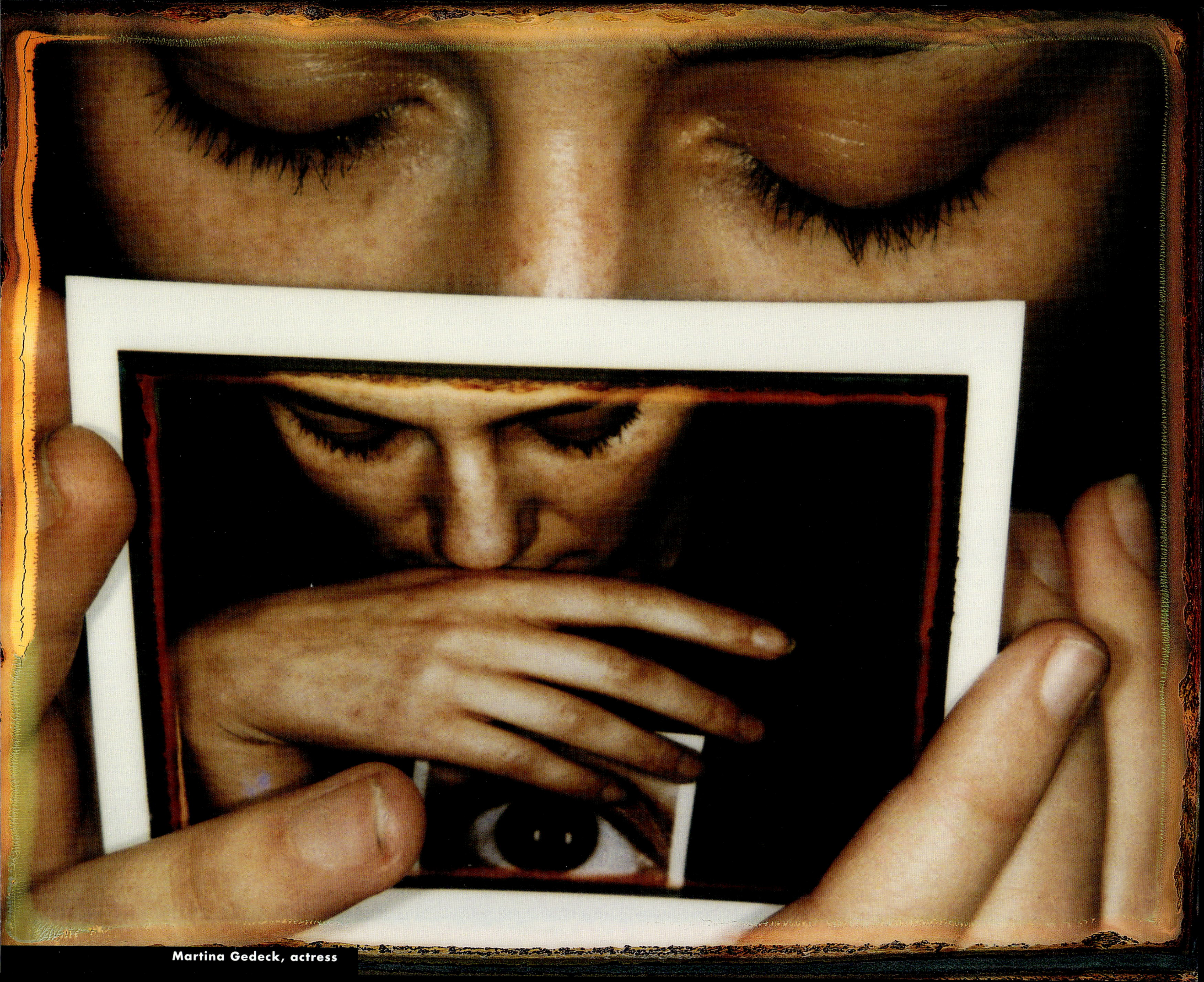

Martina Gedeck, actress

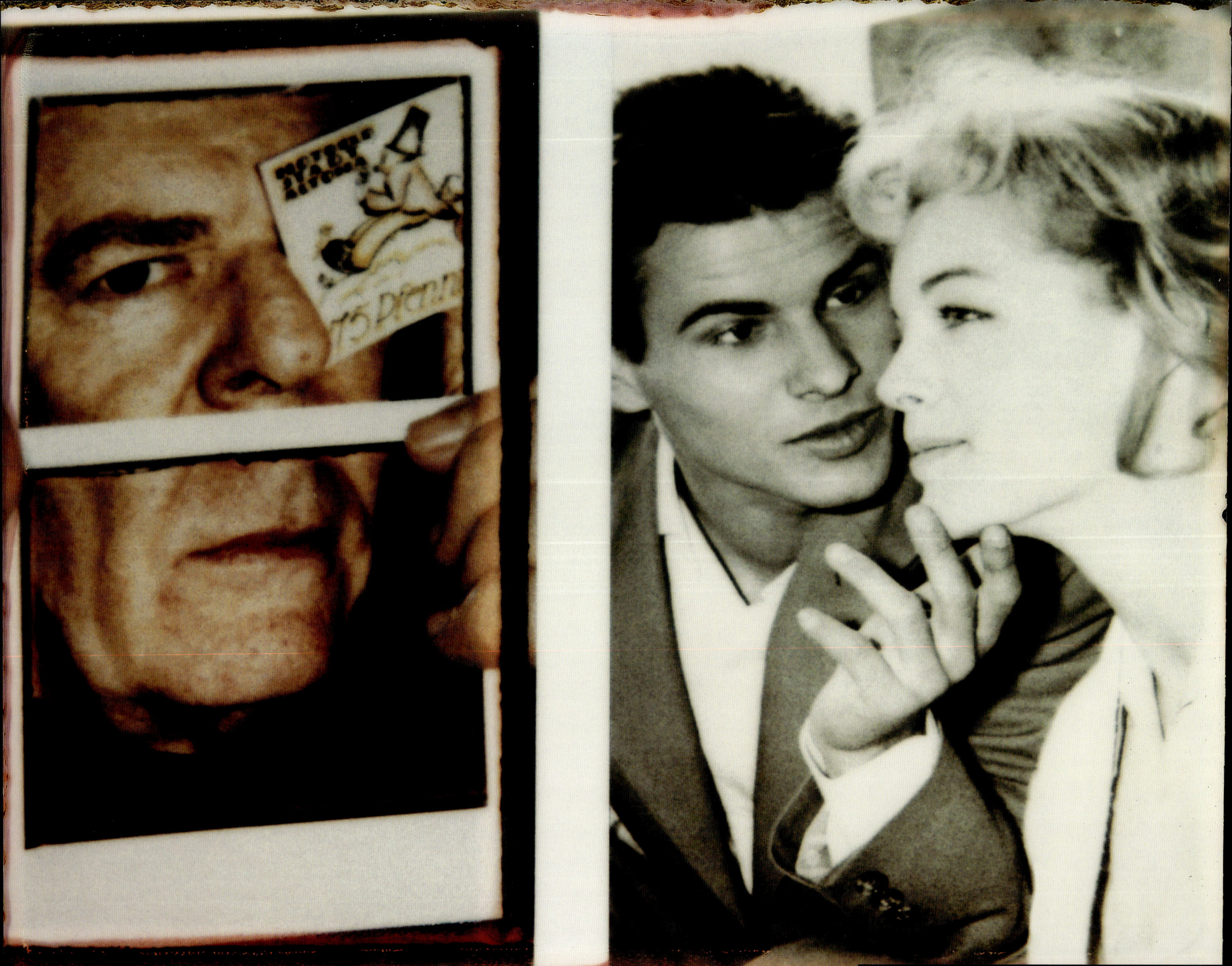

Paul Morrissey, director

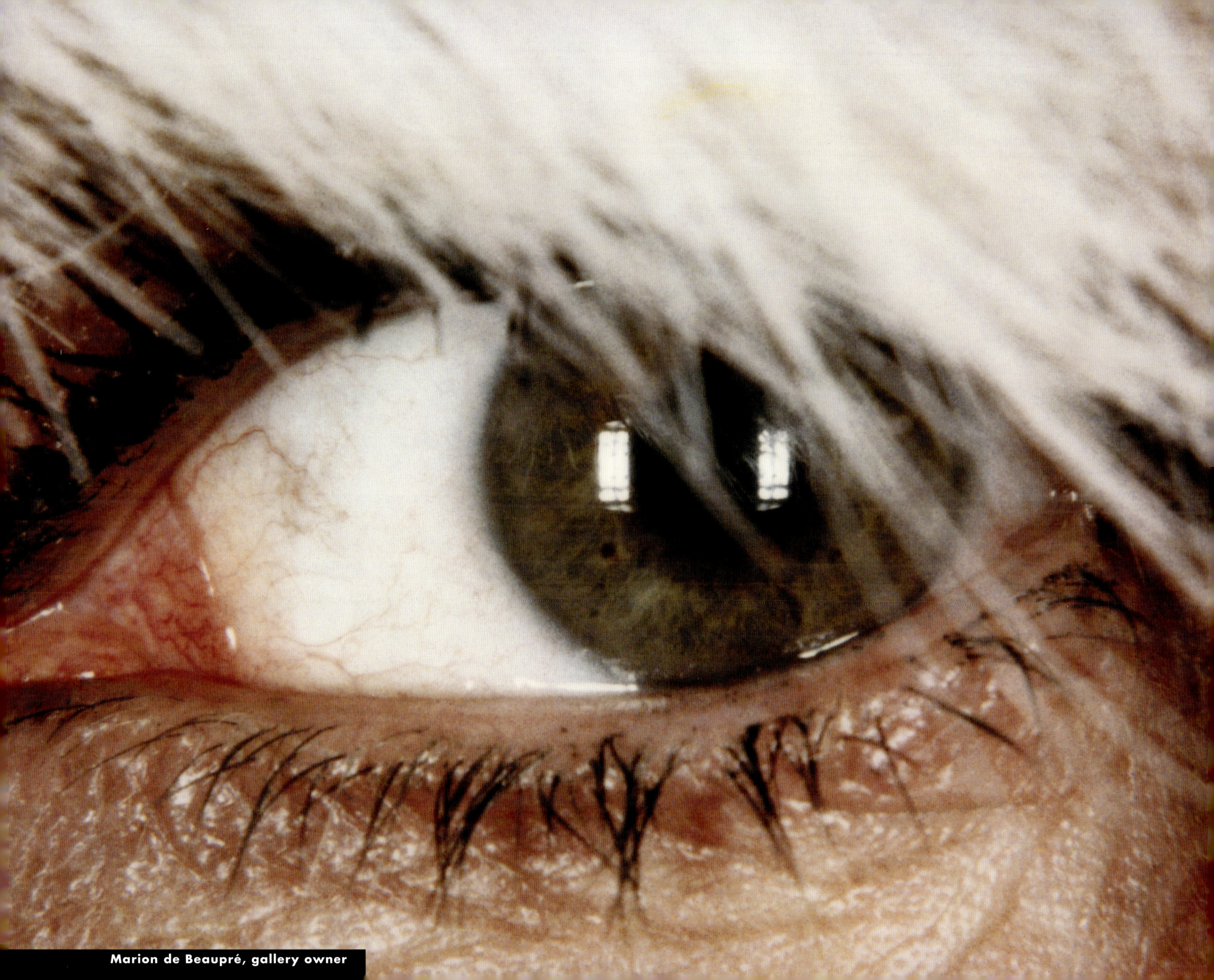

Marion de Beaupré, gallery owner

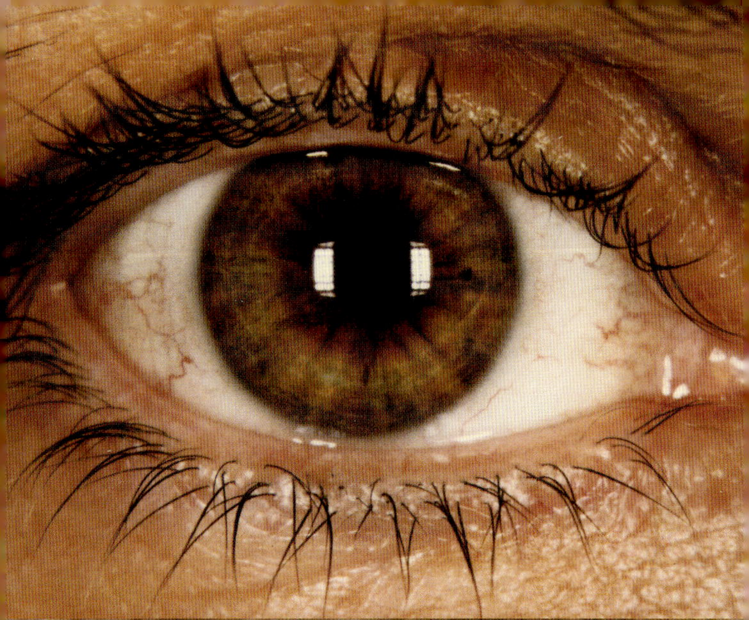
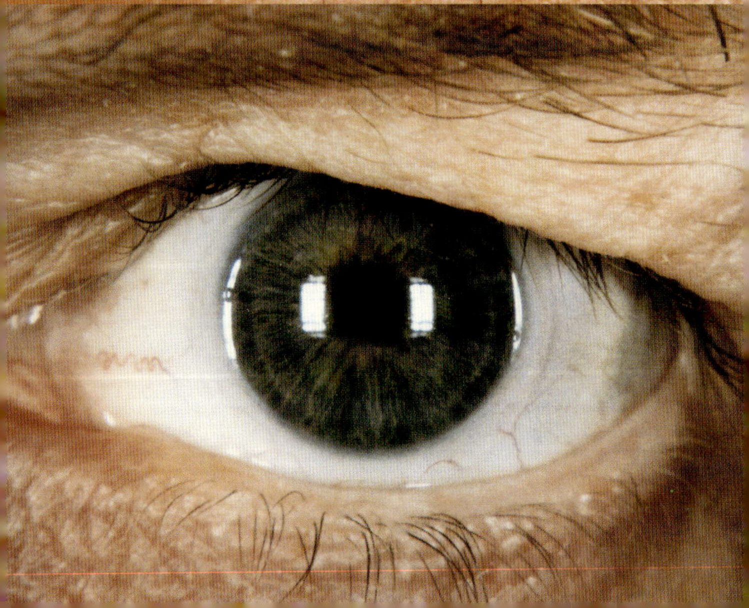

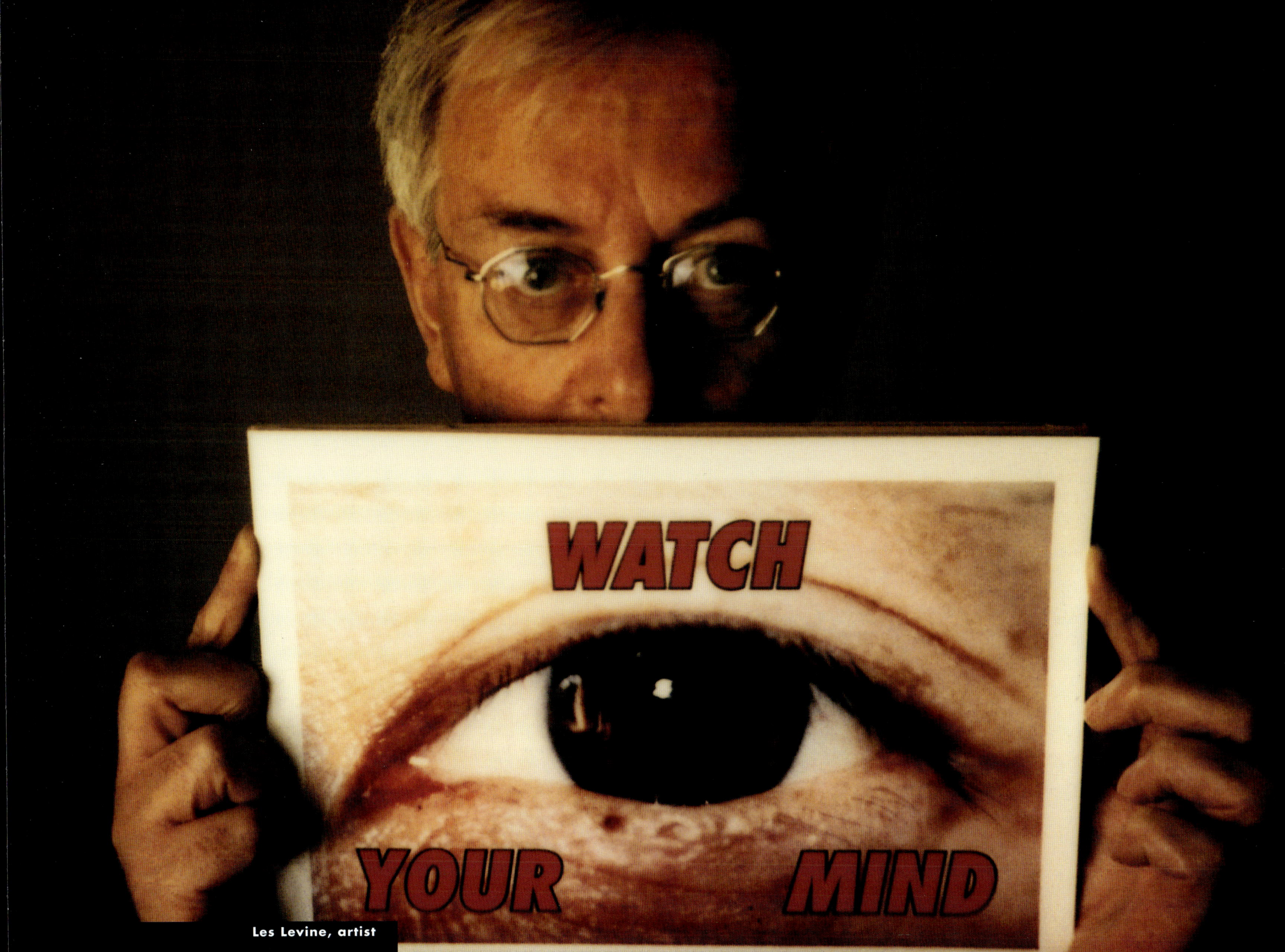

Les Levine, artist

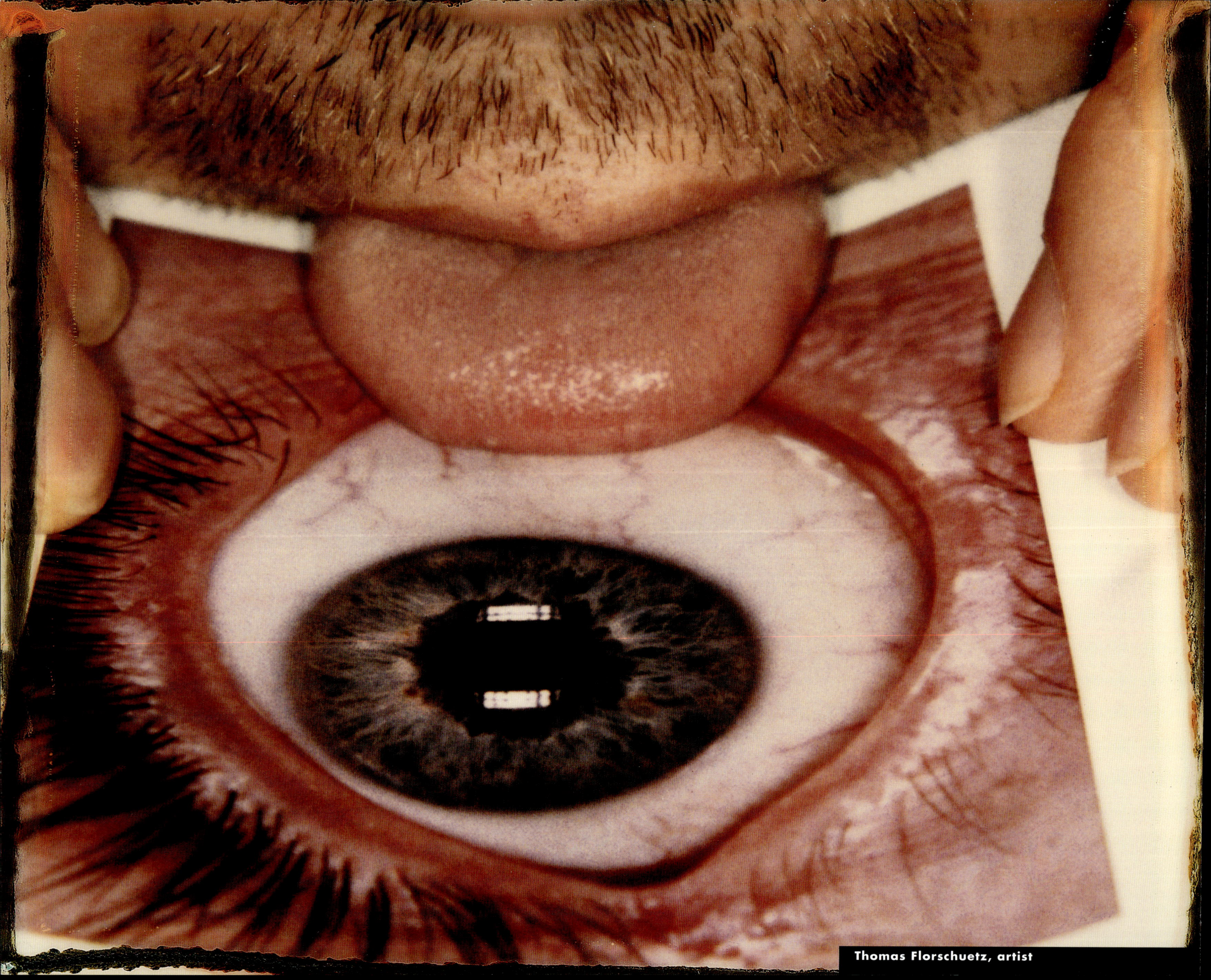

Thomas Florschuetz, artist

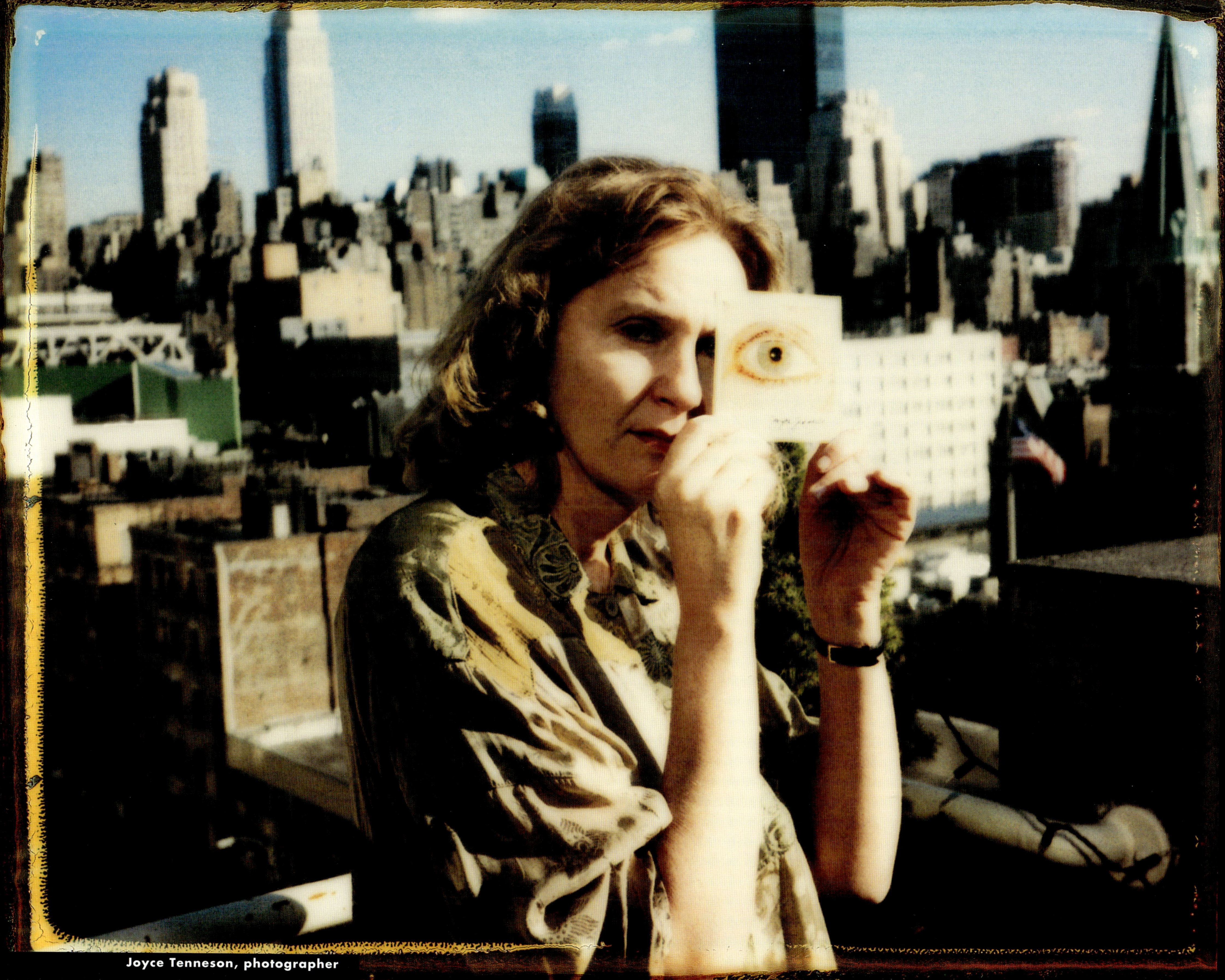

Joyce Tenneson, photographer

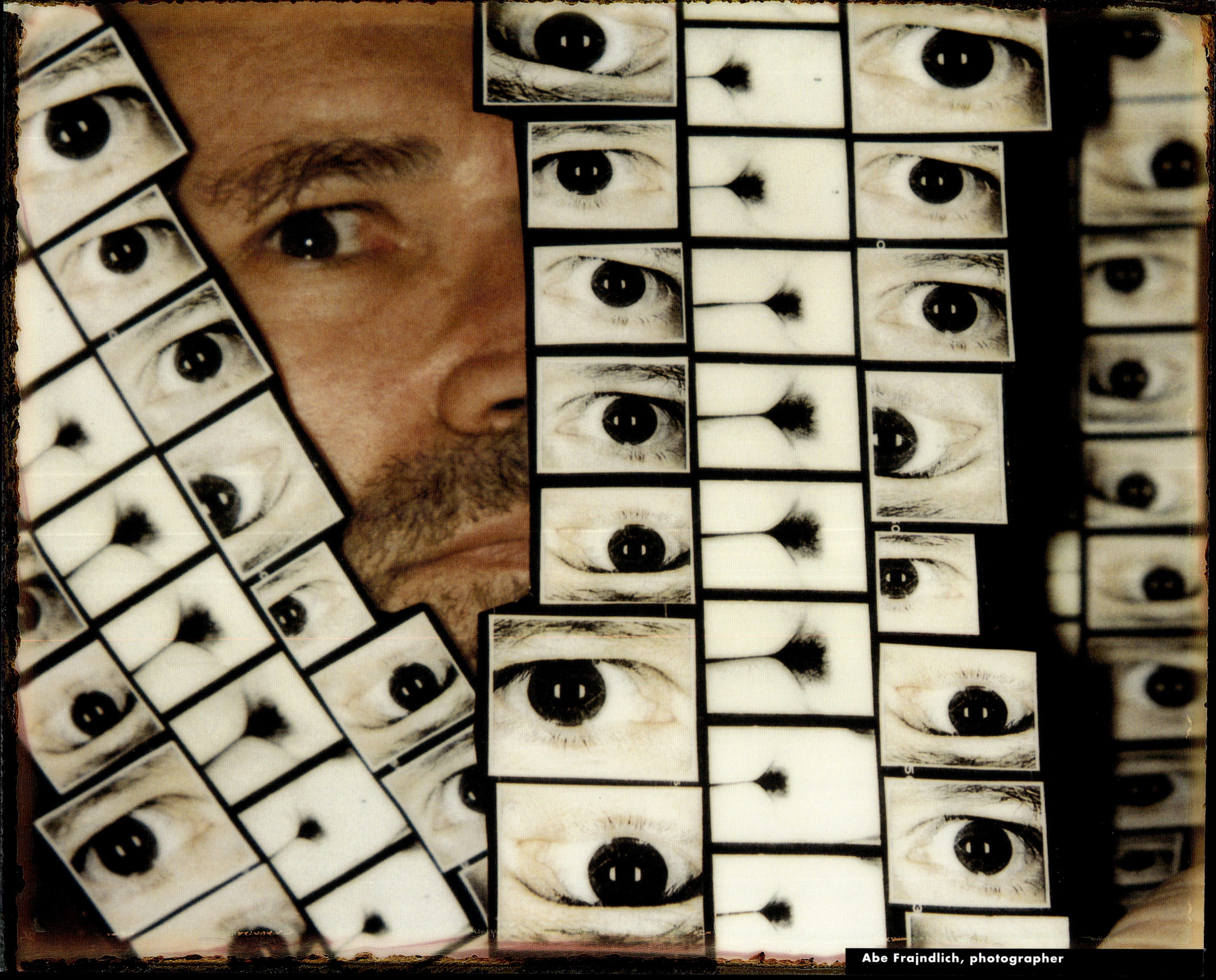

Abe Frajndlich, photographer

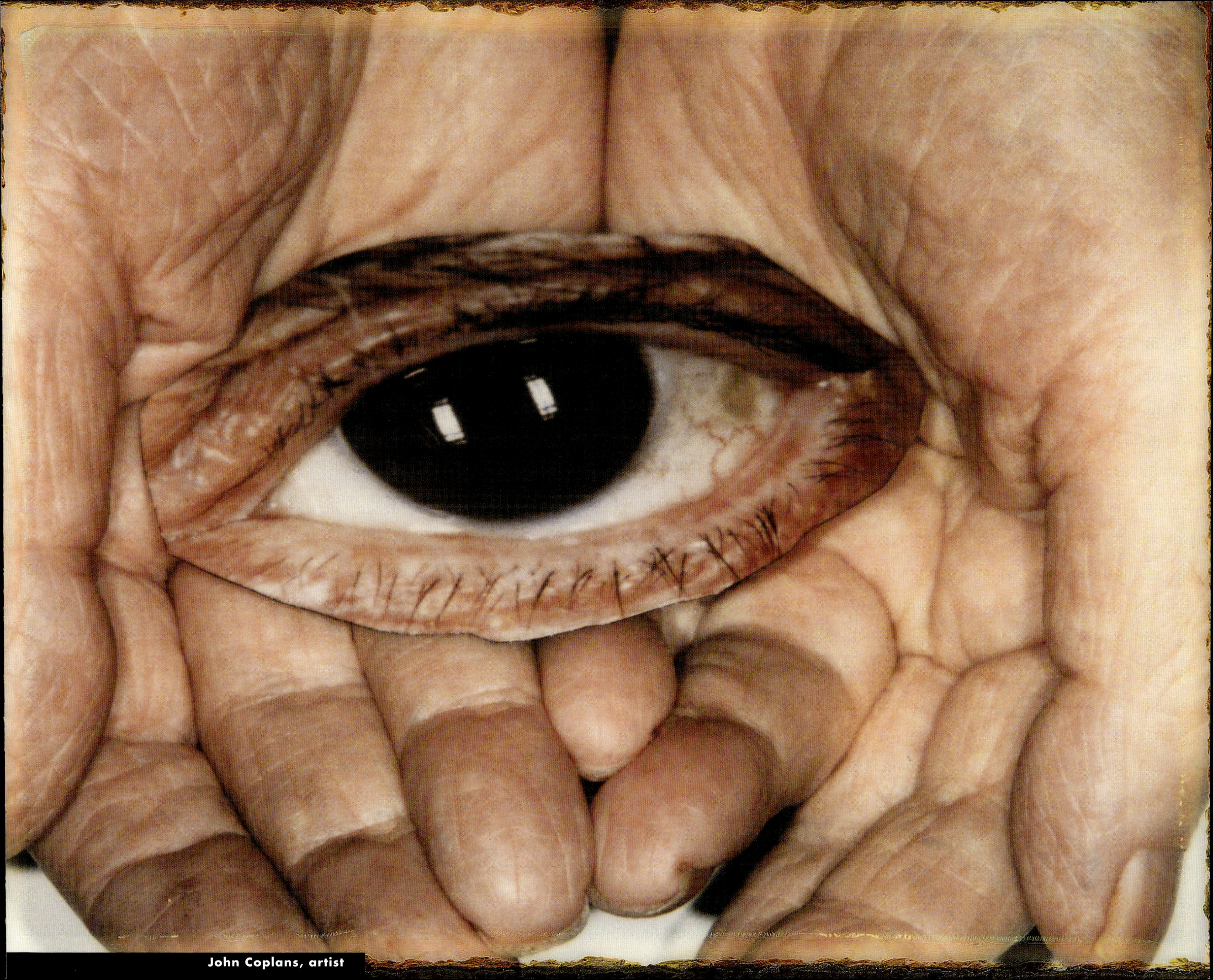
John Coplans, artist

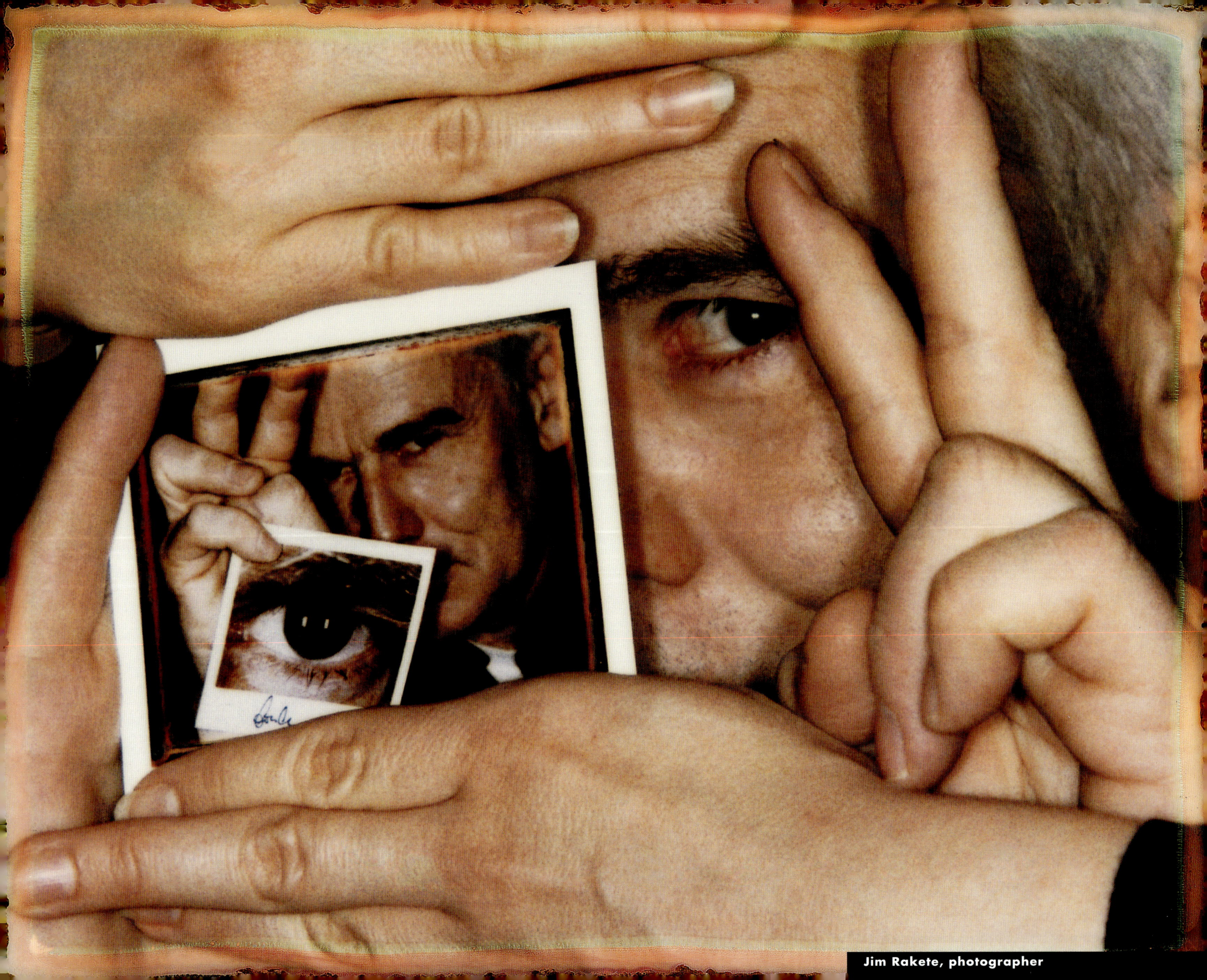

Jim Rakete, photographer

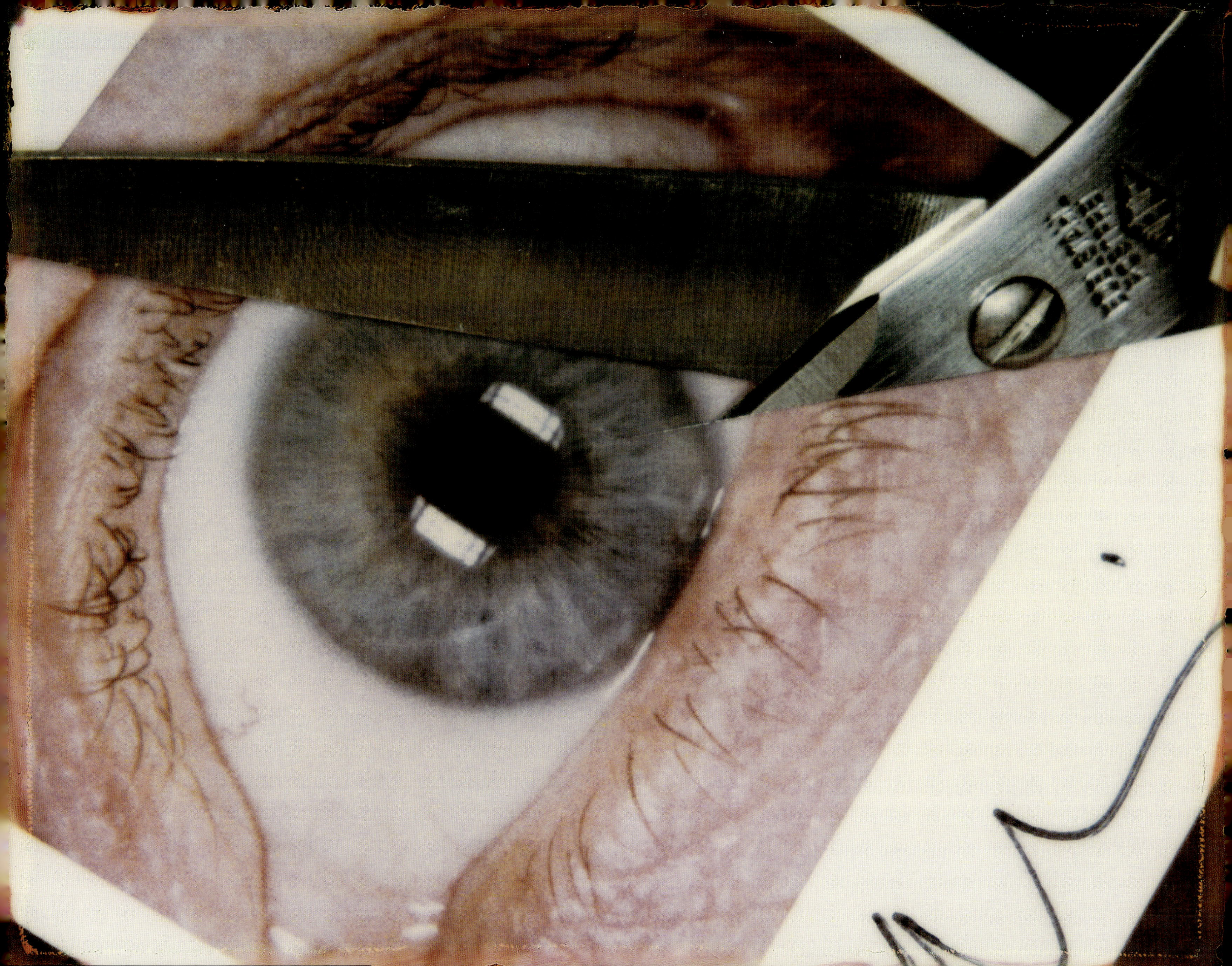

Peter Weiermair, curator

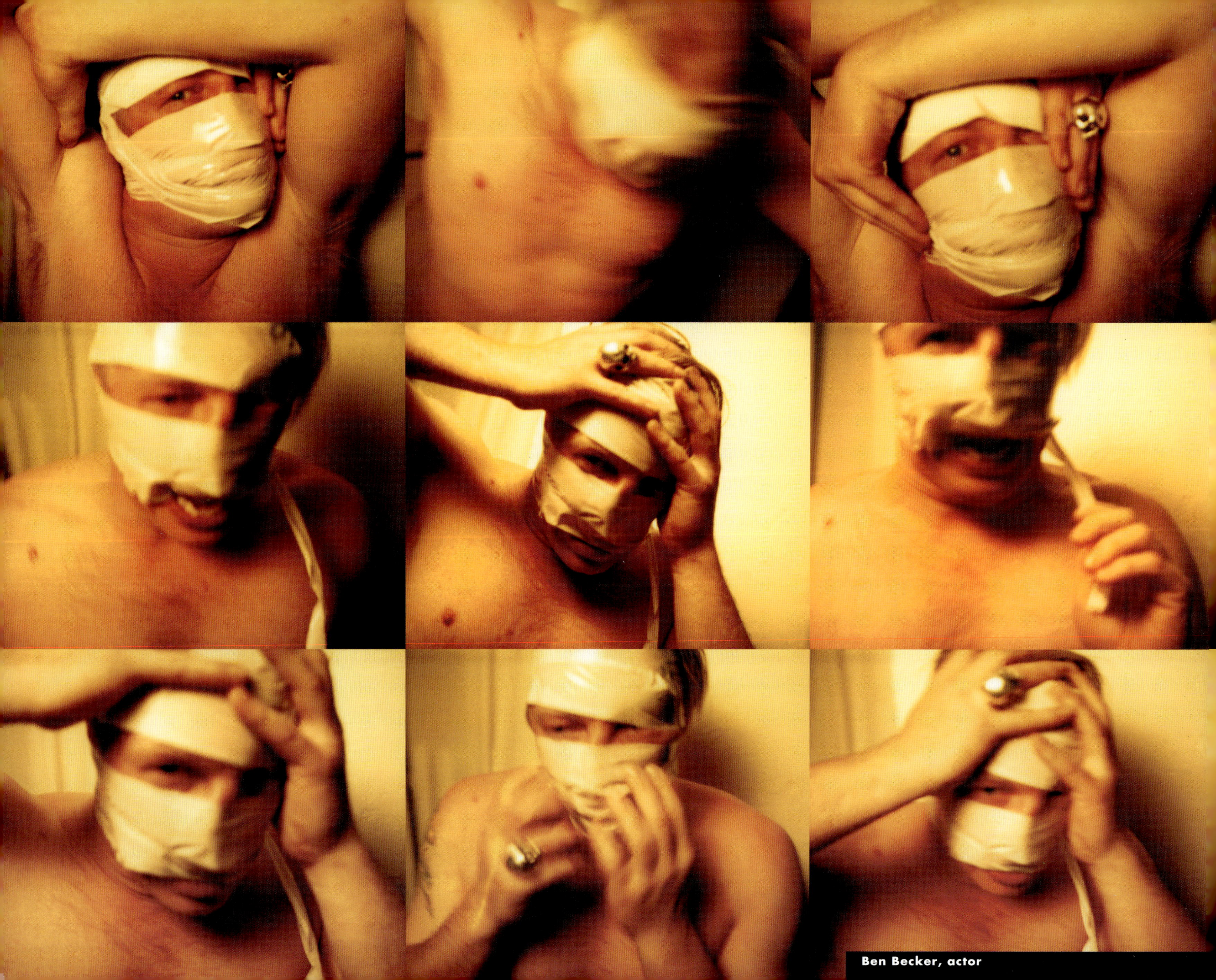

Ben Becker, actor

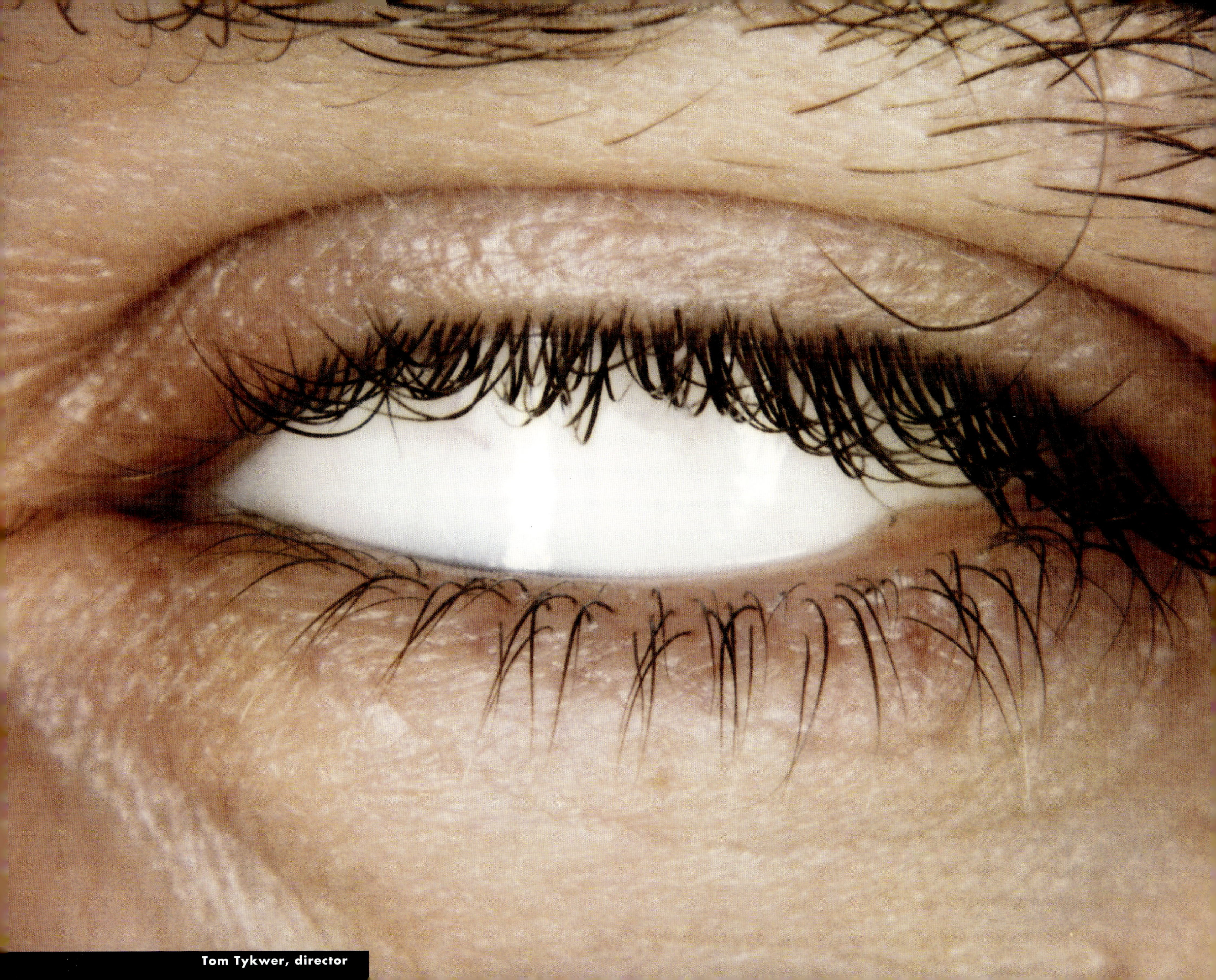

Tom Tykwer, director

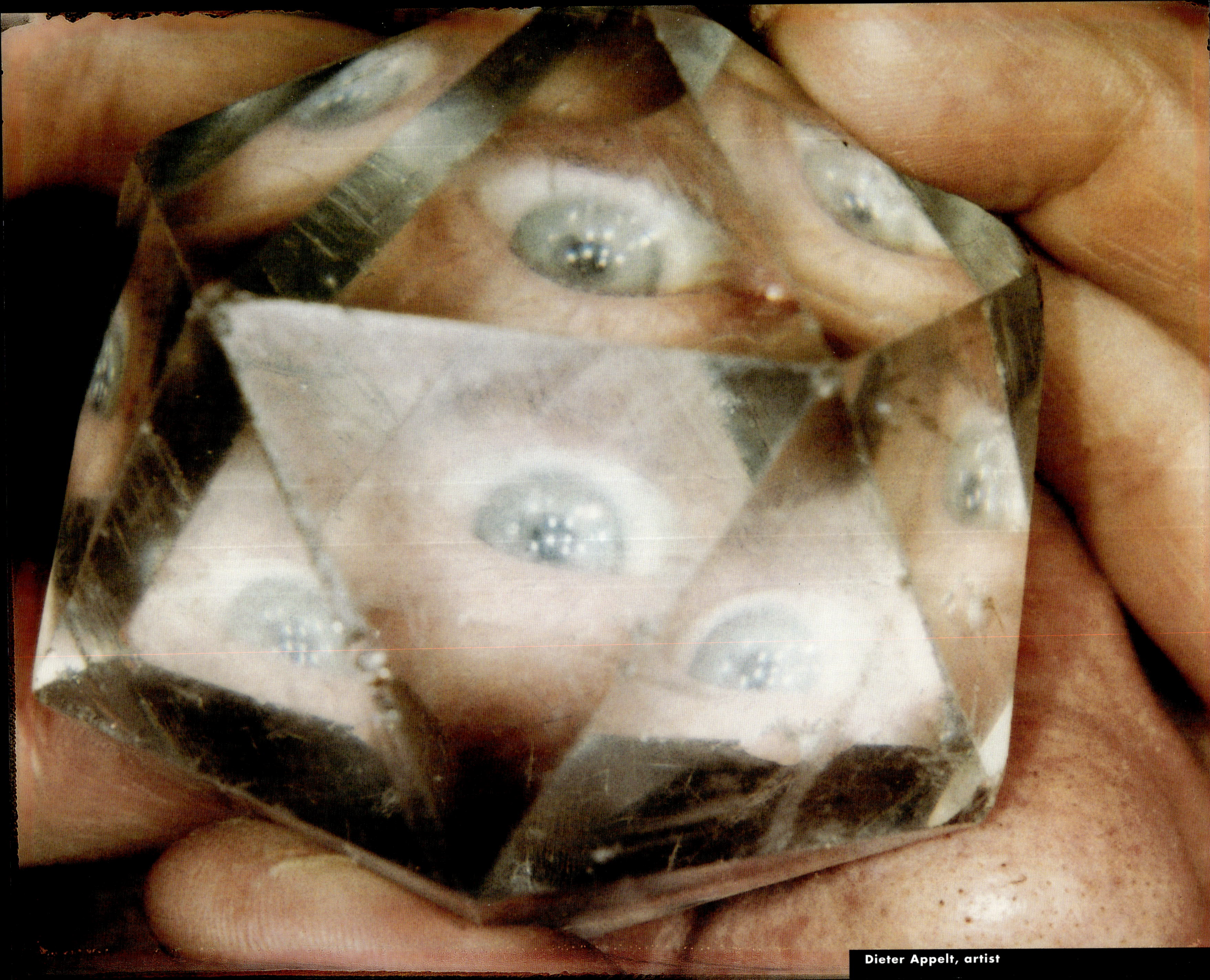

Dieter Appelt, artist

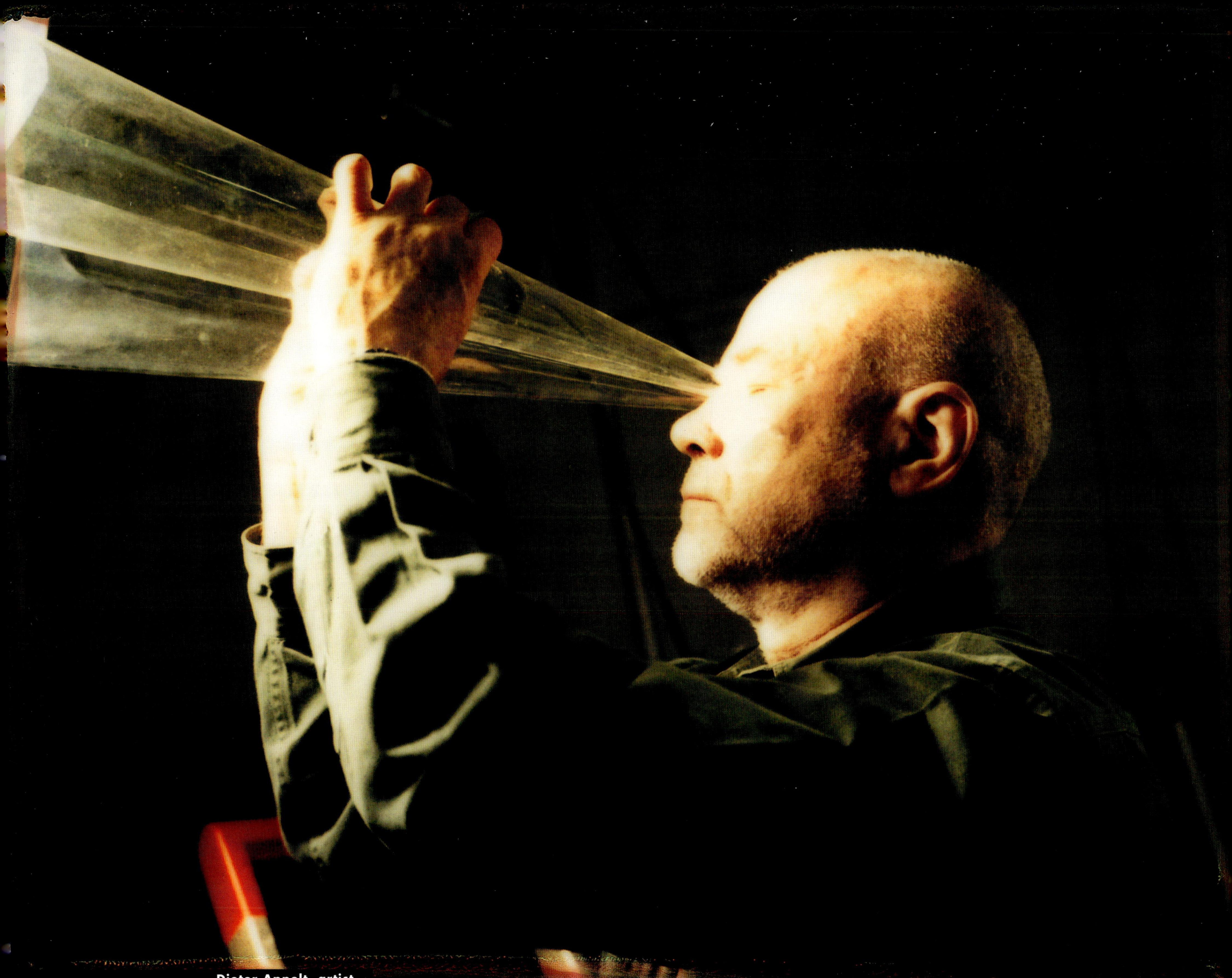

Dieter Appelt, artist

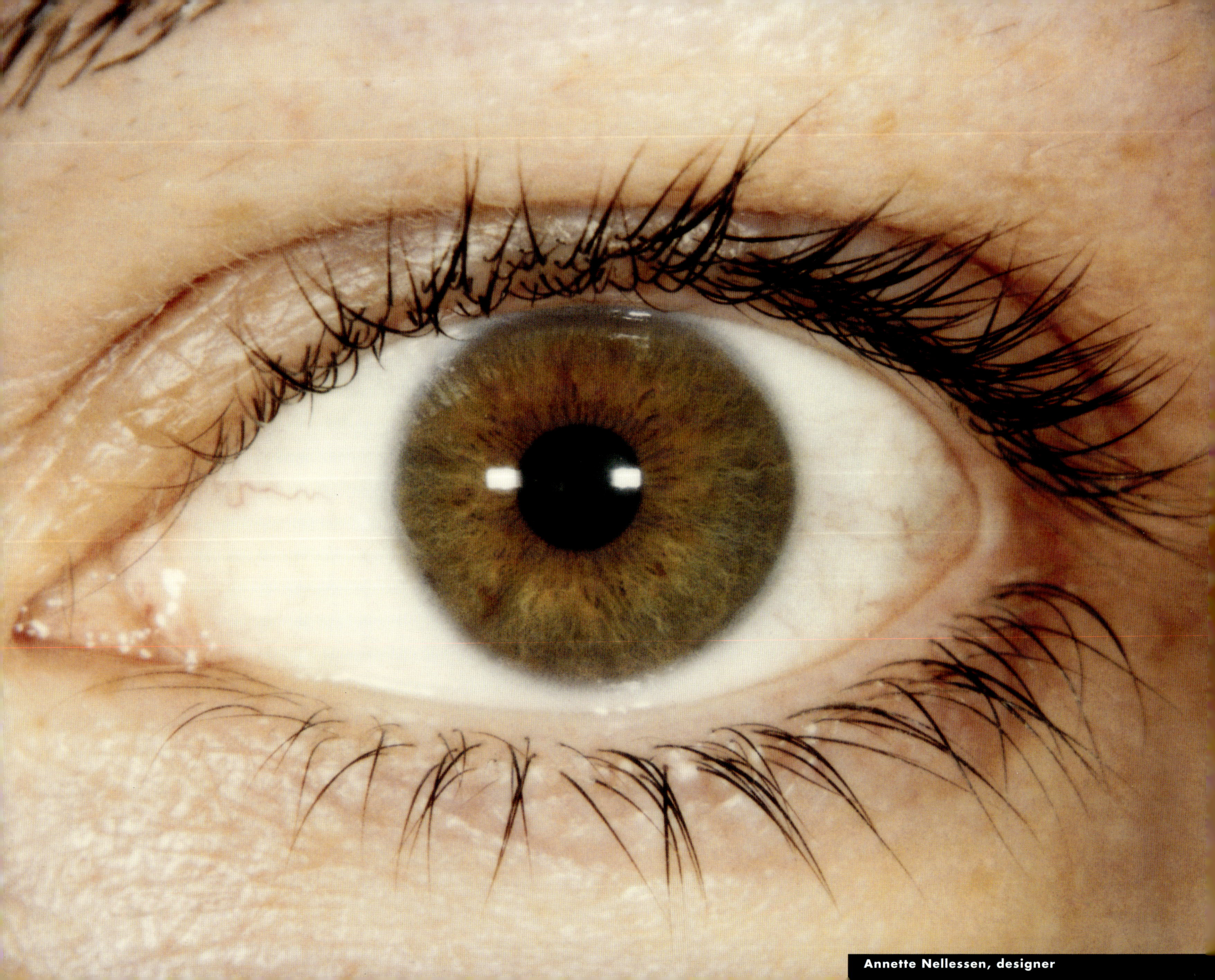

Annette Nellessen, designer

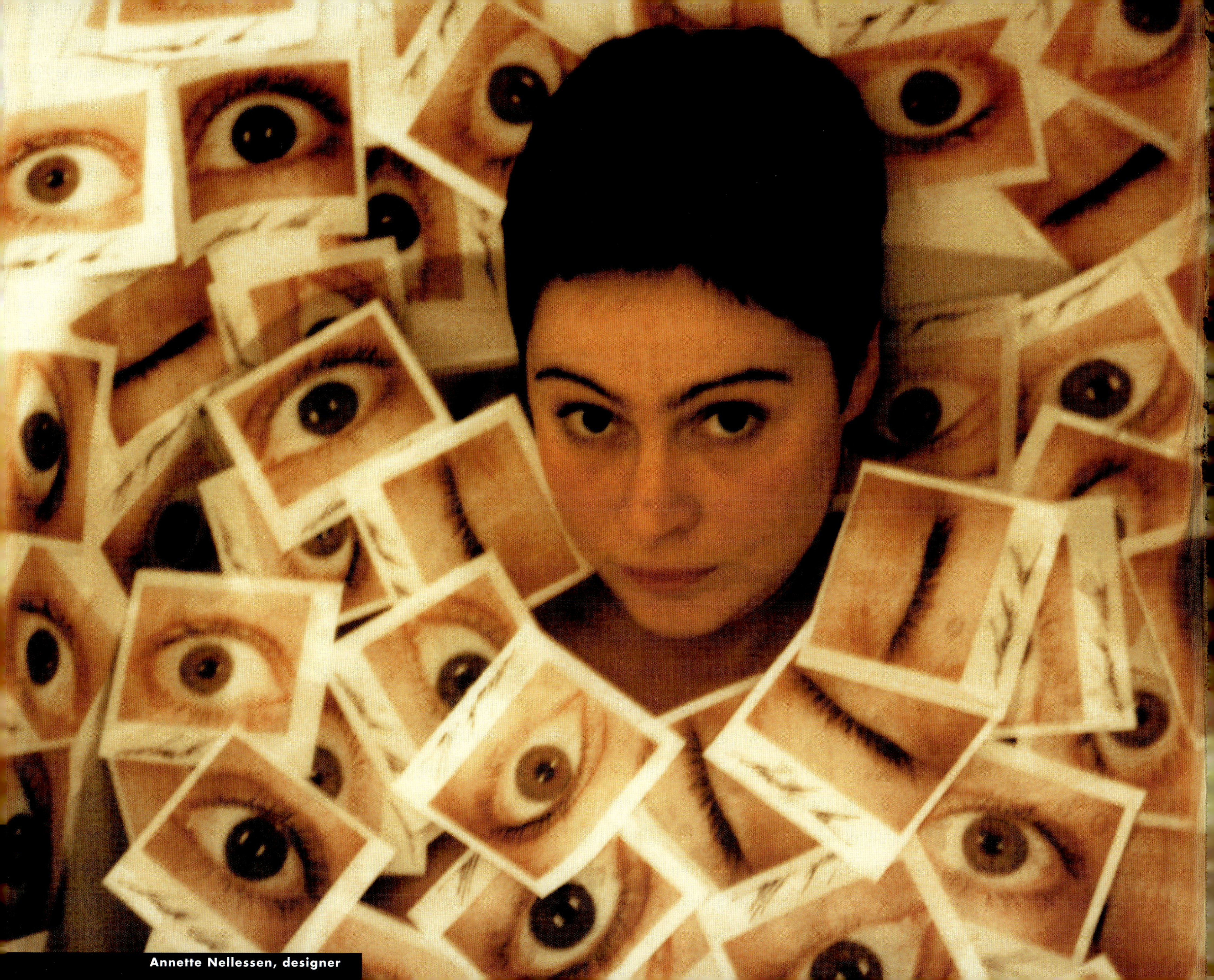

Annette Nellessen, designer

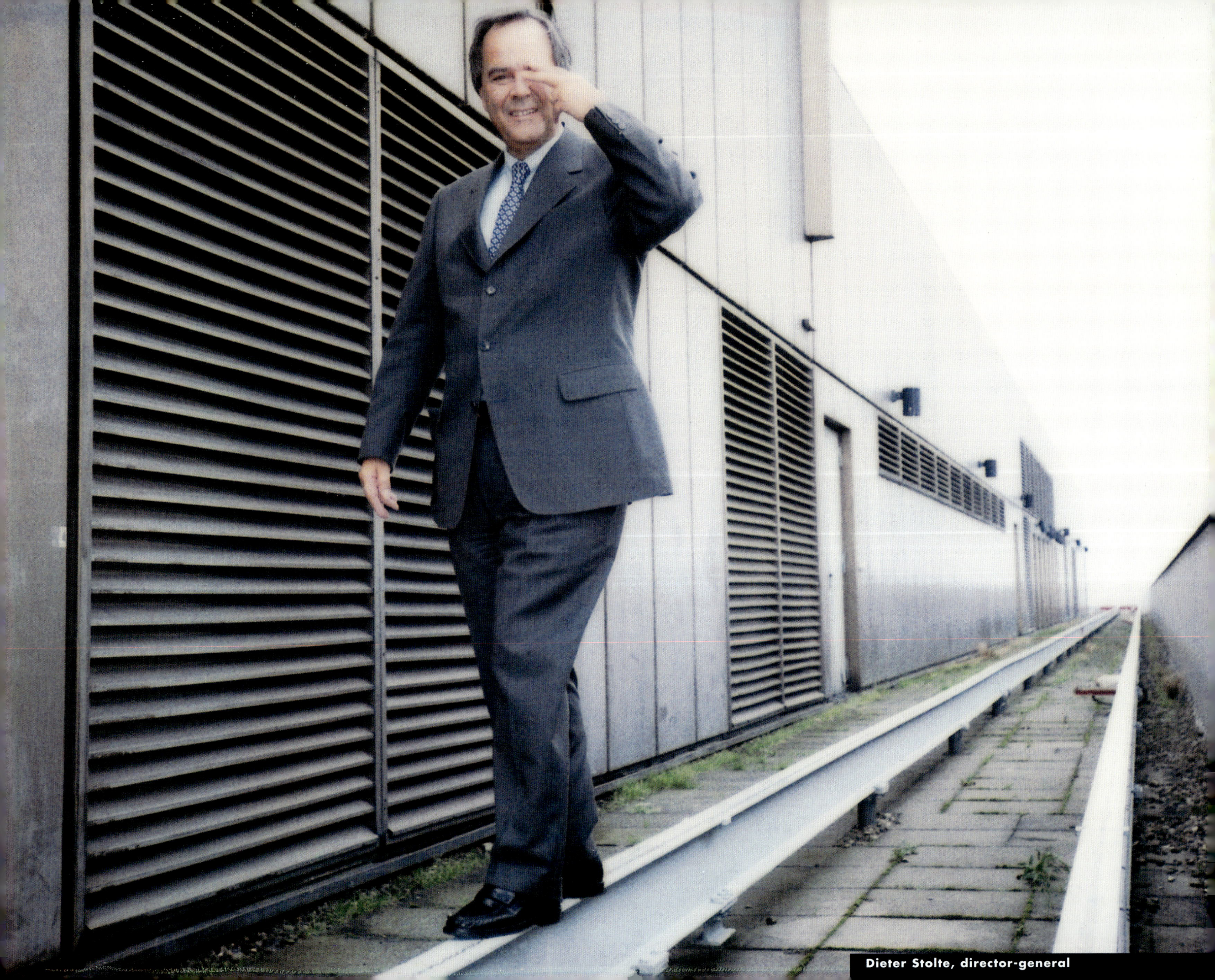

Dieter Stolte, director-general

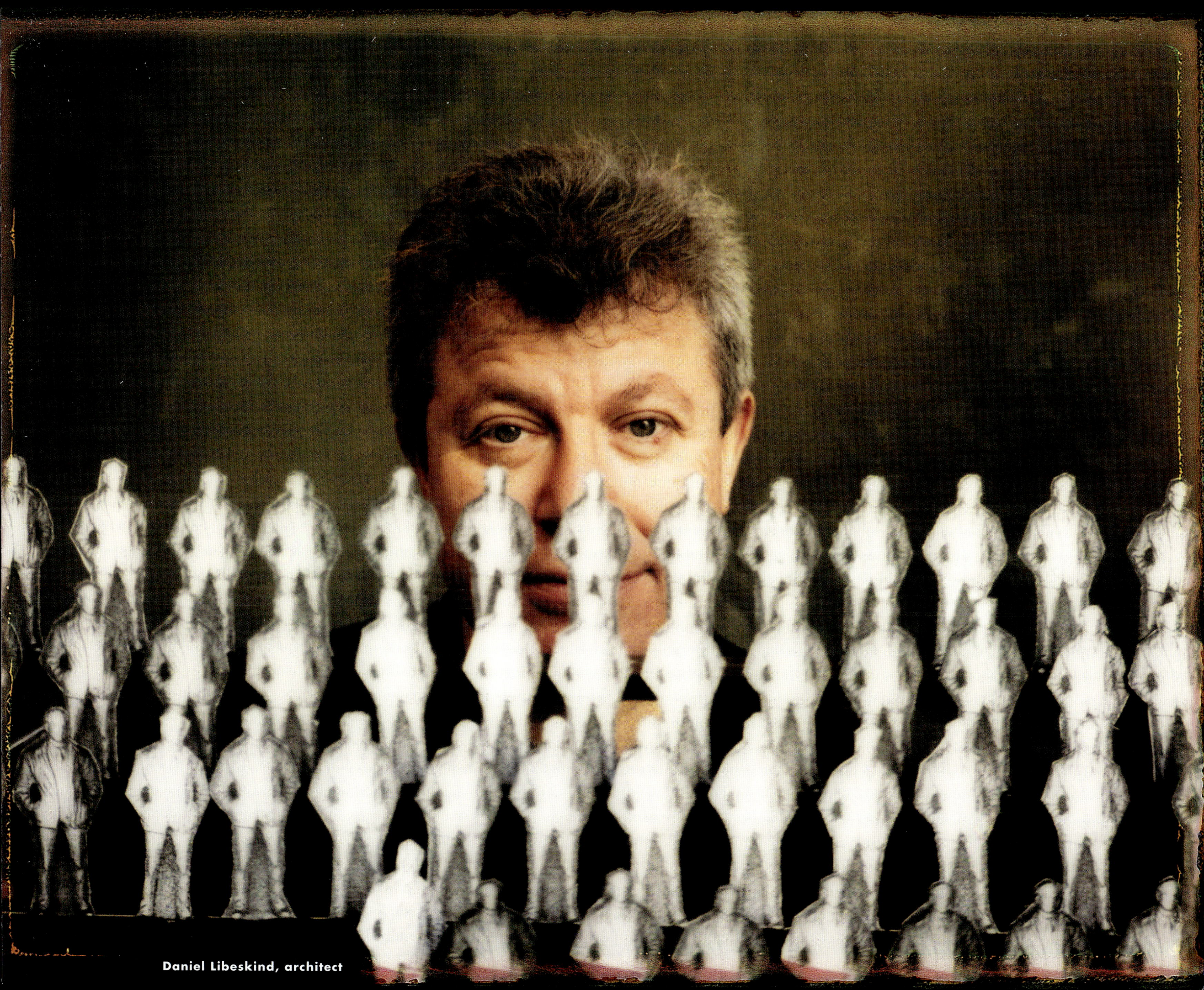

Daniel Libeskind, architect

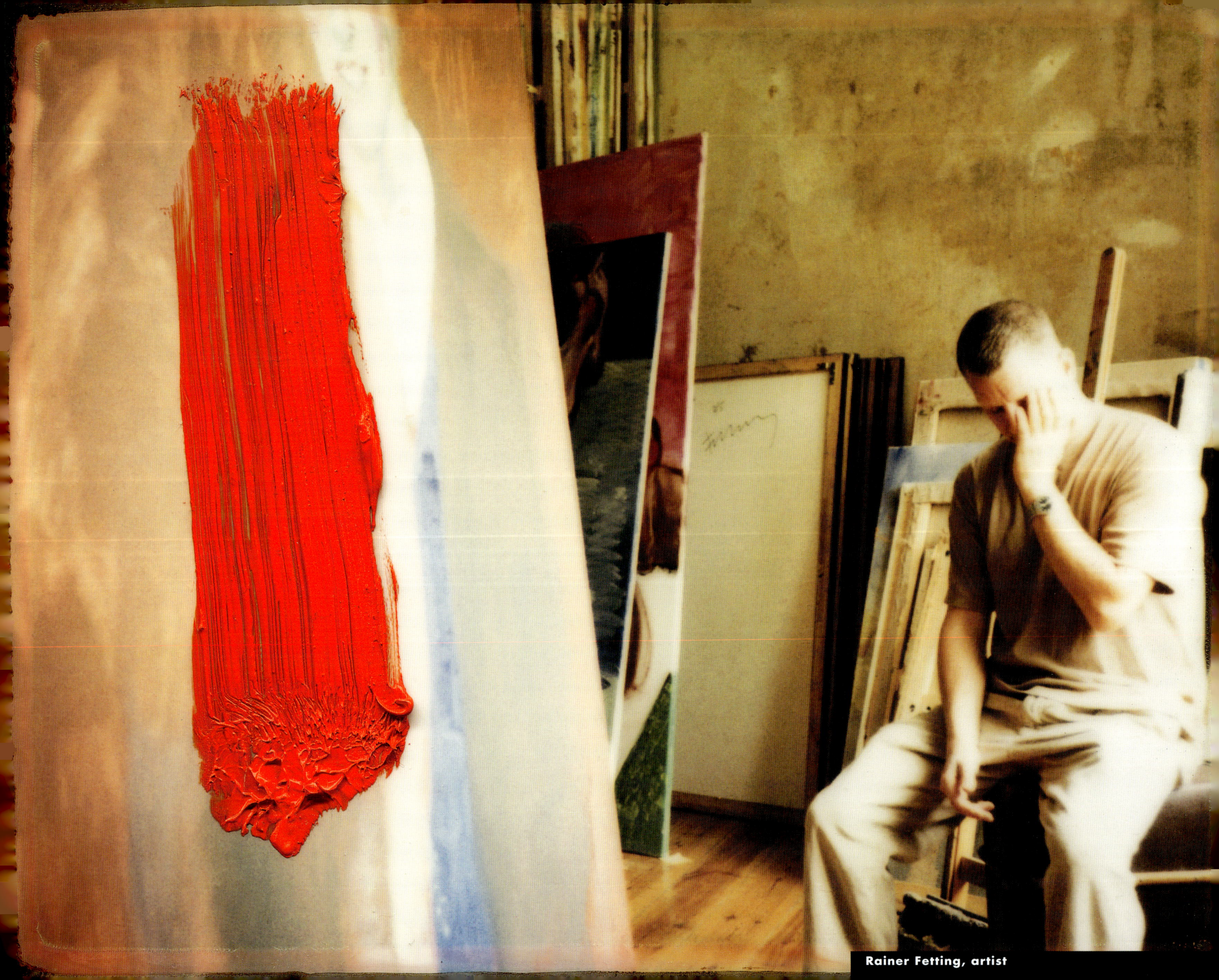

Rainer Fetting, artist

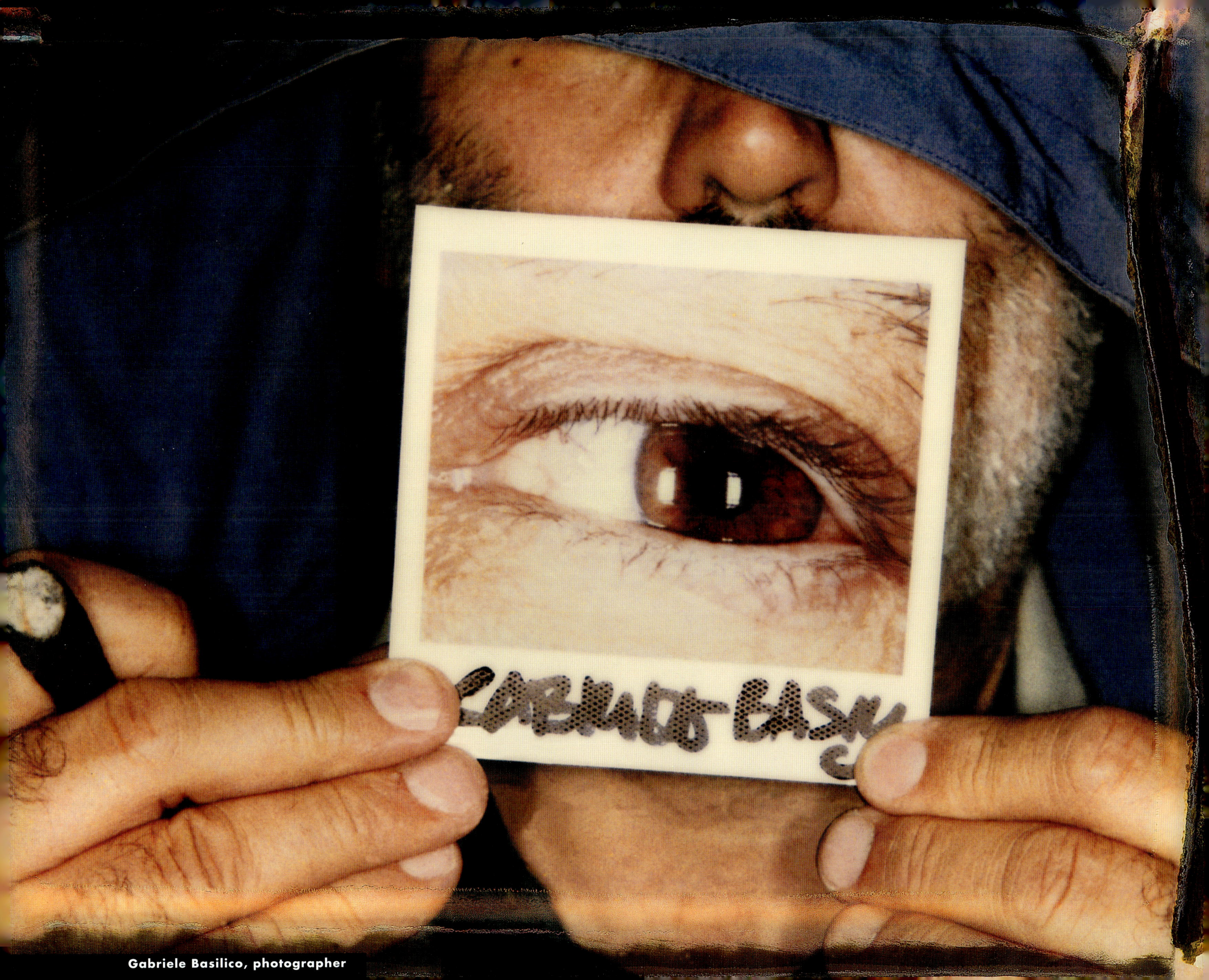

Gabriele Basilico, photographer

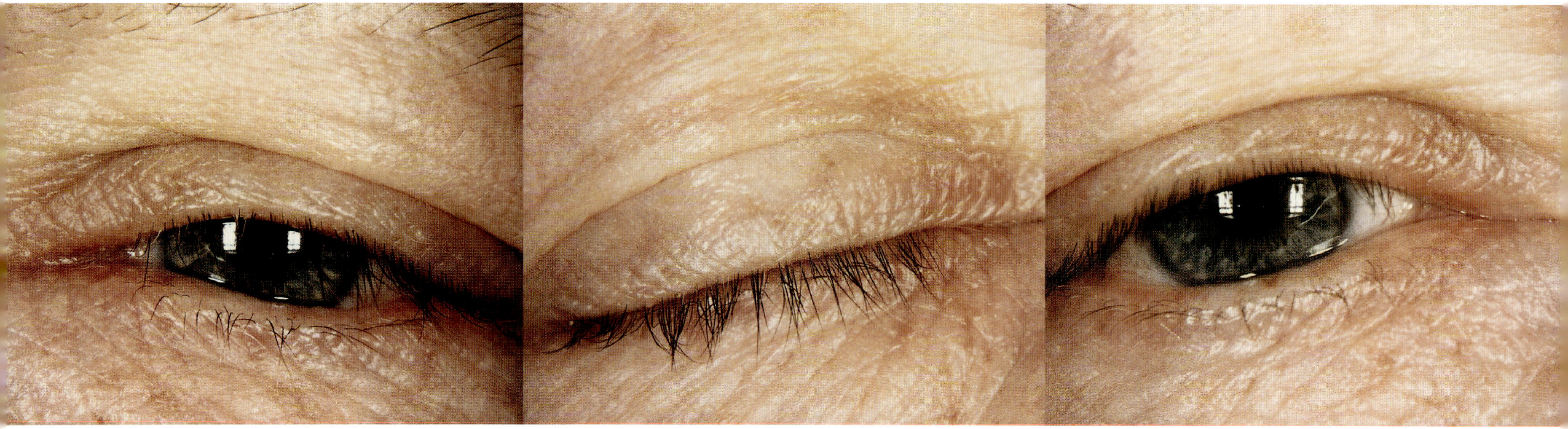

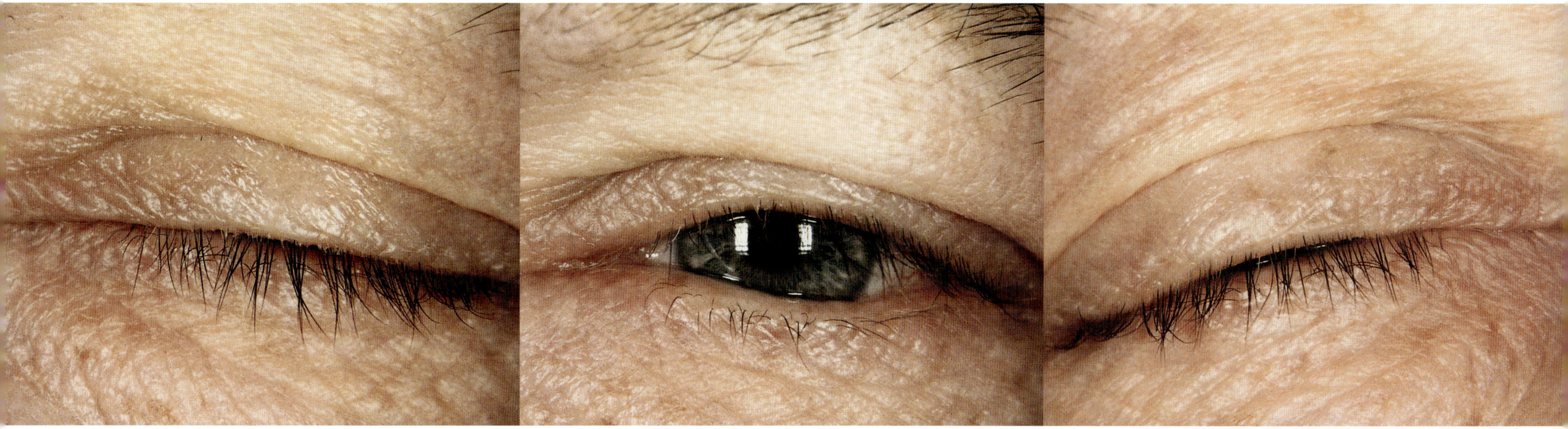

Imi Knoebel, artist

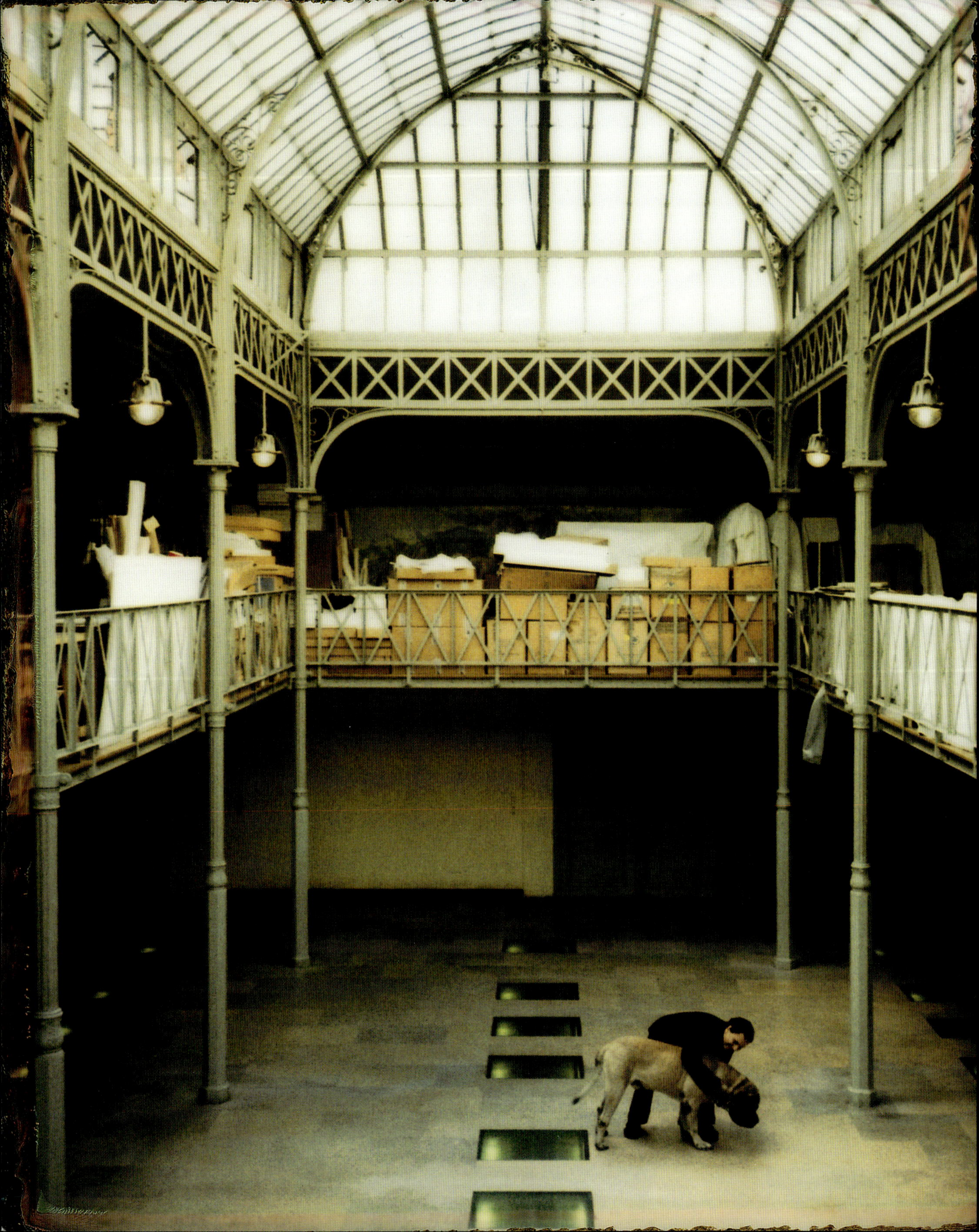

Azzedine Alaïa, designer

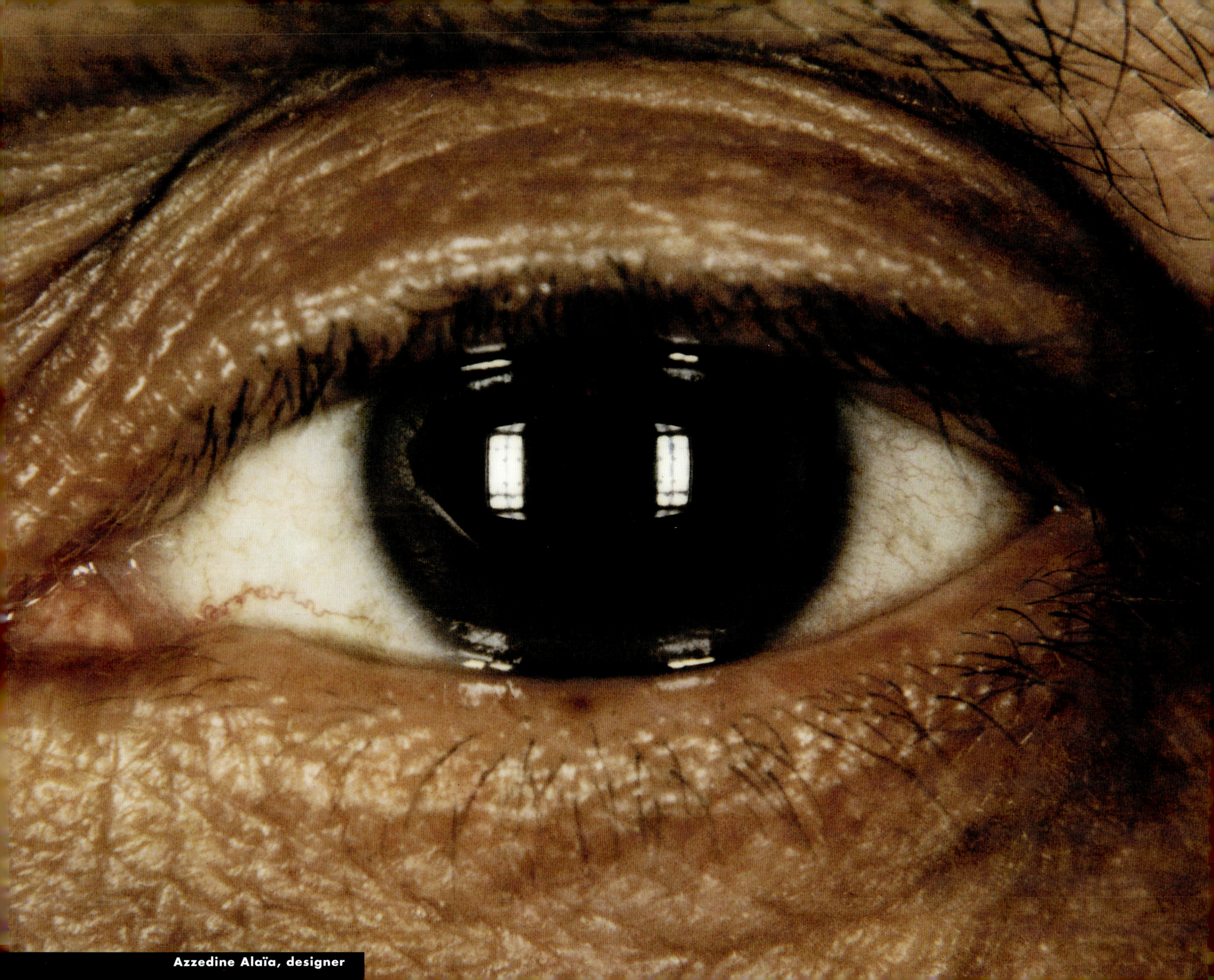

Azzedine Alaïa, designer

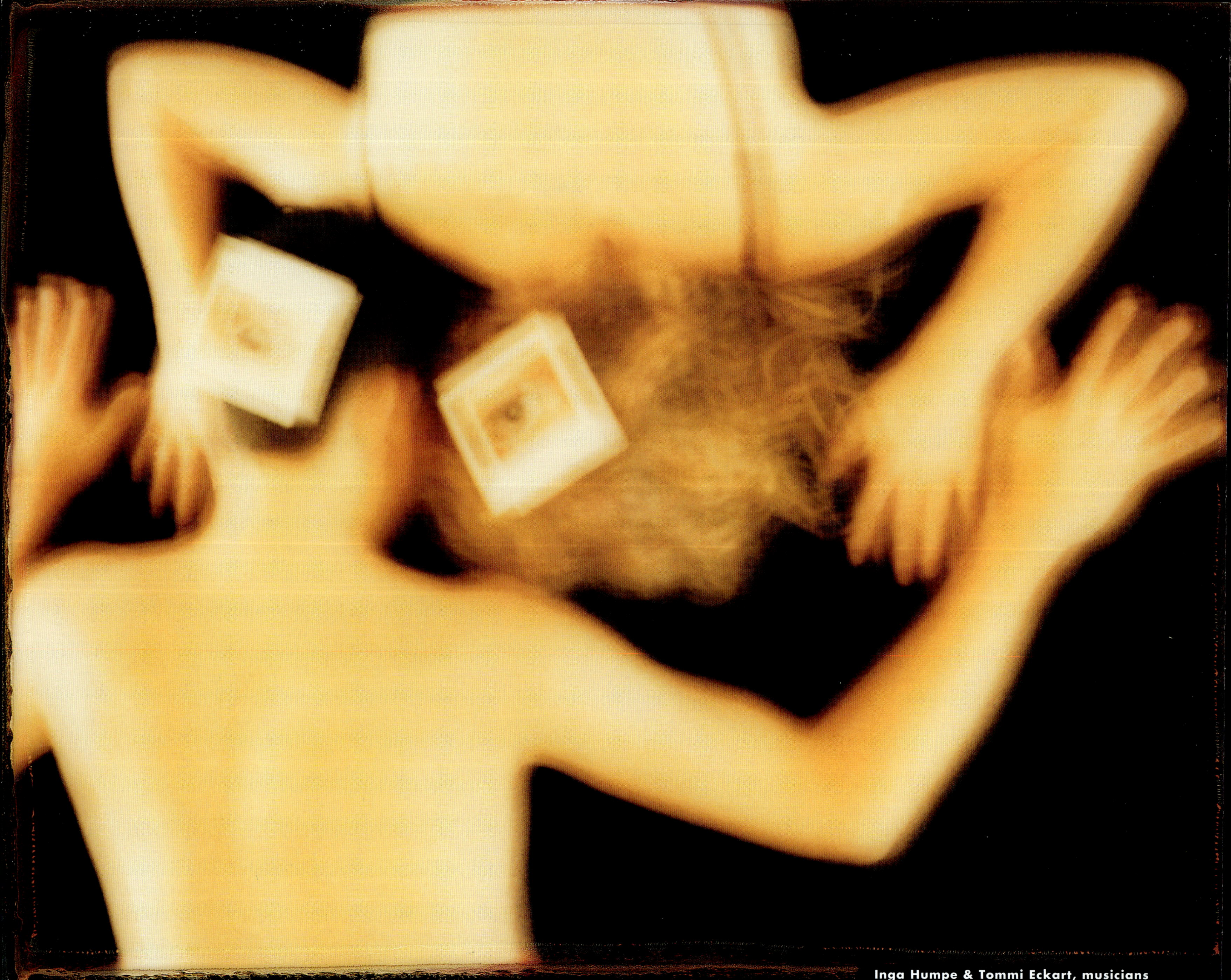

Inga Humpe & Tommi Eckart, musicians

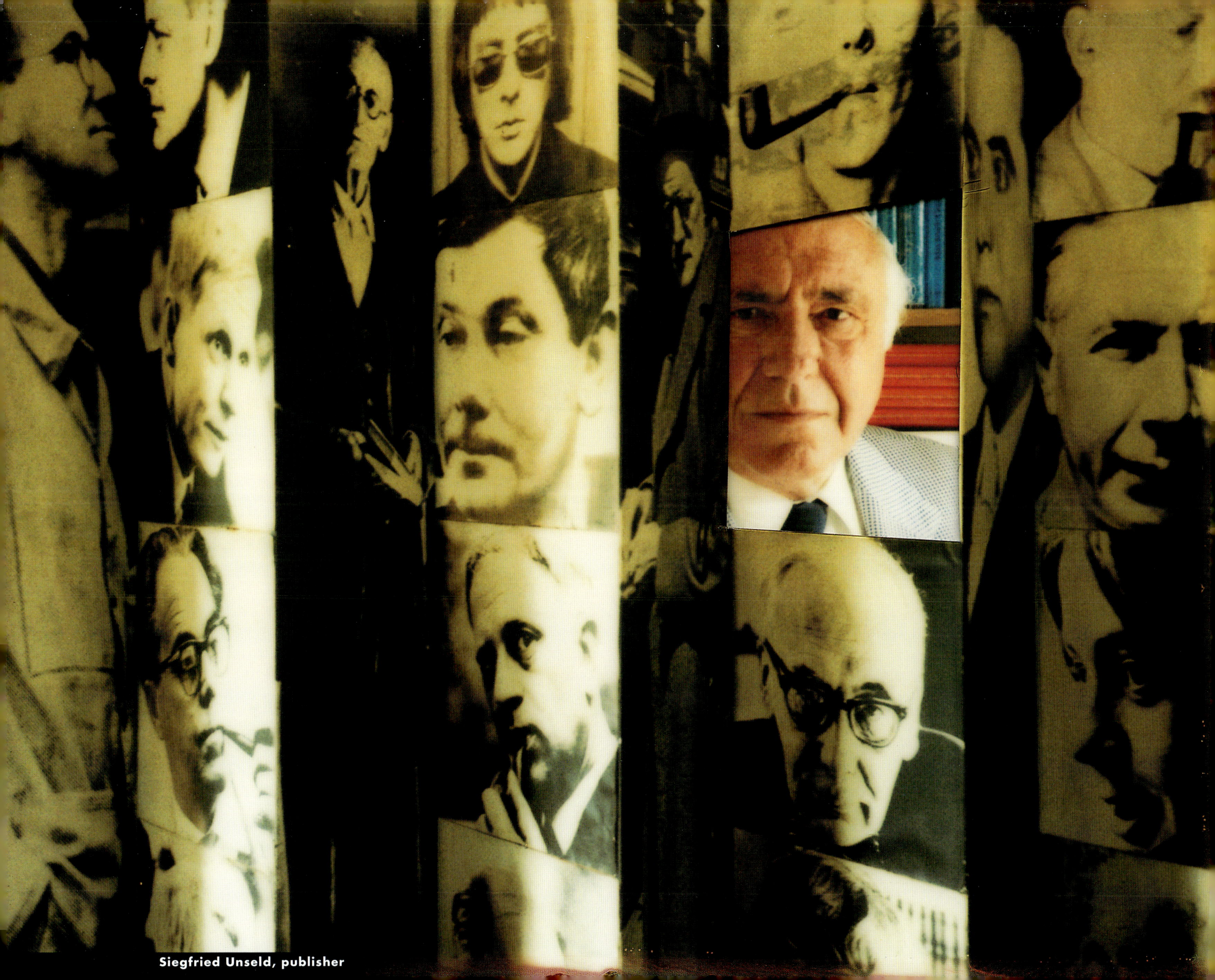

Siegfried Unseld, publisher

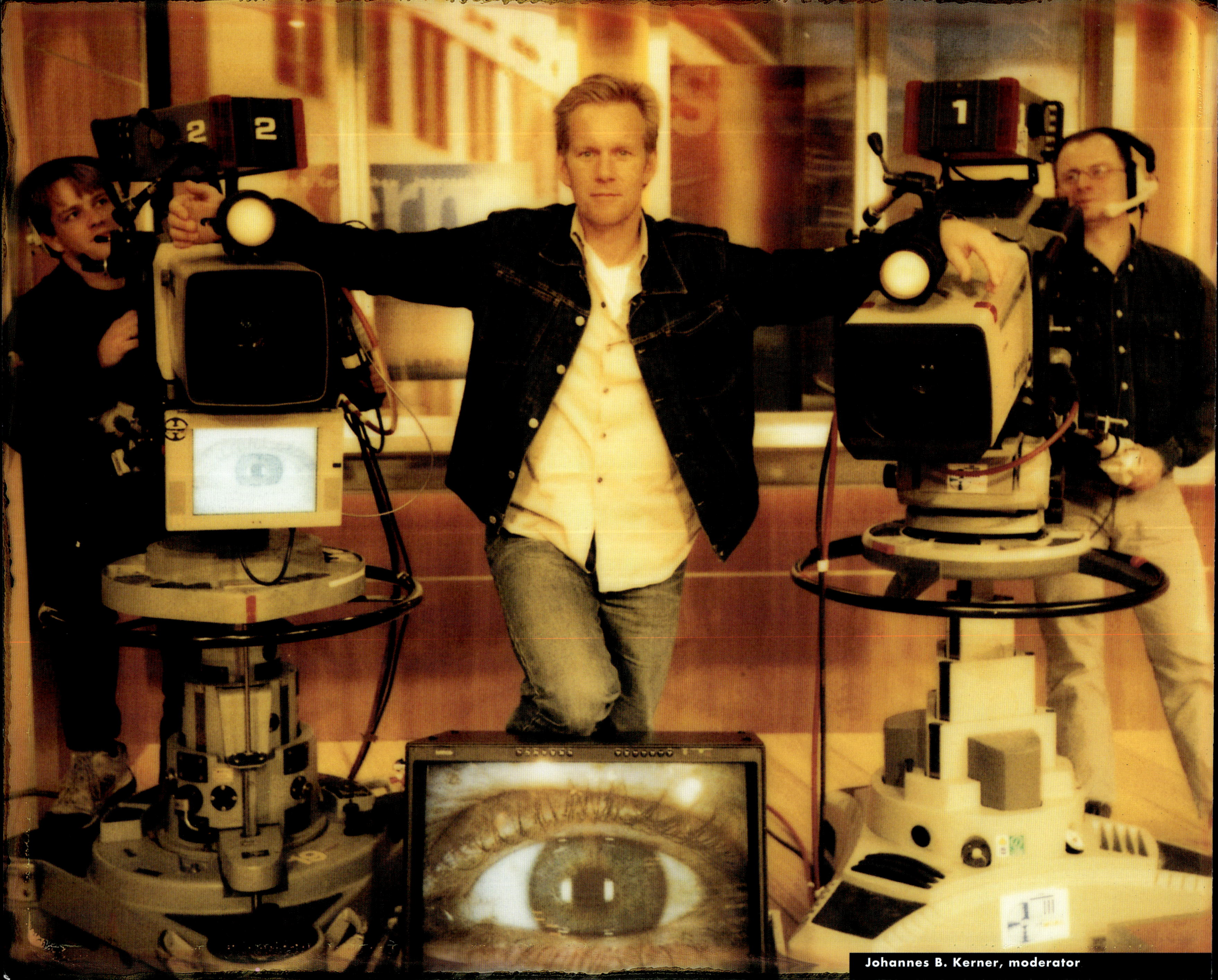

Johannes B. Kerner, moderator

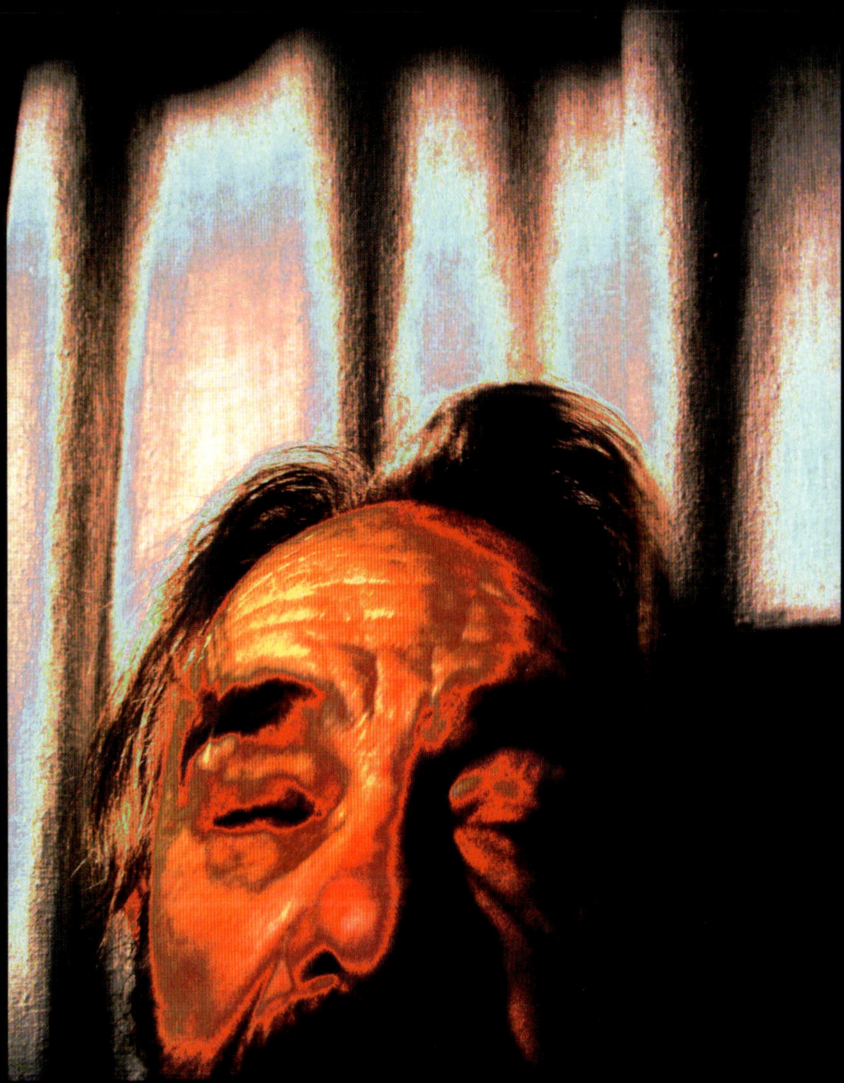

Arnulf Rainer, artist

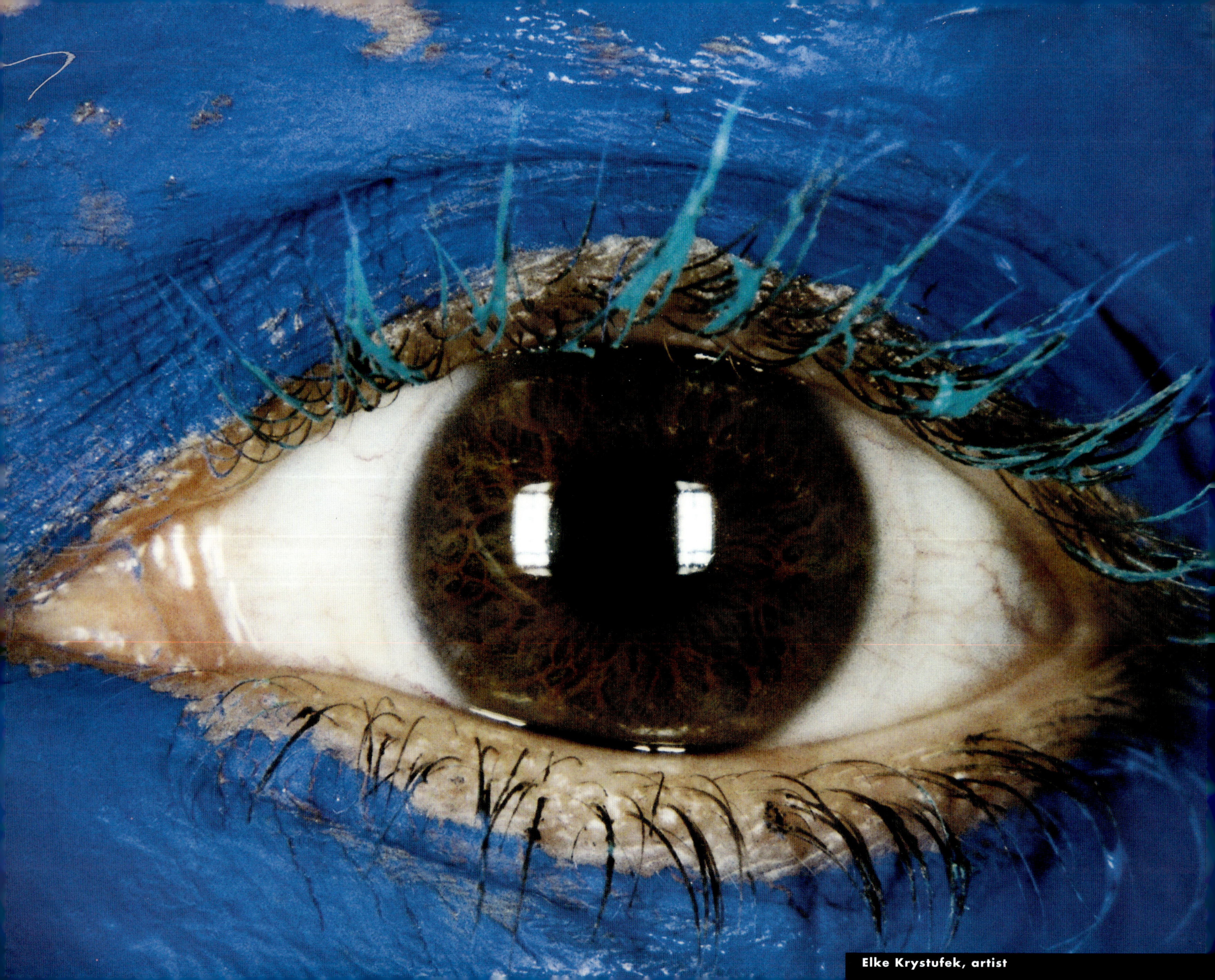

Elke Krystufek, artist

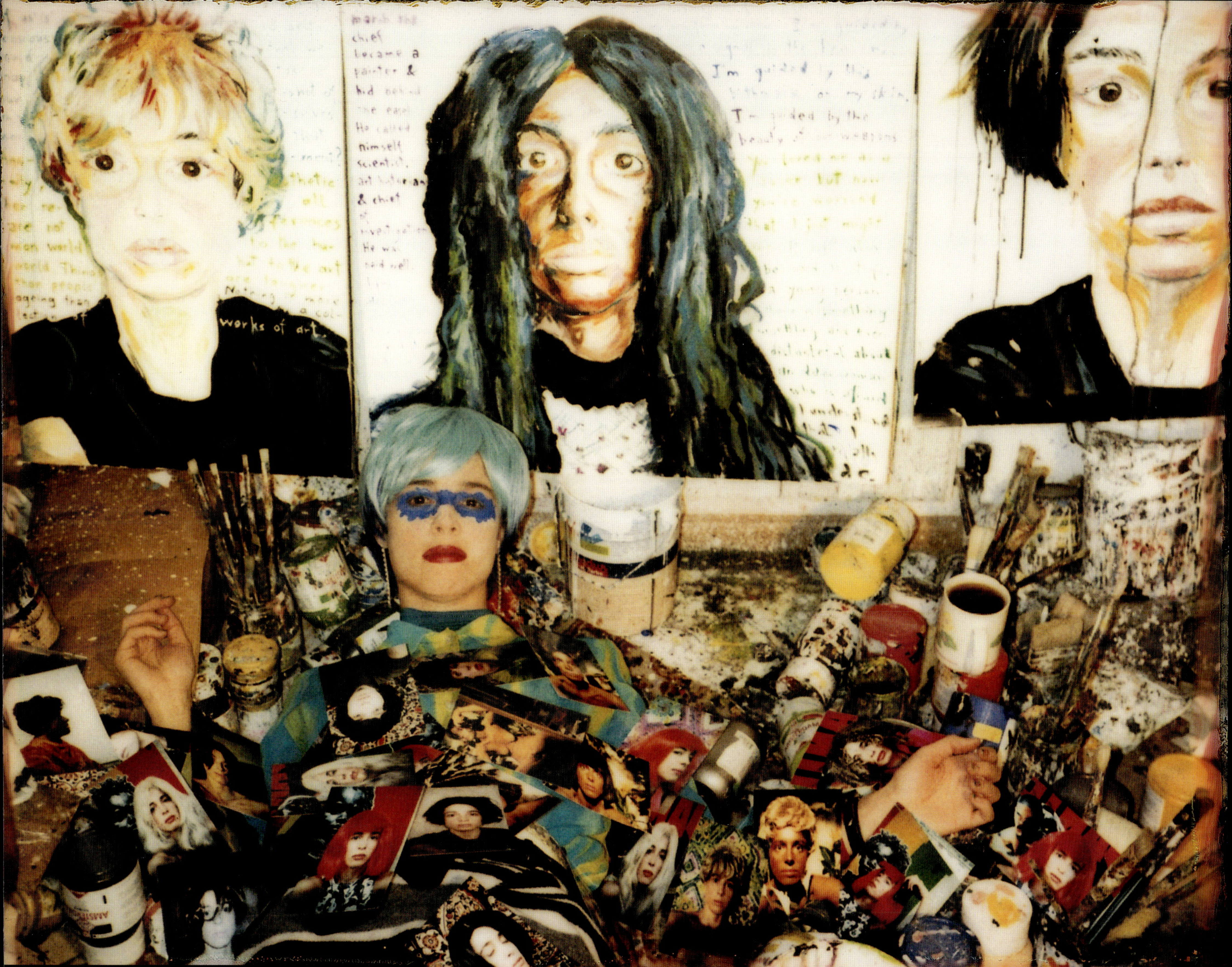

Elke Krystufek, artist

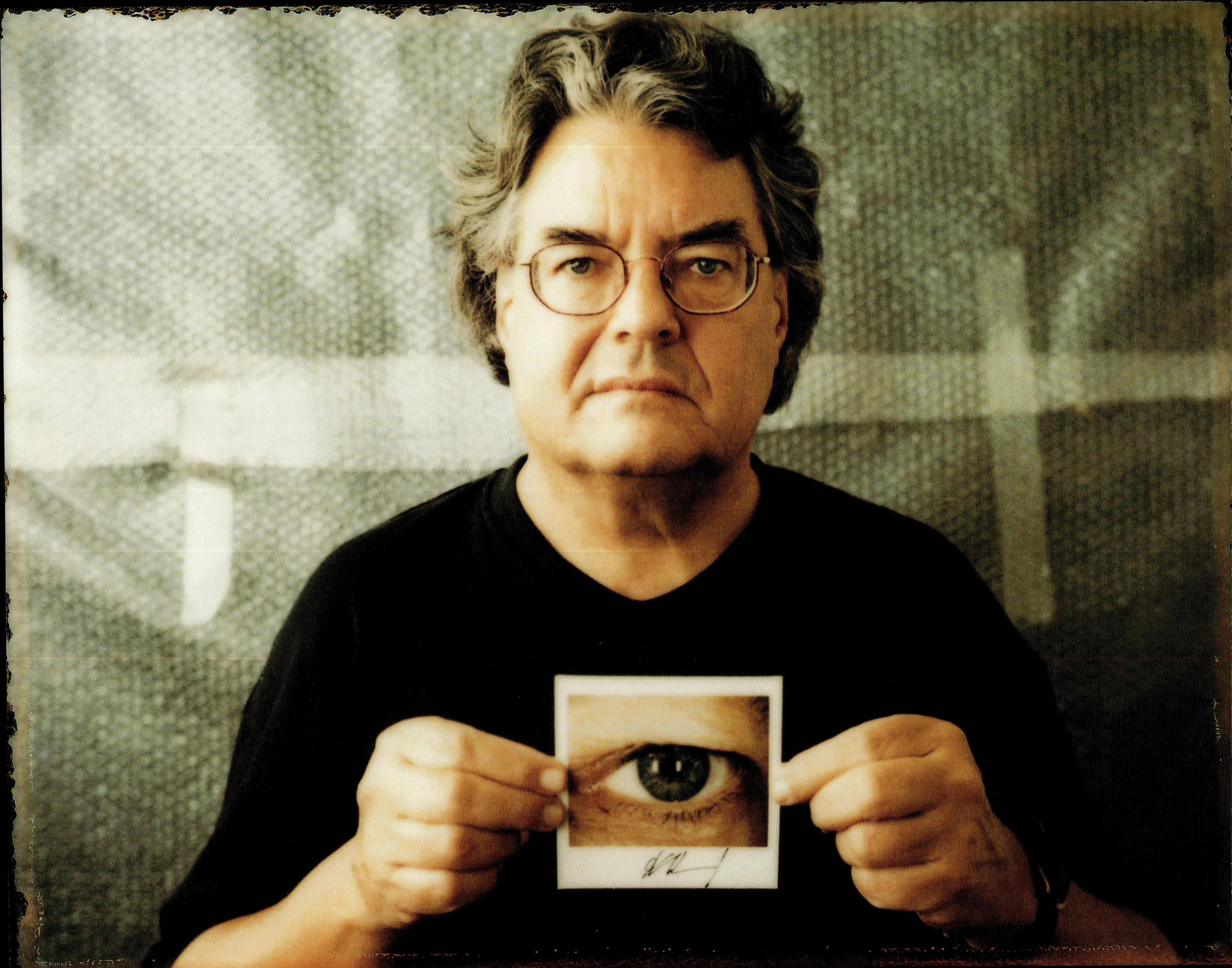

Klaus Honnef, curator

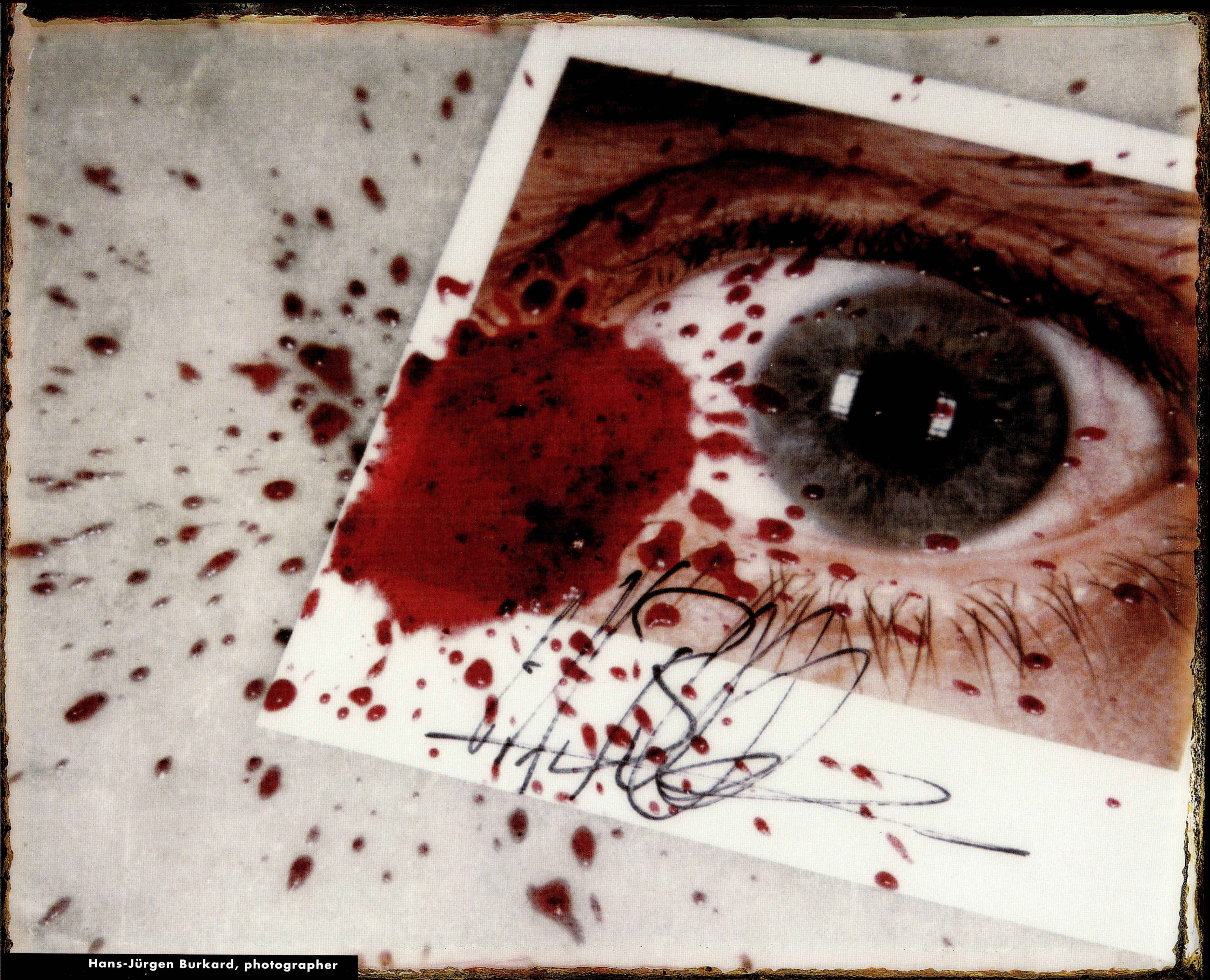

Hans-Jürgen Burkard, photographer

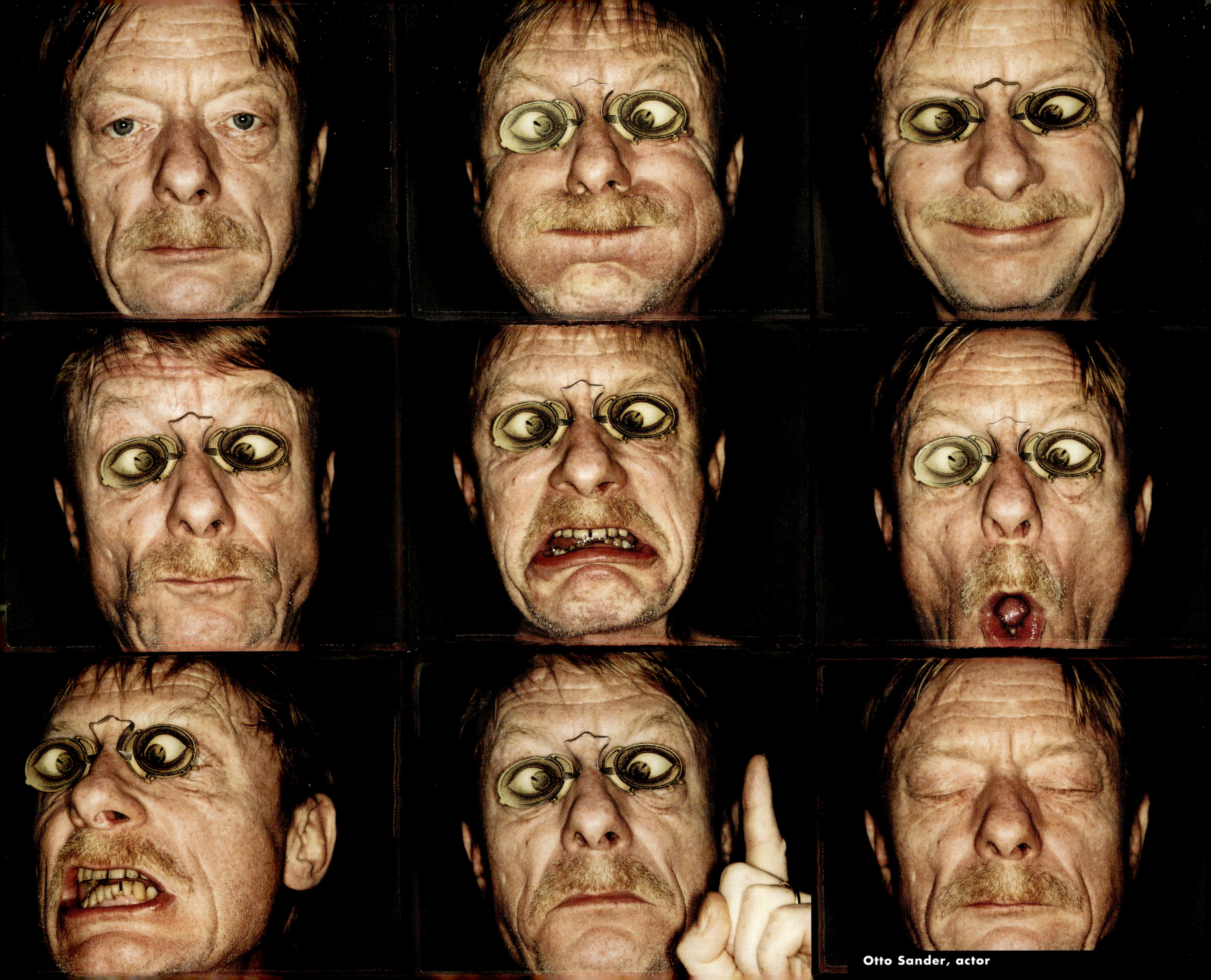

Otto Sander, actor

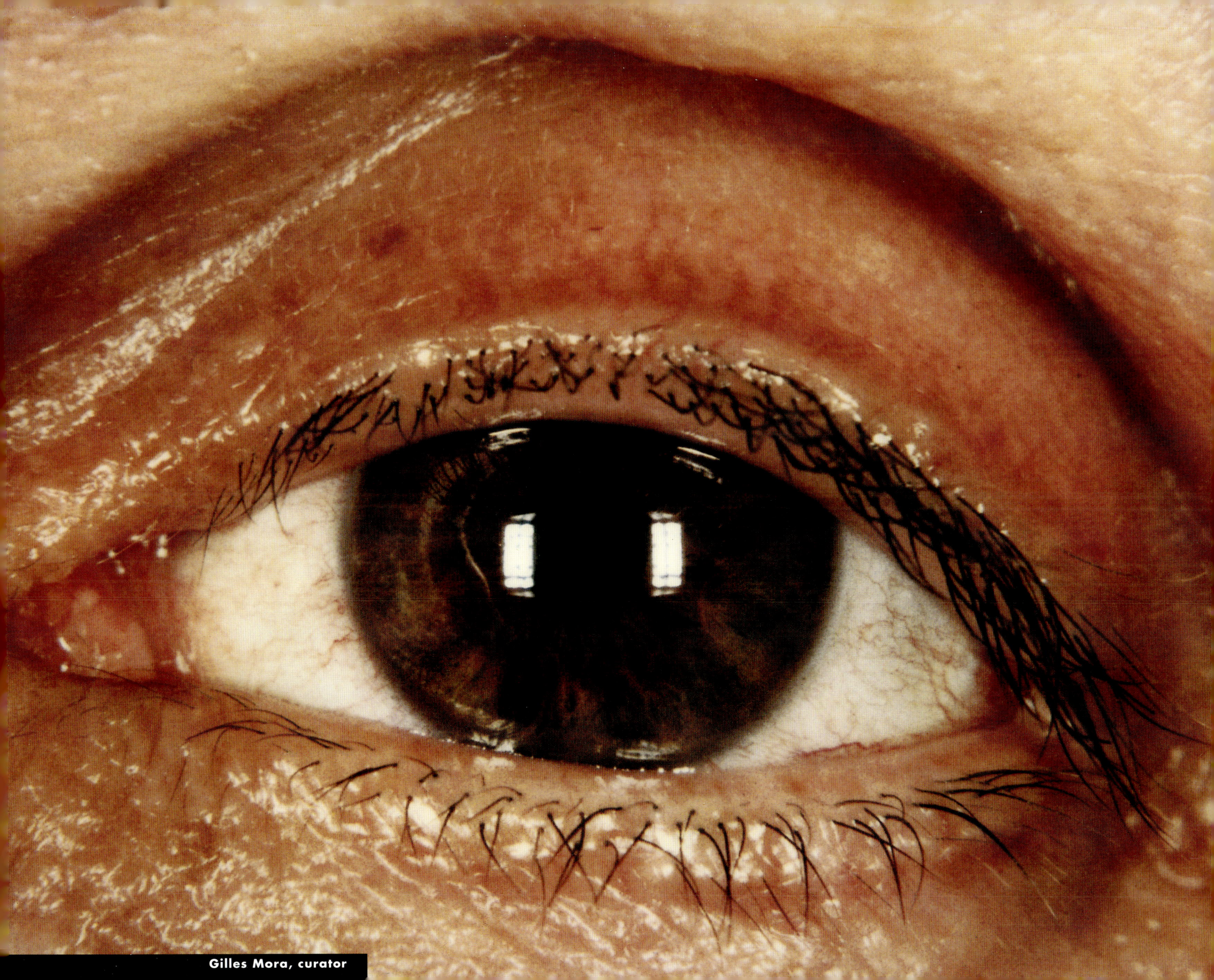

Gilles Mora, curator

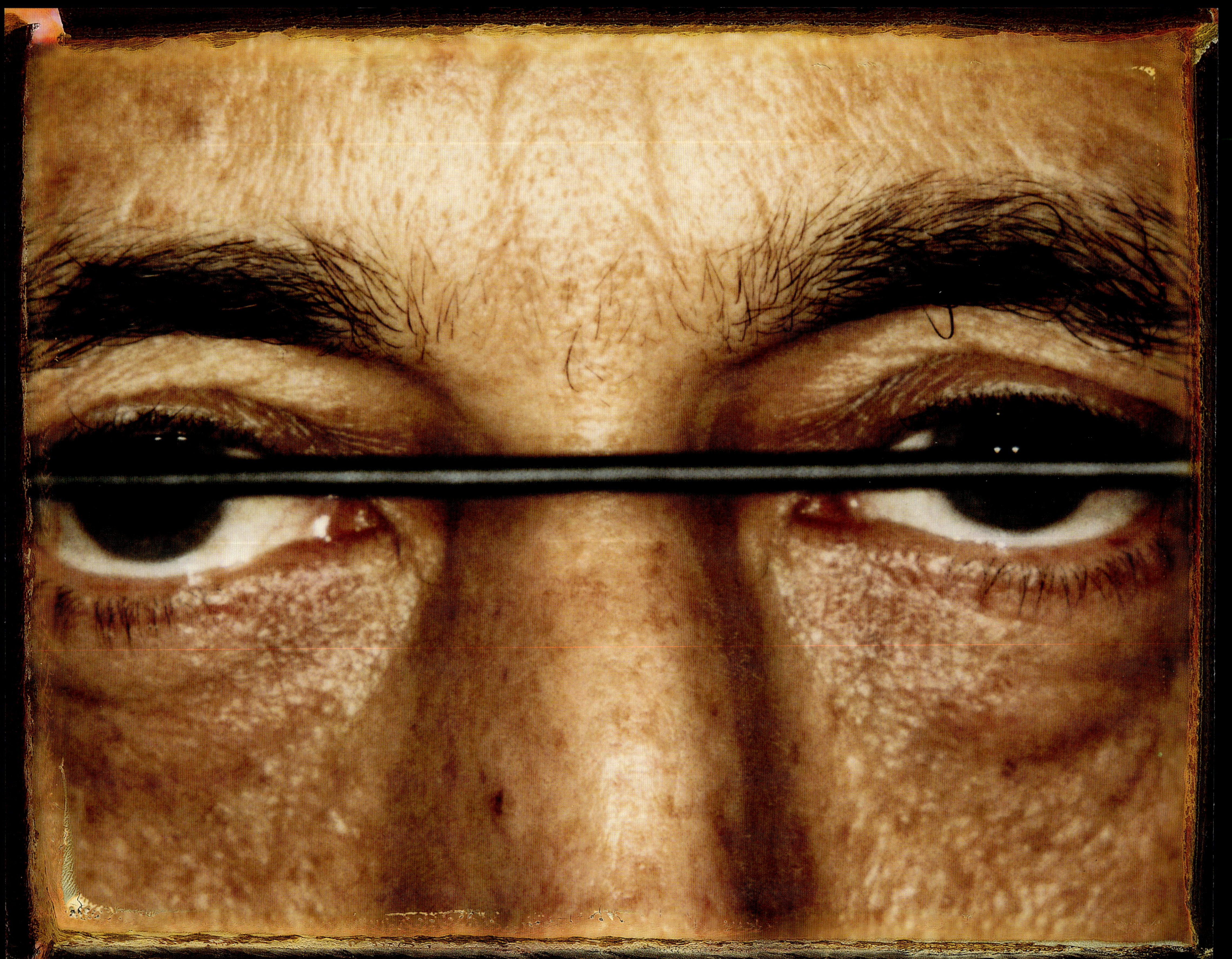

Roger F. Ballen, photographer

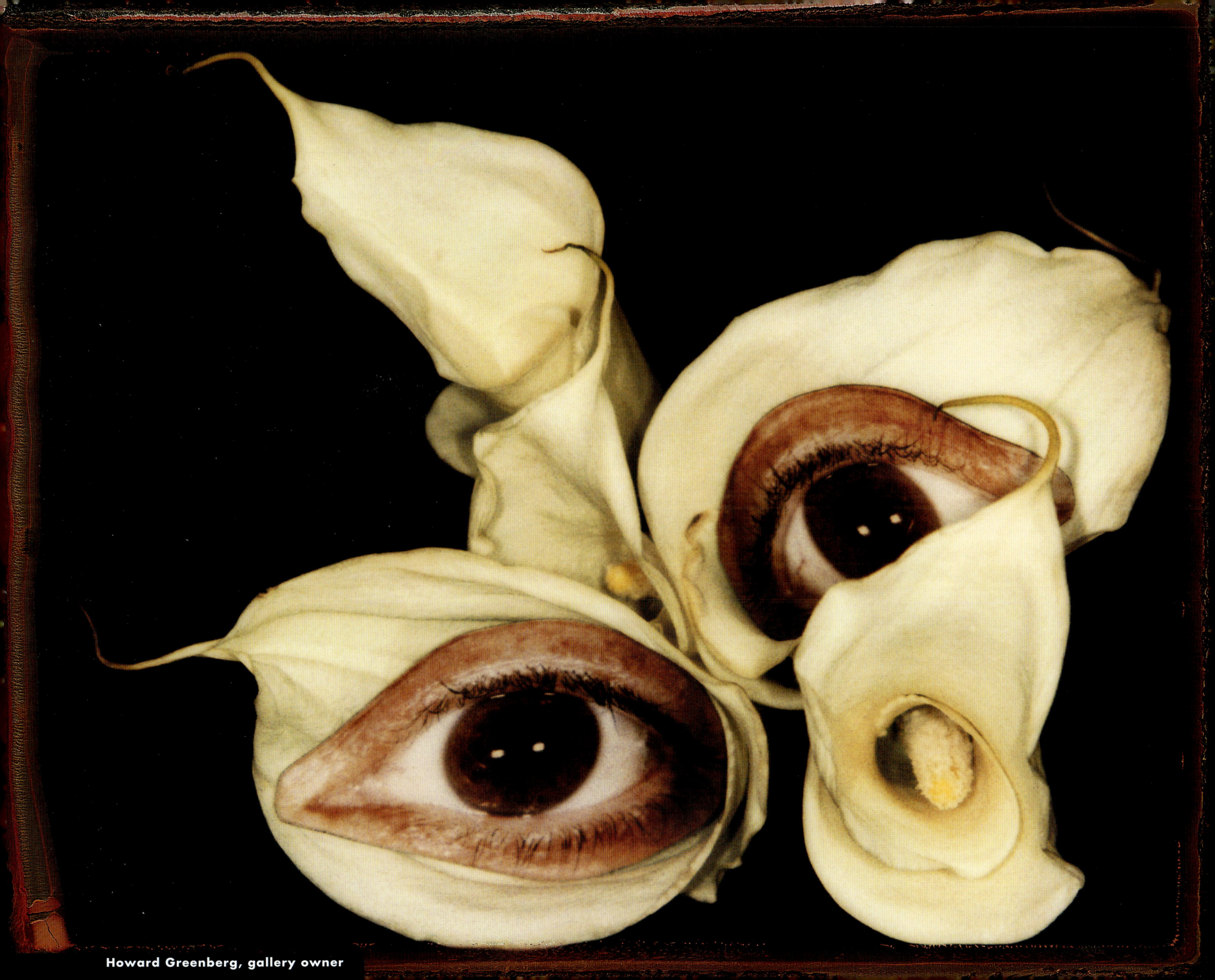

Howard Greenberg, gallery owner

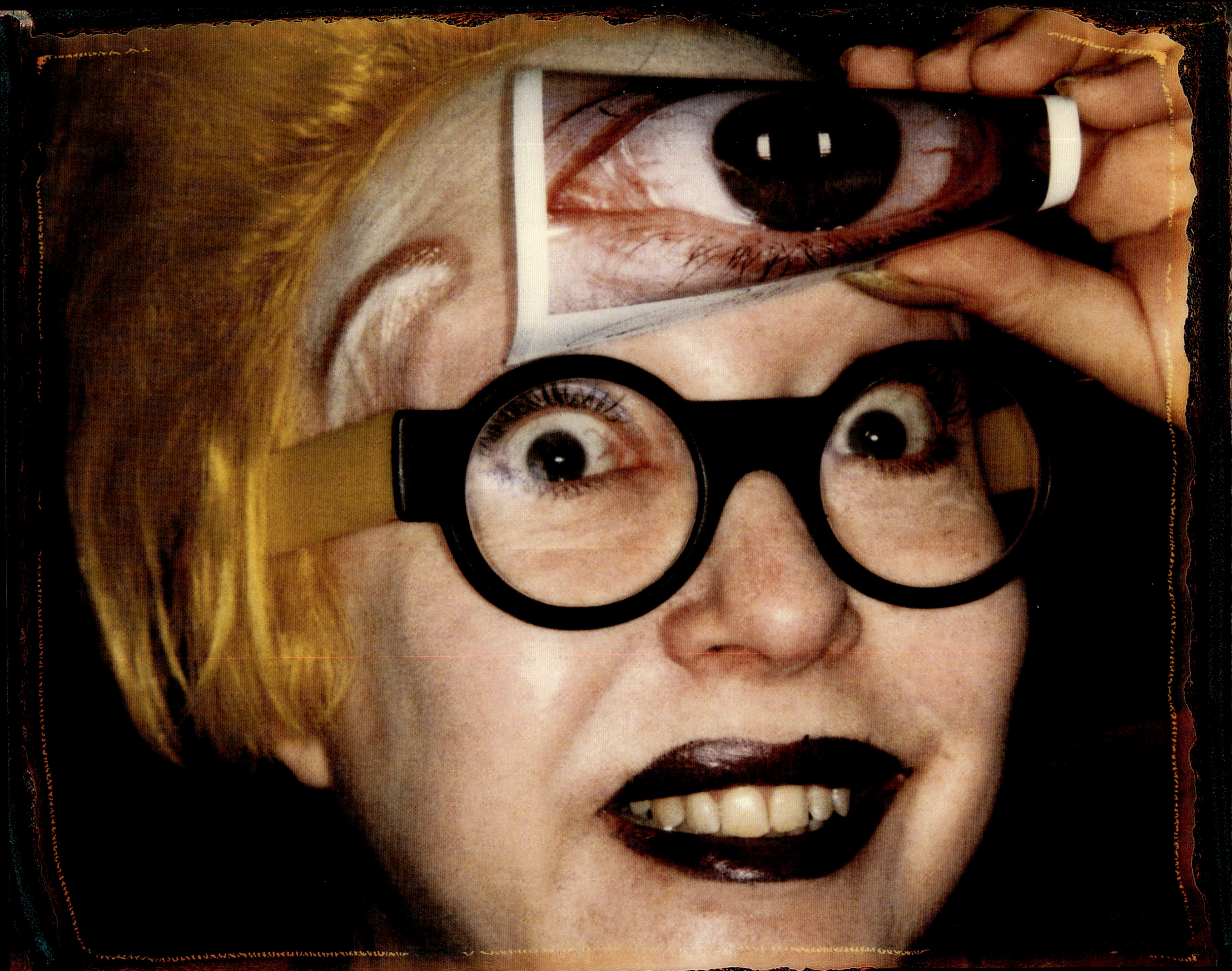

Orlan, artist

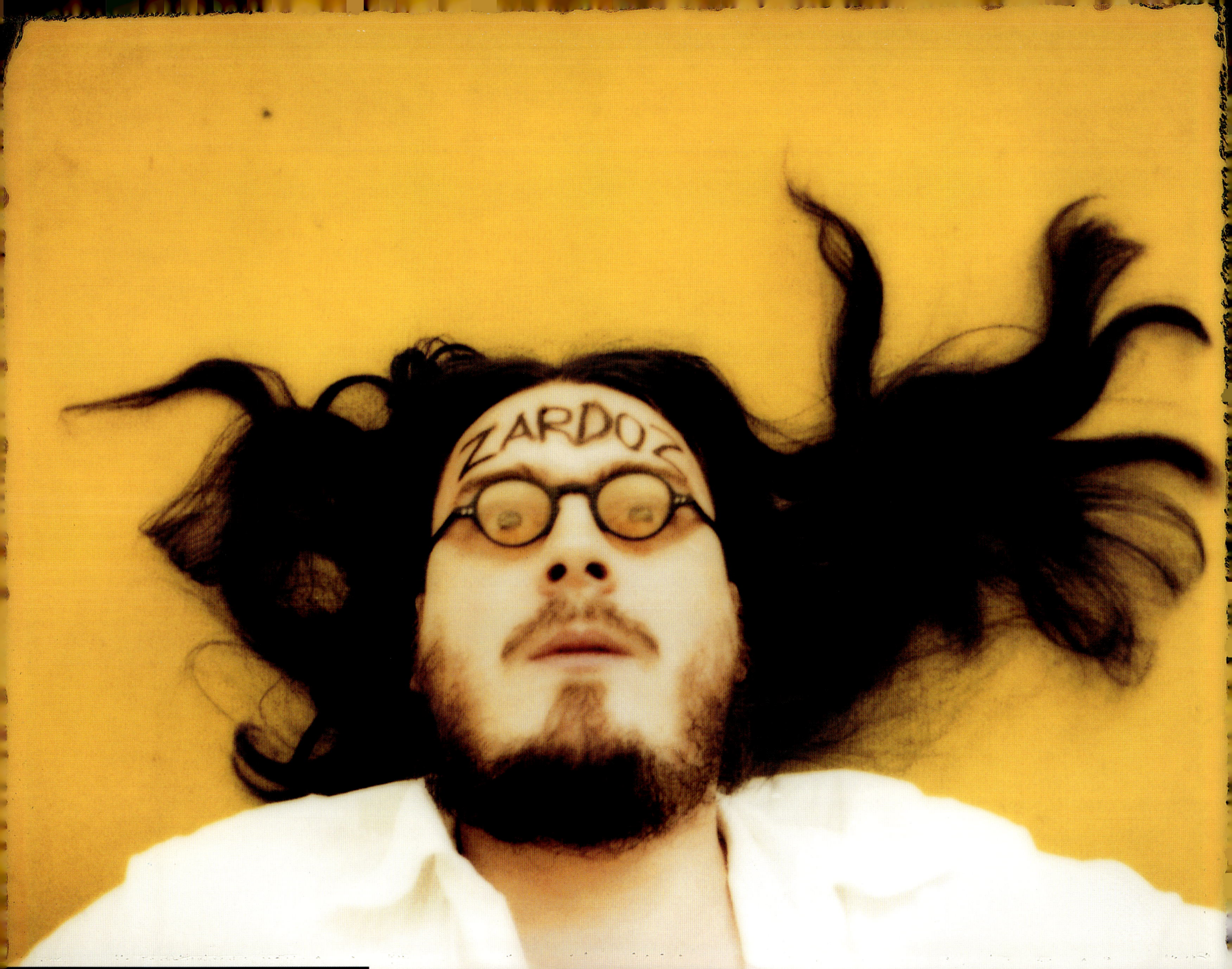

Jonathan Meese, artist

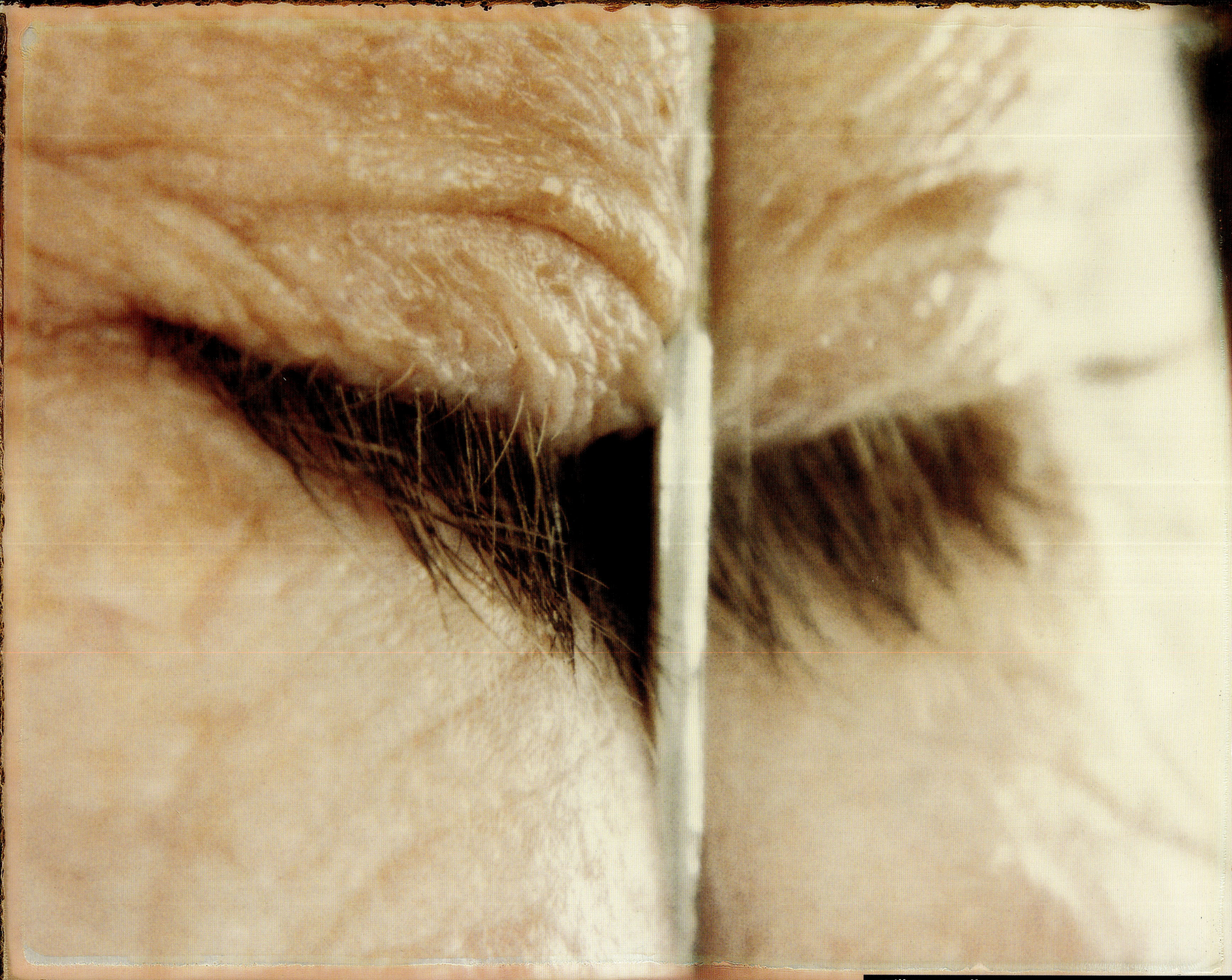

Will Hunt, collector

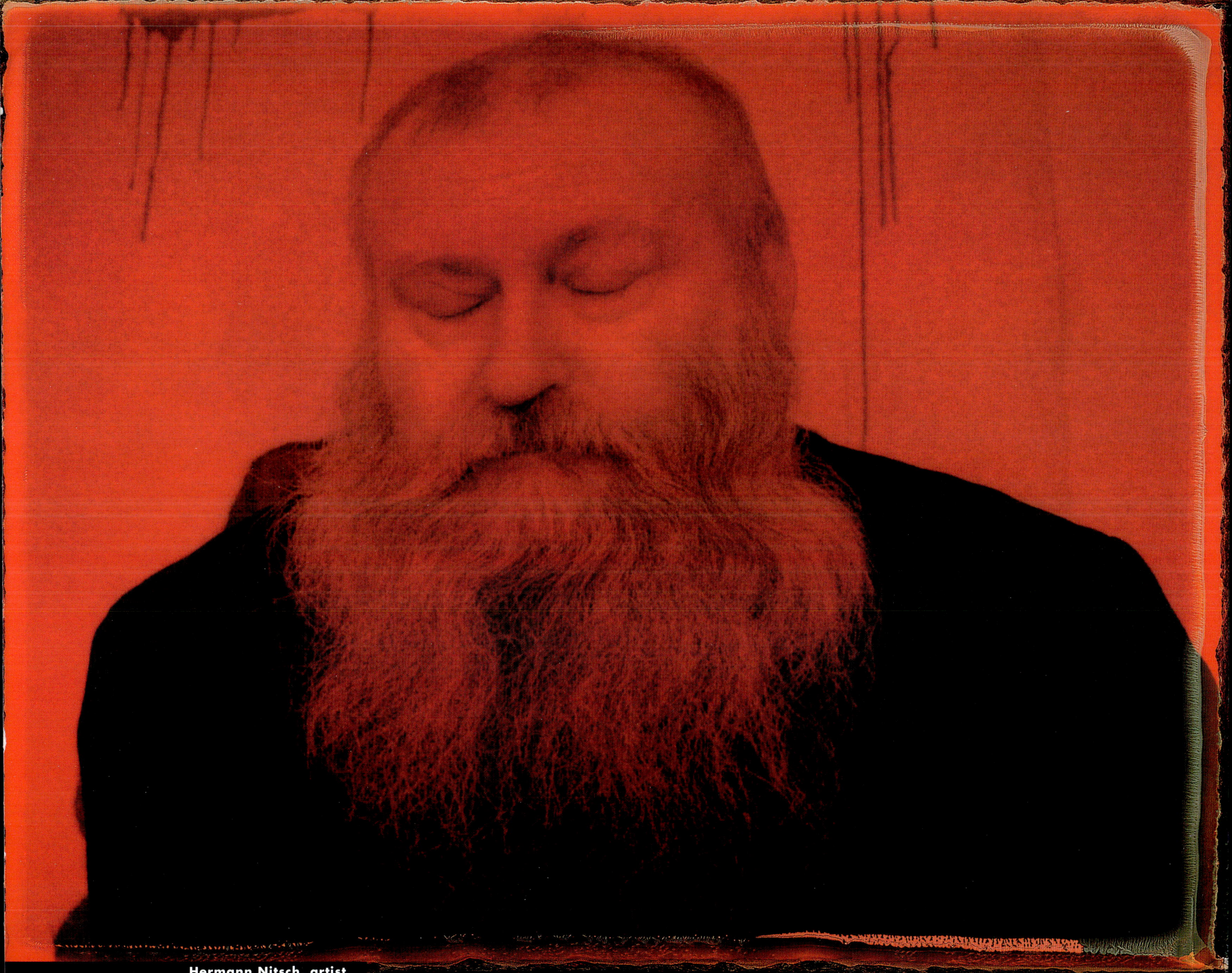

Hermann Nitsch, artist

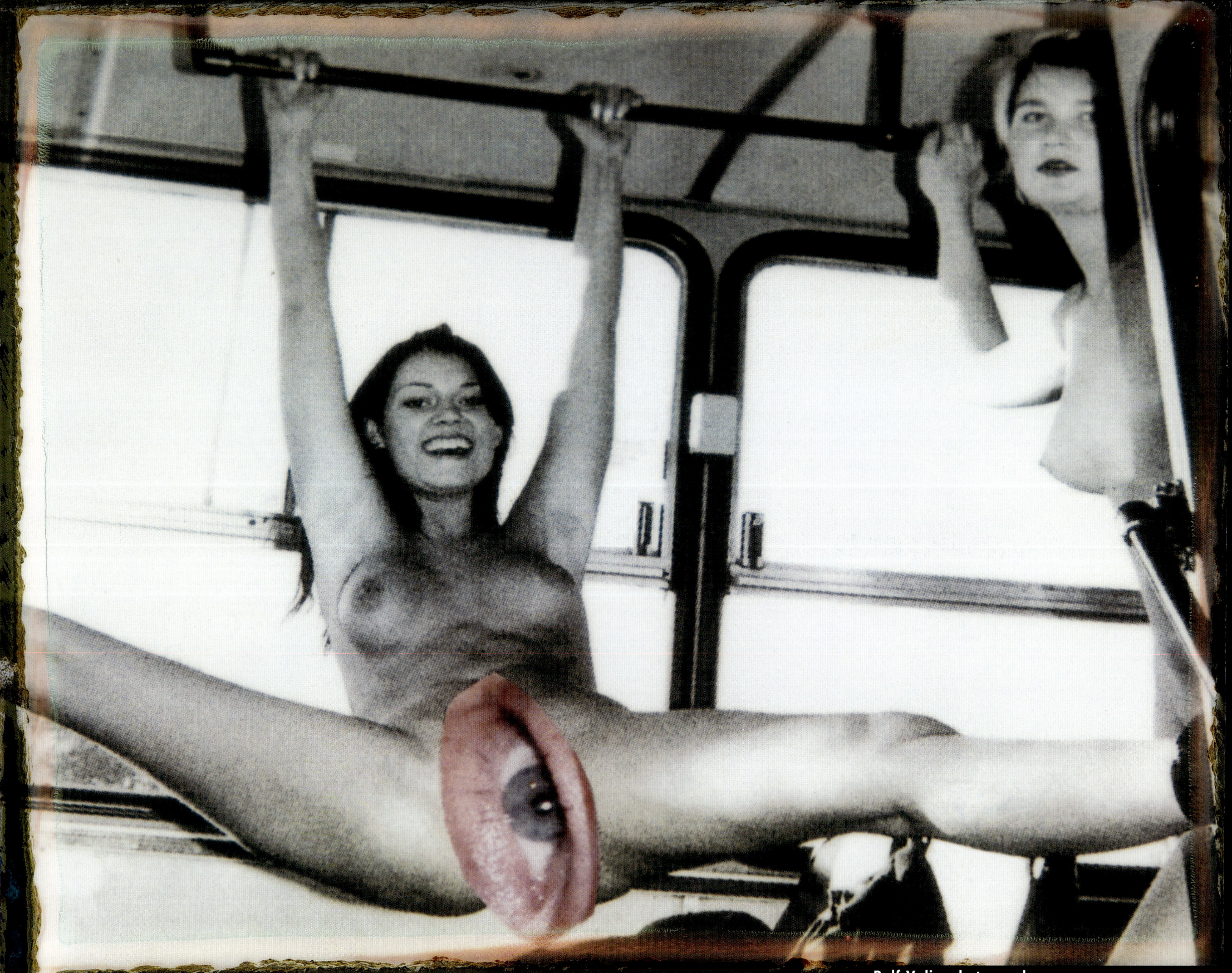

Ralf Vulis, photographer

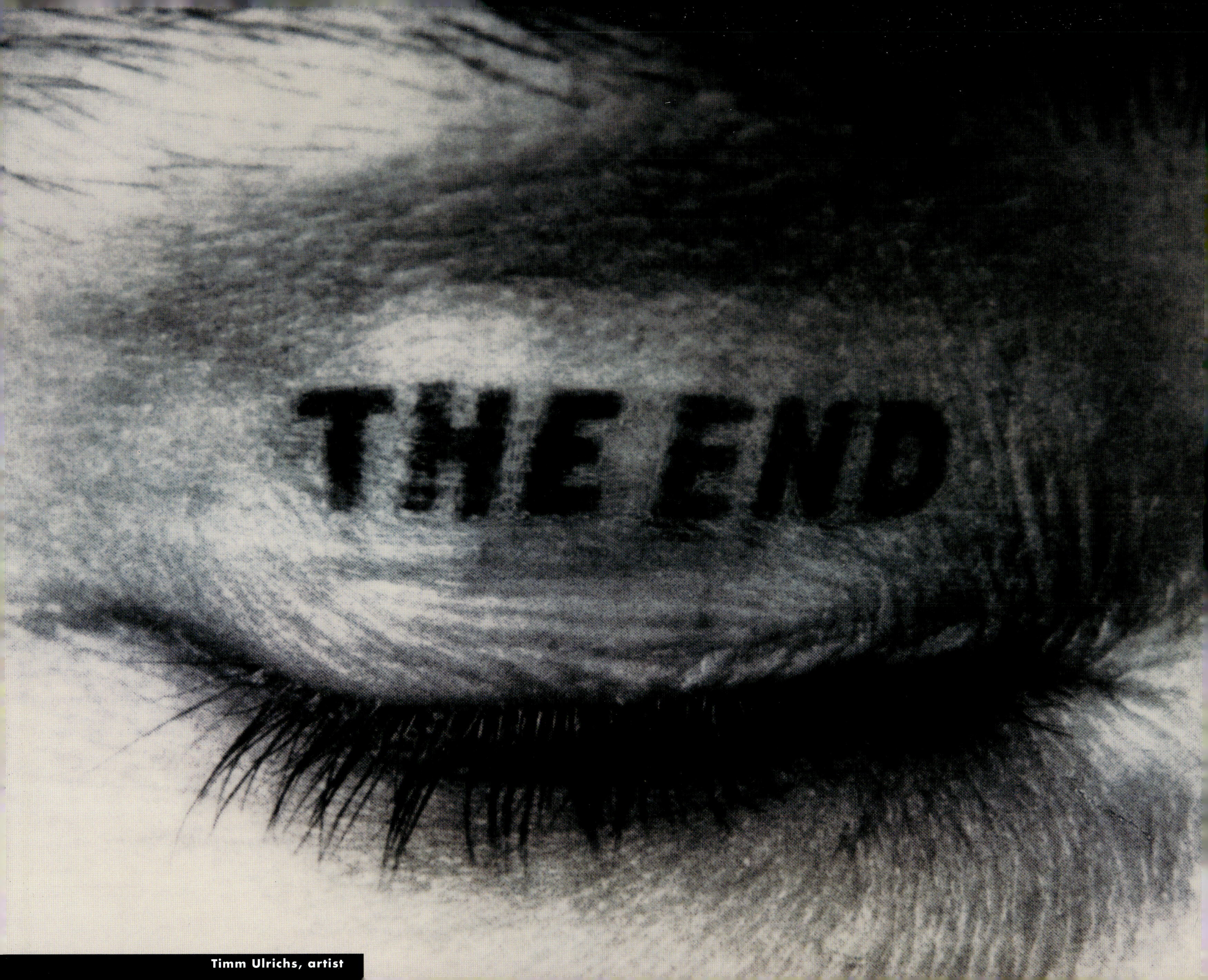

Timm Ulrichs, artist

Artistic biography

Daniel Fuchs, born in 1966

Geo Fuchs, born in 1969

Selected solo exhibitions

2002
Kunsthaus / Barlach Halle K, Hamburg, "Famous Eyes"
Camera Work, Hamburg-Berlin, "Famous Eyes"
Galerie Clairefontaine, Luxemburg, "Famous Eyes"

2001
Congress Centrum Saar, Saarbrücken, "Conserving"
Medizinhistorisches Museum, Zürich, "Conserving"
Berliner Medizinhistorisches Museum, Berlin, "Conserving"
Staatsbank Berlin, slide projection: "Conserving – Rammstein"

2000
Galerie Le Réverbère, Lyon, "Conserving"
Stephen Bulger Gallery, Toronto, "Conserving"
Rencontres internationales de la photographie, Arles, "Conserving"
Naturhistorisches Museum, Basel, "Conserving"
Fotografie Forum International, Frankfurt am Main, "Conserving"
Galerie Nei Liicht, Dudelange, "Conserving Fish"
Barlach Halle K Hamburg, "Conserving"

1999
Kunsthaus / Barlach Halle K, Hamburg, "Espada"

1998
Barlach Halle K, Hamburg, "Conserving Fish"
Galerie Nei Liicht, Dudelange, "Espada"
galerie & edition m, Leipzig, "Im falschen Körper"

1997
Schirn Kunsthalle, Frankfurt am Main, "Espada"
Neue Gesellschaft für bildende Kunst, Berlin, "Im falschen Körper"
Grundbuchhalle, Hamburg, "Menschen in der Psychiatrie"
Paul Gerhard Kirche, Cologne, "Im falschen Körper"

1996
Halle K3 Kampnagel, Hamburg, "Im falschen Körper"

1995
Gasteig, Munich, "Im falschen Körper"
Römer, Frankfurt am Main, "Im falschen Körper"
Schauspiel, Frankfurt am Main, "Die Unbehausten"

Selected group exhibitions:

2002
Muséum national d'histoire naturelle, Paris, "Histoire naturelle"
Stephen Bulger Gallery, Toronto, "Animal Logic"

2001
FIAC Paris, Galerie Le Réverbère, Lyon
Muséum d'histoire naturelle, Lyon, "Carte blanche"
AIPAD New York, Stephen Bulger Gallery, Toronto
PHOTO Los Angeles, Stephen Bulger Gallery, Toronto

2000
Paris Photo, Galerie Le Réverbère, Lyon
Göteborg Museum of Art, "New Natural History"
Art Cologne, Lipanjepuntin Artecontemporanea, Trieste
Art Brussels, Lipanjepuntin Artecontemporanea, Trieste
Arco Madrid, Lipanjepuntin Artecontemporanea, Trieste
Arte Fiera Bologna, Lipanjepuntin Artecontemporanea, Trieste
Franckesche Stiftungen, Halle/Saale, "Kinder haben Rechte"
Elektrizitätswerk Reading, Tel Aviv, Goethe Institute,"Prometheus"

1999
National Museum for Photography, Bradford, "New Natural History"
Lipanjepuntin Artecontemporanea, Trieste, "Still in motion"
Art Cologne, Lipanjepuntin Artecontemporanea, Trieste
Airport Gallery, Frankfurt, "Kinder haben Rechte"

1998
Airport Gallery, Frankfurt, "Deutsche sehen Schweizer"
Internationale Photoszene, Cologne, "Espada"
Alte Völklinger Hütte, Völklingen, "Prometheus"
Airport Gallery, Frankfurt, "Die Farbe Grün"
Art Frankfurt, galerie & edition m, Leipzig

1997
Rautenstrauch-Joest Museum, Cologne: "Sie und Er – Frauenmacht und Männerherrschaft"
Airport Gallery, Frankfurt, "Zeitgenössische Modefotografie"

Awards

Society of Publication Designers New York, gold for photography, still-life 2000
Art Directors Club Germany, silver for photography 2001
Art Directors Club Europe, distinction for photography 2001
Lead-Award Germany, first prize for photography, still-life 2000
Top Job Award Photokina 2000
Kodak Photobook Award 2000

THANKS

We are particularly grateful to all the celebrities who allowed us to photograph them for *Famous Eyes*.
We would like to thank above all Matthias Reuss and Annette Nellessen, who supported us and the idea of *Famous Eyes* from the very beginning.

For the wonderful realization of *Famous Eyes* in book form and for the sensitive collaboration with us, we would like to thank Edition Reuss and Metzger Druck Obrigheim: Wolfgang Metzger, Alejandro De Nardi, Martin Booth, Peter Teschler.

Special thanks to: Anke Degenhard, Kathy Ryan and
Michael Agel, Andrea & Christian Aeby, Shelby Lee Adams, Stephen Bulger, Marion de Beaupré, Albrecht Bangert, Luciano Bobba, Jean Back, Elisabeth Biondi, Pascal Bohlen, Hans Barlach, Martin Blau, Denis Brudna, Achim Borchardt, Agentur Baumbauer, Wolfgang Behnken, Elke Brämer, Michel Comte, Paul & Karen McCarthy, White Cube, Contemporary Fine Arts, Mariateresa Cerretelli, Volker Diehl, Klaus Doldinger, Roberta Dutto, Cathy Daly, Jeff Dunas, Ruth Eichhorn, David Fahey, Marlies Fiebig, Gertrud & Dieter Fuchs, Emanuel Fialik, Thomas Grünfeld, Greg Gorman, Manuel Gonzalez, Maurizio Galimberti, Alexander Gérard, Vicky Goldberg, Thomas & Kirstin Gerwers, F.C.Gundlach, Anna Gripp, Josefine Geier, Manfred Heiting, Jacqueline Heer, Hans-Eberhard Hess, Achim Heine, Hochkant Film, Will Hunt, Jan Hnizdo, Kurt Haas, Danielle Igniti, Norbert Jiptner, Georgine Knöll, Erwin Knöll, Doris Kloster, Margot Klingsporn, Rudolf & Annette Kicken, Francoise & Douglas Kirkland, Klaus Klemp, Georg Kargl, Robert Lebeck, Robert Longo, Celina Lunsford, Carol LeFlufy, Karin & Achim Lahrem, Edwin Lemberg, Lisson Gallery, Gudrun Muschalla, Yannick Morisot, Magnum Photos, Friedrich Meschede, Jean-Luc Monterosso, Gilles Mora, Jana Marko, Trixi Mahd Soltani, Claus Mewes, "Max" magazine Germany, Hansjoachim Nierentz, Grazia Neri, Oschatz Visuelle Medien, Daniel Oschatz, Joel Oschatz, Lello Piazza, Terence Pitts, John Reuter, Jim Rakete, Thomas Ruff, Rammstein, Rencontres internationales de la photographie Arles, Mathias Reichelt, Peter Refle, Anthony Reynolds Gallery, Jerry Sohn, Dietrich Sturm, Michael Schöbel, Donald Schneider, Nicola von Senger, Tina Schelhorn, Scumeck Sabottka, Oliviero Toscani, Michael Tafelmaier, Jürgen Teller, Tumbi, Wolfgang Volz, Val Williams, Wilfried Wiegand, Arne Weingart, Albert & Elisabeth Watson, Andreas Wellnitz, Andrea Zittel.

We thank our galleries for the good collaboration:
Galerie Camera Work, Berlin: Anke Degenhard, Karlheinz Lubojanski, Pia Plitzko, Steffi Schulze
Galerie Le Réverbère, Lyon: Catherine Dérioz, Jacques Damez
Galerie Lipanjepuntin, Trieste: Cristina Lipanje, Marco Puntin, Amanda Vertovese
Stephen Bulger Gallery, Toronto: Stephen Bulger
Galerie Clairefontaine, Luxemburg: Marita Ruiter
Polaroid International: very special thanks to Markus Mahla, and Gilles Longuève, Jacques Barthélémy, Stuart B. Strong, Marie-Christine Guyot, Annette Weber.
A special "thank you" to New Best Friend Film-Production Frankfurt am Main, which accompanied us during a two-year period and which made "Eyecatcher," a sensitive documentary film about us: Patrick Sheedy, Joss Hertle, Marc Flachmeyer, Ralph Ganswindt, Heiko Brand.

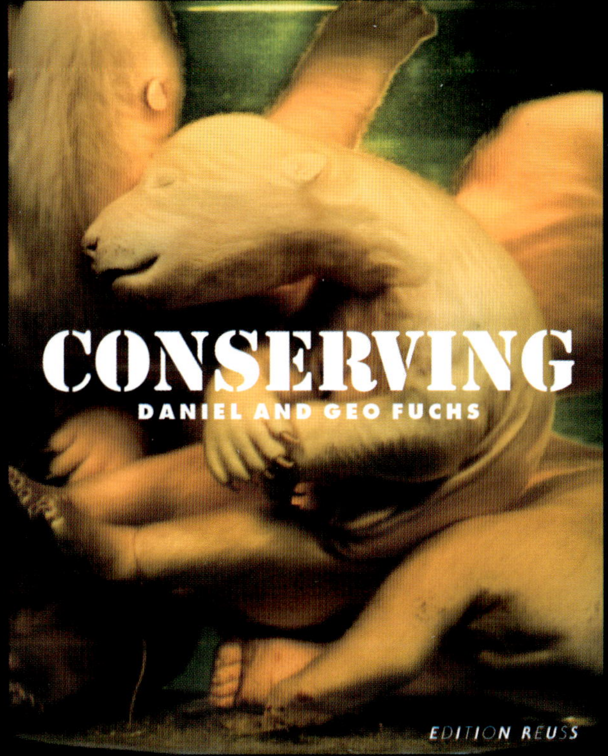

Leben und Tod sind Themen, die von jeher eine unglaubliche Faszination auf den Menschen ausüben. Die Trilogie „Conserving": „Conserving Fish" – „Conserving Animals" – „Conserving Humans" ist eine sensible Konfrontation mit einem Teil unseres Lebens, der oft verdrängt und tabuisiert wird. Daniel und Geo Fuchs gehen mit diesem fotografischen Werk Aufsehen erregende neue Wege. Mit brillanter Lichtführung gelingt es ihnen, neues Leben in die bis zu 300 Jahre alten, in Alkohol oder Formalin konservierten Exponate einzuhauchen. Ihre Bilder bestechen durch atemberaubende Schönheit der Farben und Strukturen oder sind von magischer Ausdruckskraft und erschließen dem Betrachter neue Welten. Sie erinnern an die Mysterien von Hieronymus Bosch und werden gleichsam zu Ikonen, zu fast ehrwürdigen Objekten. Weltweit erstmalig sind mit dieser künstlerischen Darstellung Fotografien von Exponaten aus naturwissenschaftlichen und anatomischen Sammlungen zu sehen, die bisher der Öffentlichkeit verschlossen blieben. Daniel und Geo Fuchs gehören zu den Neuentdeckungen der zeitgenössischen Fotografie und sind in den letzten Jahren durch ihre Projekte und Ausstellungen international bekannt geworden. „Conserving" ist in zahlreichen internationalen Ausstellungen zu sehen.

Life and death are the themes which have always exerted an incredible fascination on human beings. the trilogy entitled *Conserving* — which includes "Conserving Fish," "Conserving Animals" and "Conserving Humans" — is a sensitive confrontation with a part of our life that is frequently repressed or tabooed. Daniel and Geo Fuchs pursue exiting new paths in this photographic work. Their virtuosic use of lighting enables them to breathe new life into specimens, some of which have been conserved in alcohol or formaldehyde for as much as 300 years. The appeal of their photographs derives from the breathtaking beauty of the colors and structures and from a magical expressiveness which invites the viewer to enter hitherto unknown worlds. The images, which recall the enigmatic paintings of Hieronymus Bosch, simultaneously become icons and venerable, reverential objects. For the first time ever, these artistic photographs depict specimens from natural history museums and anatomical collections, many of which have never before been on public display. Daniel and Geo Fuchs rank among the newly discovered stars in the world of contemporary photography. In recent years, they have won international accolades for their projects and exhibitions. The photos in their *Conserving* trilogy are shown at numerous international exhibitions.

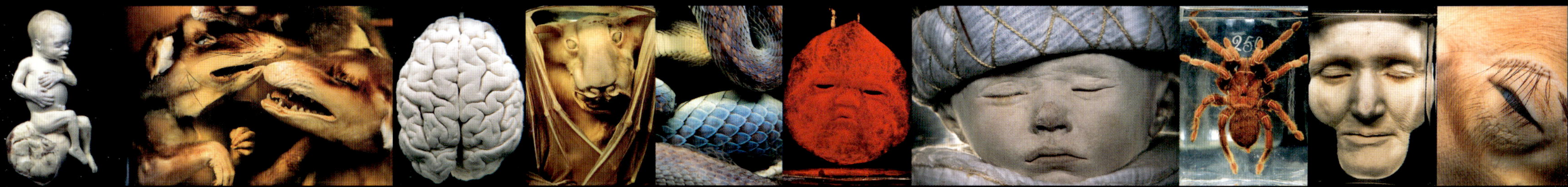

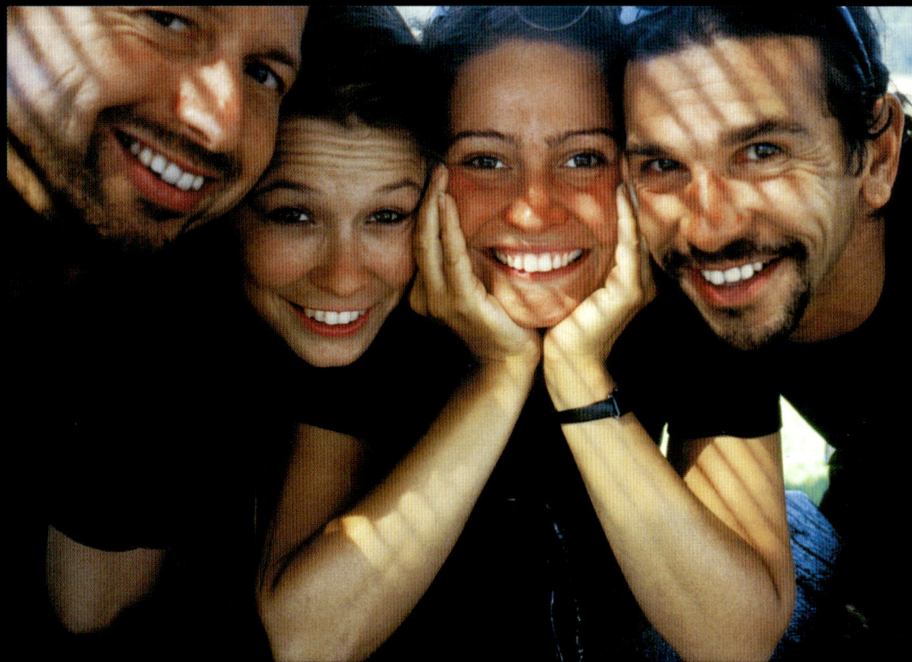

D&G Fuchs, A. Nellessen and M. Reuss at the legendary highway rest stop near Dole, France.

http://www.edition-reuss.de
photobooks@edition-reuss.de
Munich, Germany, FAX + 49 (0)89 50.13.95

Daniel & Geo Fuchs: http://www.daniel-geo-fuchs.com

English translation: Howard Fine
French translation: Sophie Vieillard

Lithography and printing by Metzger Druck GmbH, Obrigheim, Germany

ISBN 3-934020-06-2

ART PHOTOGRAPHY

CONSERVING
PHOTOGRAPHY BY DANIEL AND GEO FUCHS
ISBN 3-934020-01 1

NUDE PHOTOGRAPHY

SOLANGE ICH LIEBE
NUDE PHOTOGRAPHY AND LOVE POEMS
ISBN 3-9805017-2-8

SCHWANGER (PREGNANT)
NUDE PHOTOGRAPHY BY RALF MOHR
ISBN 3-9805017-5-2

FAMILY NUDES
NUDE PHOTOGRAPHY BY RALF MOHR
ISBN 3-934020-10-0

PUBLIC NUDISTS
NUDE PHOTOGRAPHY BY DIETER NAGEL
ISBN 3-934020-11-9

BLACK MAGIC WOMEN
NUDE PHOTOGRAPHY BY DANIEL BAUER
ISBN 3-9805017-4-4

BODYVISION
NUDE PHOTOGRAPHY BY
TEN LITHUANIAN PHOTOGRAPHERS
ISBN 3-9805017-0-1

EROTIC PHOTOGRAPHY

RUSSIAN LOLITA
EROTIC PHOTOGRAPHY BY PETTER HEGRE
ISBN 3-934020-16-X

SHAVEN BEAUTY MYSTERY
EROTIC PHOTOGRAPHY BY TORSTEN SEIDEL
ISBN 3-934020-12-7

VULIS' CRAZY SEXY PICTURES MIX
EROTIC PHOTOGRAPHY BY RALF VULIS
ISBN 3-934020-09-7

100 NAKED GIRLS ON A CHAIR
EROTIC PHOTOGRAPHY BY RALF VULIS
ISBN 3-934020-08-9

SEXY GIRLS NEXT DOOR
EROTIC PHOTOGRAPHY BY RALF VULIS
ISBN 3-934020-07-0

CRAZY SEXY GIRLS
EROTIC PHOTOGRAPHY BY RALF VULIS
ISBN 3-934020-00-3

SEX, KITSCH, GIRLS AND FUN
EROTIC PHOTOGRAPHY BY RALF VULIS
ISBN 3-9805017-3-6

SEXY SPORTS 1
EROTIC PHOTOGRAPHY BY RALF VULIS
ISBN 3-934020-02-X

SEXY SPORTS 2
EROTIC PHOTOGRAPHY BY RALF VULIS
ISBN 3-934020-03-8

WILD SKIN
EROTIC APHOTOGRAPHY BY CARLOS BATTS
ISBN 3-934020-14-3

PISSEUSES
EROTIC PHOTOGRAPHY BY CLAUDE FAUVILLE
ISBN 3-9805017-6-0

FETISH THEATRE
EROTIC PHOTOGRAPHY BY ALAN TEX
ISBN 3-9805017-7-9

SHAVEN ANGELS 1
EROTIC PHOTOGRAPHY BY PETER LORENZ
ISBN 3-9805017-8-7

SHAVEN ANGELS 2
EROTIC PHOTOGRAPHY BY PETER LORENZ
ISBN 3-934020-04-6

SHAVEN ANGELS 3
EROTIC PHOTOGRAPHY BY PETER LORENZ
ISBN 3-934020-13-5

LUST AND PAIN
EROTIC PHOTOGRAPHY BY GUY LEMAIRE
ISBN 3-9805017-05-4

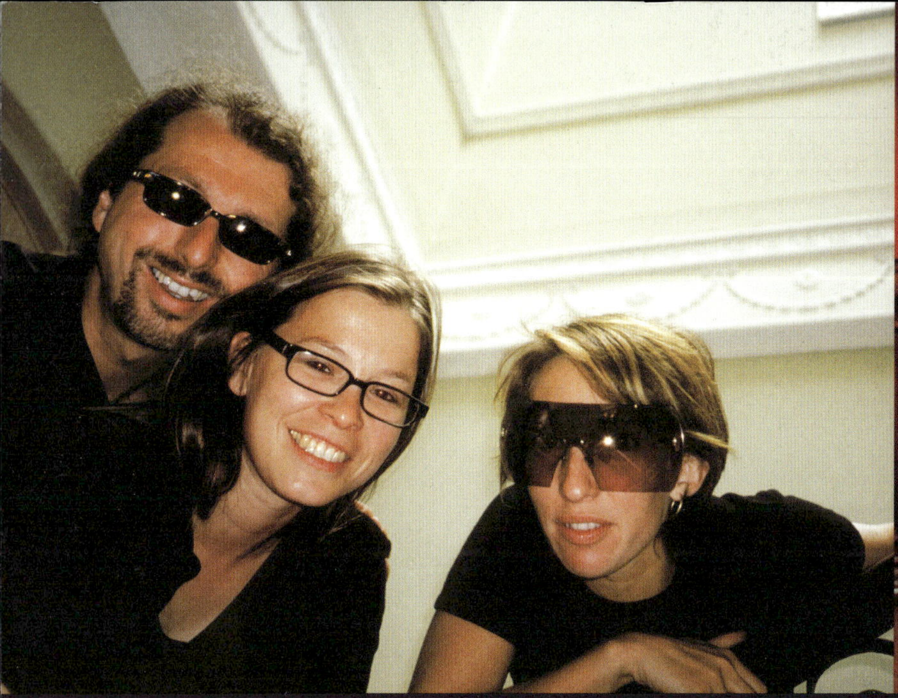
Sam Taylor-Wood, London

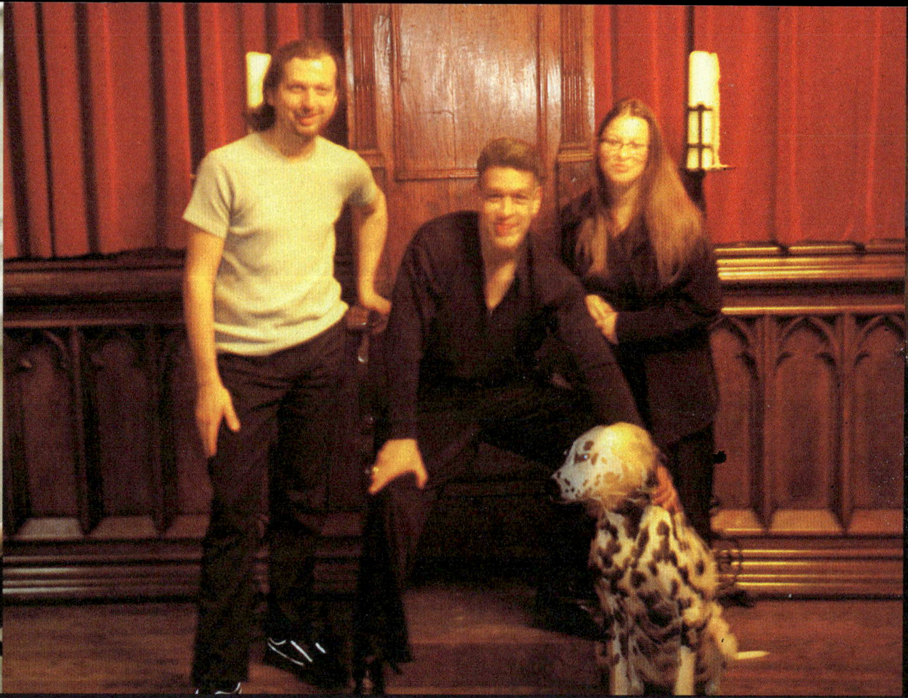
Andres Serrano, New York City

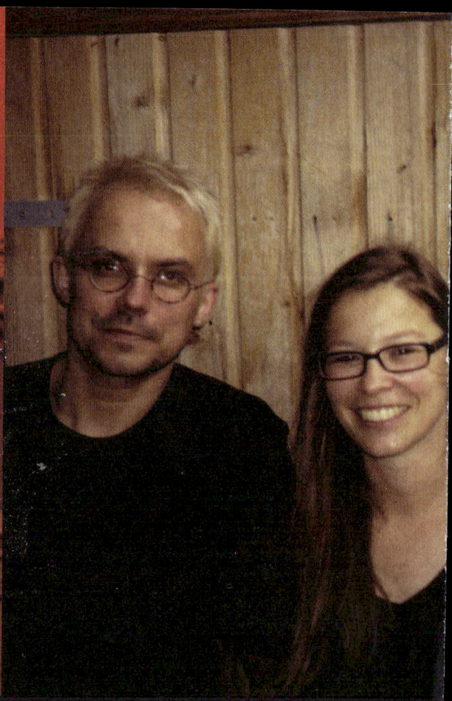
Henni & Gerd of Badesalz

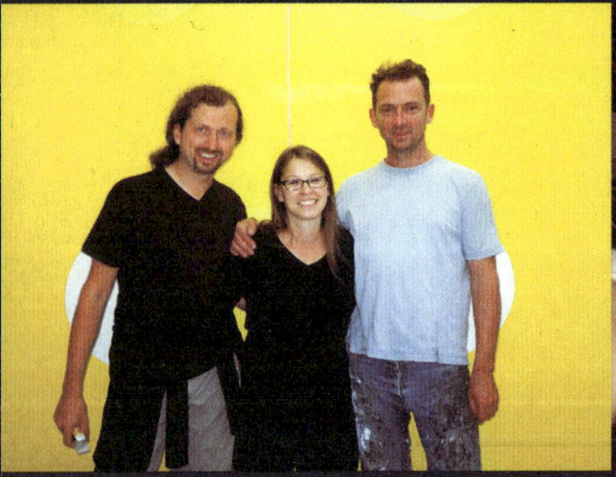
Gary Hume, London

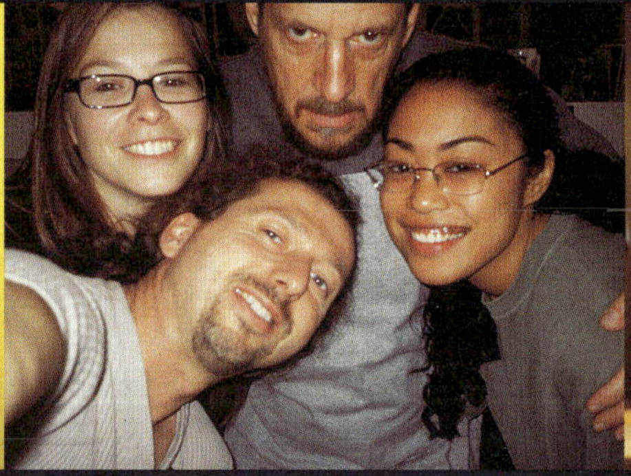
Larry Clark & Tiffany, New York City

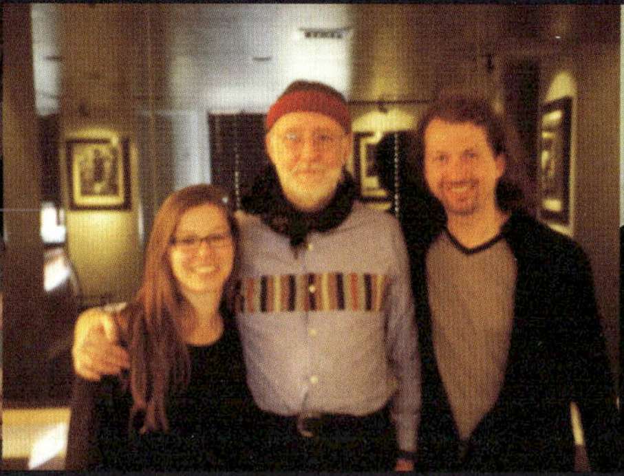
Albert Watson, New York City

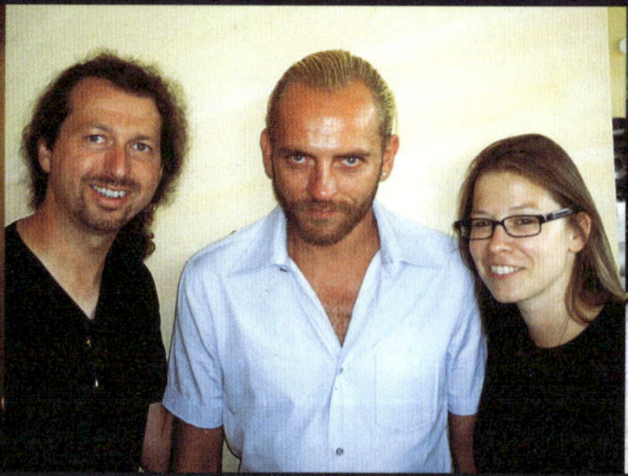
Sven Väth, Frankfurt

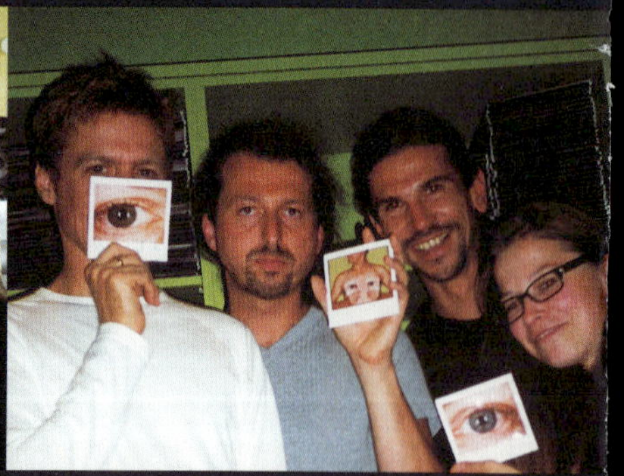
Bryan Adams, Matthias Reuss, Cologne

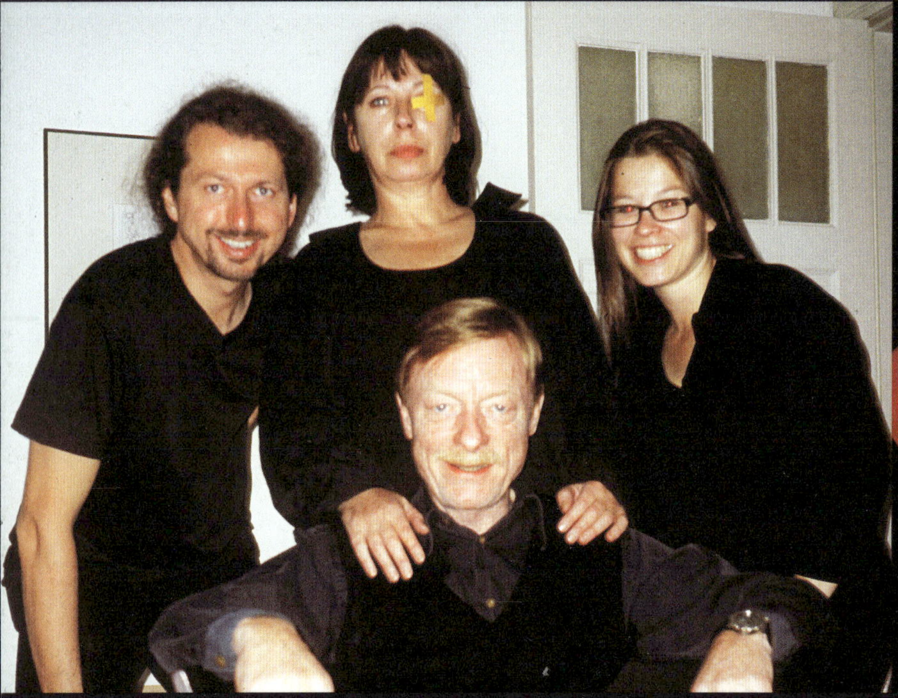
Otto Sander & Monika Hansen, Berlin

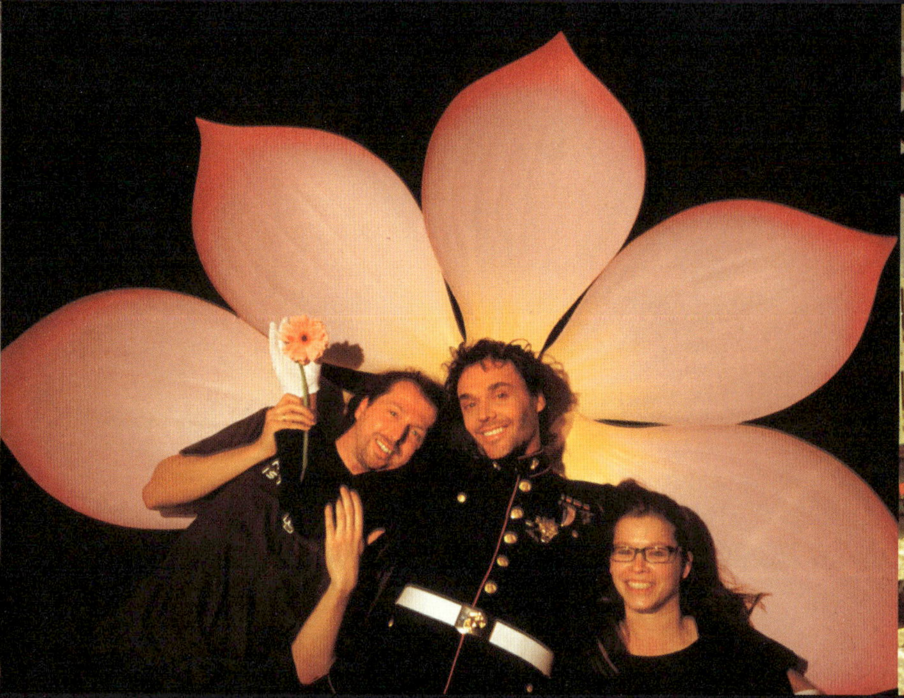
David LaChapelle, New York City

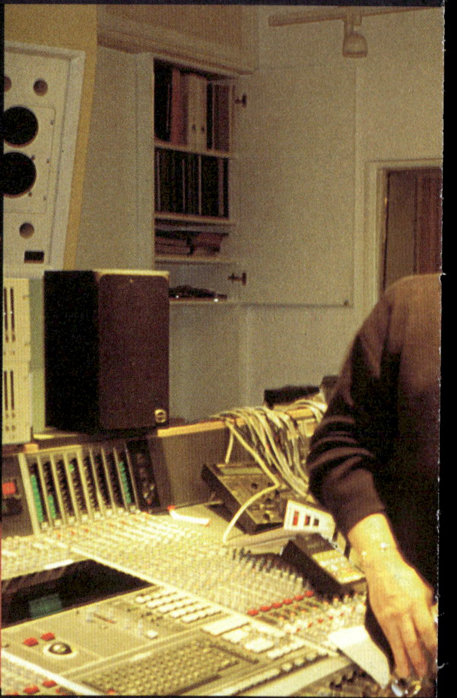
Klaus Doldinger